AIR CONTAMINANTS, VENTILATION, AND INDUSTRIAL HYGIENE ECONOMICS

The Practitioner's Toolbox and Desktop Handbook

AIR CONTAMINANTS, VENTILATION, AND INDUSTRIAL HYGIENE ECONOMICS

The Practitioner's Toolbox and Desktop Handbook

ROGER LEE WABEKE

CRC Press
Taylor & Francis Group
Boca Raton   London   New York

CRC Press is an imprint of the
Taylor & Francis Group, an **informa** business

CRC Press
Taylor & Francis Group
6000 Broken Sound Parkway NW, Suite 300
Boca Raton, FL 33487-2742

First issued in paperback 2017

Version Date: 20130227

ISBN 13: 978-1-4665-7790-9 (hbk)
ISBN 13: 978-1-138-07300-5 (pbk)

Library of Congress Cataloging-in-Publication Data

Wabeke, Roger L.
    Air contaminants, ventilation, and industrial hygiene economics : the practitioner's toolbox and desktop handbook / author, Roger Lee Wabeke.
    pages cm
    Includes bibliographical references and index.
    ISBN 978-1-4665-7790-9 (hardback)
    1. Industrial safety--Handbooks, manuals, etc. 2. Environmental engineering--Handbooks, manuals, etc. 3. Industrial hygiene--Handbooks, manuals, etc. 4. Air--Purification--Handbooks, manuals, etc. 5. Ventilation--Handbooks, manuals, etc. 6. Industrial toxicology--Handbooks, manuals, etc. 7. Work environment--Economic aspects--Handbooks, manuals, etc. I. Title.

T55.W335 2013
658.2'8--dc23   2013003656

Visit the Taylor & Francis Web site at
http://www.taylorandfrancis.com

and the CRC Press Web site at
http://www.crcpress.com

I dedicate this book to:

Mary, my best friend, wife, and life companion; our wonderful children: Lisa, Lori, and Michael and their spouses, Richard, John, and Andrea, respectively; and our beautiful grandchildren: Abigail, Charlotte, Grace, Kettler, and Veronica;

present and former colleagues and my students who, through support and guidance, offered constructive criticism, shared their many solved problems, and gave insights without which this book would not have sufficient richness and value;

my manuscript reviewers: Drs. Patricia A. Brogan, MSc, PhD, CIH, ROH, and Ernest P. Chiodo, MD, JD, MPH, MSc, MBA, CIH,

all who strived, currently endeavor, as well as future practitioners who work to improve our environmental air quality applying fundamental industrial hygiene practices, and

my publisher, Taylor & Francis/CRC Press, who with their endorsement and oversight, nudged me to embark on this project.

Contents

Disclaimer

Conscientious effort was made to ensure the contents of this book and every problem, in particular, are technically accurate, complete, and useful in the day-to-day practice of contemporary industrial hygiene and air pollution control engineering. All calculations and problems were critically peer reviewed. However, when thousands of informational items are entered into a published work, a few typographical errors may result even with the best efforts of everyone involved in the process. To ensure completeness and accuracy of these calculations, users are encouraged to send any corrections, additions, and comments that enhance usefulness of this handbook to the author or to the publisher.

Roger Lee Wabeke
Detroit, Michigan

Man is here for the sake of other men—and for all unknown souls with whose fate we are connected by a bond of sympathy.

Albert Einstein

Author

The author spent 47 years in full-time professional practice as a chemical safety engineer, industrial hygienist, and occupational toxicologist. Throughout a richly diverse practice, he encountered many challenging situations that he wished to share with those who follow. This is the overarching purpose of this book.

Air Contaminants and Industrial Hygiene Ventilation, published in 1998 by Lewis Publishers/Taylor & Francis/CRC Press, was his first book. Included here are 450 solved problems with comments from the book. Added were 275 problems, and the original 450 were expanded and modified to reflect current ventilation engineering and industrial hygiene practices plus advances in knowledge of atmospheric toxicants and control by elimination, substitution, process changes, ventilation, personal protective equipment, warnings, wet methods, and other time-tested interventional and prevention strategies.

Someone once jokingly said that every equation in a book cuts potential sales by one half. If so, this book, replete with numerous, but simple, equations, will never be a best seller. Regardless, the publisher and author disagree and expect ample acceptance.

Preface

A few problems and calculations in this book were initially developed to assist industrial hygienists preparing for their board-certification examinations. After the author passed (albeit probably marginally) two-day board-certification examinations in 1973 in Boston, he and a few other certified industrial hygienists from the Detroit area organized a one-day—admittedly brief, review course for others preparing for their examinations. Now, almost four decades later, that process is the primary launch point for this book.

The course was offered semi-annually, on Saturdays, at Wayne State University School of Medicine in the Department of Occupational and Environmental Health. The Michigan Industrial Hygiene Society provided ample refreshments. The registrants and instructors brought "brown bag" lunches. There were no fees; instructors provided *pro bono* lectures. Nine discussion rubrics covered air sampling and analysis, ventilation, toxicology, calculations, radiation, respiratory protection, industrial hygiene chemistry, noise, and heat stress and strain. With 50 minutes for each subject, only the industrial hygiene "pearls" could be presented. After the didactic lectures, many met afterwards for pizza, libations, and to discuss questions. Today, of course, there are several high-quality one-week review courses offered around the country for a reasonable fee. The depth and scope of these courses are far more encompassing and preparative than our early initiative one-day course.

This is the author's second book addressing air quality, ventilation, toxicology, and other industrial hygiene issues. The first book, *Air Contaminants and Industrial Hygiene Ventilation* (1998), was widely received throughout the world. Several persons encouraged the author to write a second book expanding upon the first publication. Over recent years, he developed more lectures to assist his students in professional development, provided case studies that incorporate the business model and enterprise economics, and tips to prepare for the Certified Industrial Hygienist board examinations. In some aspects, this book could partly be considered the second edition of the first book because every problem in that book is included here, but expanded. Additional text was added to further clarify initial existing problems, and 275 new problems were added. In part, over a long career with many potholes, the author hopes that this book removes some bumps and rough spots for those who follow. The author, now late in his career and at the 47th year of full-time professional practice in 18 countries and in every state except Hawaii and Idaho is still learning and desires to share these quantitative and practical experiences and tips from his journey.

Many problems in this book were acquired during the author's career of resolving industrial hygiene exposures of workers and others in different environments. These included evaluations of numerous workplace poisonings, fatalities, chemical spills, and catastrophic air emissions. A few were provided over the years by present and former colleagues, mentors, and graduate students far too numerous to mention.

Some were air modeling calculations. Still others served as the foundation for home-work assignments and examination questions for graduate students in classes the author presented at Wayne State University Schools of Medicine and Pharmacy and Health Sciences, the University of Michigan School of Public Health, and under-graduate students at Henry Ford Community College.

How to Use This Book

This handbook is intended primarily for use by

- Industrial and occupational hygienists
- Heating, ventilation, and air conditioning (HVAC) system engineers
- Air pollution control engineers
- Chemical safety engineers
- Hazardous material managers
- Inhalation and application toxicologists
- Air contaminant emergency responders
- Health care and public health professionals concerned with air quality
- Environmental evaluation and control engineers
- Atmospheric scientists and meteorologists
- Professors and teachers of these subjects
- Graduate and undergraduate students of these subjects
- Risk managers

Several typical plus some uncommon industrial hygiene problems that require mathematical solutions are covered. Tips are given to help one prepare for and take the board-certification examinations. Common formulae, equations, conversions, and other information worthy of committing to memory and practice are also included.

This book was prepared to be browsed. To simply read this book offers little. The problems must be studied. Once the ramifications of a problem are explored, the underlying concepts must be encoded to help ensure future relevance to practitioners.

Those preparing for board-certification examinations should master introductory sections and, at a minimum, the following 14 problems: 140, 304–306, 308–314, 271, 276, and 277. Once these are successfully handled, problems 1–7, 10–14, 16–20, 27, 29, 30, 32, 70, 79, 88, 93, 101, 102, 108, 297, 316, 321, and 406 should be understood. Those in the best position to achieve a high score on these aspects in the certification examinations will begin preparing at least one year in advance. If only six problems are mastered daily for four months during the examination preparatory year, every problem and exercise in this book will be covered.

The teacher of these subjects can extract selected, relevant course problems for student assignments and homework. With only slight modifications, each problem can be custom "tailored" to make it unique with pedagogic relevance to the course materials and content. Many problems can serve as a launch point to discuss industrial hygiene control methods and consequences to workers' health if preventive steps are not taken. The author has assigned several problems to be solved by his students with supporting industrial hygiene and environmental control methods. Students defend their selections of control methods incorporating cost savings and technical solutions when feasible.

The seasoned and experienced professional is encouraged to browse this book as well. Colleagues have suggested that solving these problems might be used to help maintain board-certification maintenance points. This concept could be raised with the American Board of Industrial Hygiene. Moreover, the Board could select some key problems from this casebook to be included in the revolving examinations.

The problems in this workbook are not grouped into categories. Few problems in industrial hygiene fall neatly into a specified "type." Rather, problems faced by today's industrial hygienists embrace several diverse topics. For example, a complex chemical spill issue could entail compound calculation sets of evaporation rates, air contaminant concentrations, worker's dose determinations, additive mixtures, community evacuation parameters, dilution ventilation requirements, and chemical contaminant half-lives and environmental fate. Several problems encompass such a broad scope and are of this nature (e.g., see Problem 316, referred to by a colleague as the "mother of all industrial hygiene problems"). While daunting, Problem 316 is not as challenging as a few others. Some problems, understandably, are far more rigorous and robust than others. Certain key types of problems are presented in various ways.

Since the problems in this handbook cannot be easily sorted into groups, it is the author's hope that industrial hygienists, chemical safety engineers, students and their teachers, atmospheric toxicant scientists, ventilation engineers, and all others will frequently peruse the problems. Only in this way will users begin to develop calculation methods and their "mental index" of the broad scope and diversity of a plethora of problems and their own systematic schema to easily define and solve them.

Another purpose of this handbook is to provide an assortment, a repository, and a reference set of calculations to assist industrial hygienists throughout their careers. Calculations to arrive at solutions for many of these types of problems are performed frequently by industrial hygiene practitioners. Others are done less often, some rarely, but the examples are included to assist resolution of uncommon problems. The author hopes that students of these problems will not only be challenged, but will also see the diversity of issues industrial hygienists and chemical safety engineers encounter, and that they will be stirred to regularly return to this handbook and engage themselves by these problems. In so doing, those responsible for conserving health and providing for the comfort of workers and the public's health and safety might see solutions applicable to situations they regularly, or perhaps even rarely, encounter during their professional practice.

Many problems can be solved by a variety of methods. The calculations are not always necessarily the "best" or the quickest way; however, the author was comfortable with the approach that was taken and fully recognized that stated solution steps might not be the most expedient method to arrive at the answers. Most problems, hopefully, balance theory and practical applications. A tad of levity is injected into some problems to help break the tedium and chore of some of the more daunting, lengthy calculations.

These problems do not require mathematical skills beyond college algebra and elementary statistics. Most calculations are simple arithmetic, but all require humility and critical, logical thinking based on sound understanding of basic industrial hygiene principles relating to evaluation of exposures to air contaminants. Since the science aspects of industrial hygiene (as contrasted with intuitive "art" parts) are quantitative in nature, those who are adept at number "crunching" and mathematical logic should have no difficulty solving these problems. Once the principles of a problem are understood, then it becomes, as the engineers say, simply a matter to methodically "plug and chug" the often ponderous numerical arithmetic parts. Several problems in this workbook are variations on a common theme. The author believes that, to fully understand some of the key concepts, one must be able to see a multifaceted problem from all aspects and be able to solve for the different variables.

Finally, these problems are a work practice module. For those preparing for the board-certification examinations, little will be gained by sitting down and simply reading these problems and skipping the exercises. One will reap only nominal benefit from the problems unless they are systematically analyzed, comprehended, and completed.

Remember that the certification examination questions are highly quantitative in nature. Over half of the core aspects examination questions may involve questions that require calculations, whereas the comprehensive practice portion may be 20% or more questions requiring calculations in air pollution, noise, radiation, heat strain, chemistry, ventilation, statistics, toxicology, safety engineering, ergonomics, and other rubrics.

Note: Throughout the book, unless otherwise stated, the concentrations of vapors and gases are expressed as parts per million by volume (ppm = ppm_v), not parts per million by mass or weight (ppm_m). Likewise, listed concentrations of air contaminants in high concentrations may be expressed as percent by volume ($\%_v$), or just %. Assume NTP (25°C, 760 mm Hg, dry air) unless otherwise stated. With knowledge of geography, we assume that port and sea level cities have a barometric pressure of 760 mm Hg unless otherwise stated. Although the author is knowledgeable in significant numbers, you will encounter calculations where this principle was not applied because application will not offer more insight.

CONTENTS

1. 725 practical (and some unusual, but helpful), solved problems with relevant, timely comments and helpful application tips covering:
   * Air contaminants, toxicants, and toxins released from our mobile and stationary sources that can adversely affect the health and comfort of people at worksites and those in residential, public, and ambient atmospheric environments with a major emphasis on industrial hygiene practice
   * Industrial ventilation system design engineering, testing, and intervention.
   * Inhaled, dermally absorbed, and ingested doses of toxicants and toxins
   * Air-sampling statistics and probability
2. 154 win–win business economic case studies demonstrating how to preserve your clients' financial resources, promote industrial hygiene, foster worksite safety, learn the financial ropes of business economics, and help control your

clients' potential adverse environmental impact and, in so doing, greatly enhancing career progress

3. Tips on preparing for, and passing, Certified Industrial Hygienist board examinations

4. Keys to professional development and future success for every industrial hygiene student, early-career professionals, and those launching careers as consultants in workplace safety, health, risk management, and environmental engineering

Introduction

TIPS FOR THE AMERICAN BOARD OF INDUSTRIAL HYGIENE CERTIFICATION EXAMINATIONS

1. Bring a scientific calculator with fresh batteries. Be able to apply all major functions. The calculator should have scientific notation, $\log_{10}$ and natural logarithms, common conversions (e.g., gallons to liters, lb to kg, °F to °C, and so on), exponential notation, and basic statistical functions.
2. Bring sharp pencils and new ball point pens to the exams. You might consider hard candy, mints, and gum (nonbubble type). How about a canteen of cold juice? Or a Thermos® of coffee? Cans of caffeinated soda (Mountain Dew®, Coke®, Pepsi®)? An eye moisturizer (e.g., Visine®) might provide ocular relief as needed.
3. Do not "cram" on the nights before the examinations. Kick back. Consider going to a light air pollution opera or to a nice movie—preferably a good industrial hygiene comedy or a documentary on the correct application of Pitot tubes and velometers.
4. Get a good night's sleep. Arrive refreshed and confident and fearlessly, boldly stroll into the examination room. Intimidate those about you in these competitive tests by sneering and insinuating that these examinations are just a "walk in the park." The examination is difficult. Some have sat for the examination as many as four times before passing. Having interviewed several who sat at least twice, I quickly learned they initially failed to take the time and apply hard work to prepare.
5. Wear comfortable clothes, for example, big, soft, over-sized shoes are nice at examinations. Sit in the center of the examination room to avoid any cold drafty walls and windows, solar heat, and excessive glare and contrasty shadows. Select a comfortable chair seat. Consider bringing a chair seat pan cushion.
6. Eat a light, well-balanced, nourishing breakfast and a similar lunch. Avoid heavy pancakes, fatty food, highly fibrous food, greasy donut "sinkers," and so on. Sugars from fresh fruits and a couple of proteins might not be a bad idea. Active brains need amino acids. Bring chunks of cheese to the examinations. Studies show we reason better when well hydrated. Taking a laxative the night before is not prudent. Be mindful that the residence time of food in the gastrointestinal tract can be 24 or more hours. Third helpings of stewed prunes after a beans, cabbage, and turnips dinner the day before examination are foolish menu choices. Pray that those seated around you did not eat such a meal.
7. Your first examination calculation must be:

$$\frac{\text{examination duration (in minutes)}}{\text{total number of questions}} = \frac{\text{average number of minutes allocated}}{\text{question}}$$

Be mindful of this average, and do not spend too much time on a single question. Try to pace yourself. Questions and problems involving calculations will normally require more time than others and might be weighted more heavily. Wear a wrist watch to keep track of the time and your mental pacing schedule.

8. Answer every question. There are no penalties for guessing. A chimpanzee will get about 20% correct answers if there are five choices and the multiple answers and his or her guesses are truly randomized. But, knowing absolutely nothing, you must do at least three to four times better than the chimp to be considered for a passing grade. Yet, three times nothing is still zero.

9. There is an ancient bromide: If you must guess at the answer, stick with your first hunch. If you erase it and replace it with a second guess, odds are you will be farther from the scientific truth. And the Board is searching for the scientific truth.

   Having stated that, while scientific principles are critical in the practice of industrial hygiene, professional judgment is as well. We often encounter exposure scenes where controls are obviously indicated, but a standard or guideline has not been exceeded. This must never deter us from intervening on behalf of workers at the risk of the six Ds: Discomfort, health Disorders, reversible Diseases, irreversible Diseases, Deaths, and significant property Damages.

10. Disregard previous answers; that is, if you guessed (or correctly answered) choice "C" on the previous five questions, do not think "C" for the next answer is incorrect if you must guess. There is a 20% chance it is correct.

11. Always ask yourself once you have selected an answer: "*Does my answer make sense?*" I have seen some absurd answers from students who, when rushed, did not take time to ask this simple question, yet they could otherwise correctly solve the problem. Wild answers included 154.7×106 ppm$_v$, a TWAE of 879 gram of dust/m^3, and a 30-minute exposure of a worker to 2300 ppm$_v$ HCN, followed by a 7-$\frac{1}{2}$ hour exposure to the same gas at 0.01 ppm! But then she/he will no longer be working after an inhalation or two. Where is it written that one cannot evaluate the exposure of a corpse?

12. Memorize equations, constants, formulae, atomic weights, conversions, and so on given on the following introductory pages. Equations and constants in boxes are especially important and are "must knows."

13. Watch decimal points and orders of magnitude! Watch units! Ensure that they are consistent—10 thimbles do not a gallon make. Five cm ≠ 11'. Be able to perform *dimensional analysis* to convert to other units, for example: from 100 fpm to mph, from 173 mg/sec to tons/day, and 2.3 mg/m^3 to lb/ft^3:

$$\frac{100\,\text{feet}}{\text{minute}} \times \frac{60\,\text{minutes}}{\text{hour}} \times \frac{\text{mile}}{5280\,\text{feet}} = \frac{1.14\,\text{miles}}{\text{hour}}, \text{barely a light breeze}$$

$$\frac{173\,mg}{sec} \times \frac{60\,sec}{min} \times \frac{60\,min}{hr} \times \frac{24\,hr}{day} \times \frac{g}{1000\,mg}$$

$$\times \frac{lb}{453.59\,g} \times \frac{ton}{2000\,lb} = \frac{0.0165\,ton}{day}$$

$$\frac{2.3\,mg}{m^3} \times \frac{m^3}{35.315\,ft^3} \times \frac{g}{1000\,mg} \times \frac{0.00220\,lb}{g} = \frac{1.43 \times 10^{-7}\,lb}{ft^3}$$

Note how fractions are arranged so identical units cancel each other. Two final practices: convert 1.8 mcg/m^3/second into lb/ft^3/year (= 0.00354 lb/ ft^3/year), and if men's facial hair growth rate is an average of 0.5 cm/day, calculate a beard-second (a standard unit for "slow," a counterpart to light-year) (5.79 × 10^{-9} cm/sec). The speed of light is 29,979,245,800 cm/sec. So, the ratio of beard-second to the speed of light is (5.79 × 10^{-9} cm/sec)/ 2.9979 × 10^{10} cm/sec = 1.93 × 10^{-19}—of highly redeeming social and scientific value. OK, let us try the inverse of a beard-second to the speed of light to arrive at values closer to which we can understand: (1/5.79 × 10^{-9} cm/sec)/ 2.9979 × 10^{10} cm/s = 0.00576. Now, most can deal easily with that number. The inverse of beard-second is an in-grown whisker which divided by speed of light is 0.00576 centimeter/second.

An excellent Website for conversions is www.onlineconversion.com where you can convert almost any metric to virtually any reasonable other—some amusingly silly.

14. Finally, *PREPARE, DON'T PRAY!* If you want to successfully solve the problems, *PRACTICE, PRACTICE, and PRACTICE* some more! If you work with other industrial hygienists, ask to solve *their* problems! Ask them to share examples of problems they might have in their notes and professional repertoire.

Dr. Steven Levine (professor emeritus, University of Michigan's School of Public Health) offered a few of the following examination tips. Heed Dr. Levine's sage advice. He is certified in both the *Comprehensive Practice and Chemical Aspects of Industrial Hygiene.* The author added to and augmented Dr. Levine's tips.

AT LEAST SIX MONTHS BEFORE THE EXAMINATION

1. Take a comprehensive review course. Refrain from sight-seeing and going "out on the town" while attending this course; instead, study every night. Review the notes for the next day's lectures. Prepare questions for instructors in areas where your concepts and skills are fuzzy.
2. Outline the notes in the book from the comprehensive review course.
3. Condense the outline on 3" × 5" flip cards. The outlining and condensing "process" will be a valuable learning tool.

4. Buy and use a computerized study and simulated examination program. These programs give practice in answering multiple choice questions.

5. Practice every type of calculation you can find. Ten of the most important types of calculations in the area of air contaminants, risk assessment, and ventilation are
   • "Dr. Clum Z. Chemist" who spilled a bottle of a volatile solvent with known vapor pressure in a room of a given volume with described ventilation parameters
   • Exponential relationships (e.g., radioactive decay, half-value thickness shielding, dilution ventilation, and half-life concentrations)
   • Inverse square law
   • Converting air contaminant concentrations (e.g., converting mg/m^3 into ppm_v)
   • Vapor pressure calculations; saturation concentrations in confined, unventilated spaces; gas and vapor migration
   • Ventilation air volume and hood capture velocity and duct velocity calculations
   • Additive mixture rule for multiple airborne toxicants
   • Time-weighted average dose calculations including consideration for overtime
   • TLV®s and OSHA PELs for air contaminant mixtures
   • The three fan laws; ventilation hoods ventilation capture velocities, characteristics of mechanical local exhaust hoods and systems, and dilution ventilation

If you understand these basic types of problems, you will be able to do a significant number, perhaps all, air sampling and ventilation types of examination calculations.

6. Exercise during your study breaks. Being physically fit will help you mentally and emotionally and gives you the stamina needed for this rigorous examination.

7. Do not reward yourself for wasting study time. For example, if you have a full day to study, and find yourself unable to focus, do "nothing" until you can. If, instead, you do other productive work, you will feel good about your alternative productivity, but you will not have accomplished any studying. Every now and then, that is OK.

8. Keep an honest record of concepts you do, and do not, know. Never hesitate to ask for help—an earmark of professionals who know their limitations, weaknesses.

9. Know how to operate your calculators quickly and accurately.

10. Bring an extra calculator—same type (and extra batteries).

11. Study diligently daily. For example, if only six problems are studied from this book every day, 120 days ($\cong$ 4 months) are needed to study all but three of them. If handled this way, these study tasks are not so daunting.

IMMEDIATELY BEFORE AND DURING THE EXAMINATION

1. Stay in a comfortable, nearby quiet hotel. Indulge yourself. Consider ordering only high-quality, nutritionally balanced room service meals. Tip generously. But do not "hole up" in your room. Get outside once in a while. Take a brisk walk once or twice a day while making your final preparations.
2. Do not stay with relatives.
3. Do not bring your family, especially any young, whining children. Substitute with their charming pictures.
4. Arrive at the examination location at least one full day early.
5. Eat safe food before the examinations. Ask yourself if that taco or cheesecake that you want to eat the night before will cost you a few percentage points on the tests. Consuming alcoholic beverages the night before the examinations would be foolish; all of your central nervous system neurons must function at warp speed.
6. Wear comfortably soft, clean, loose-fitting clothing to the examinations. Bring a light sweater in the event the examination room becomes chilly. The author wore a baseball cap when he sat for the examinations. Not only did this help to keep his brain warm, but glare was reduced, and he was able to focus more clearly on the question. Whatever works.
7. Arrive before registration starts, get a good seat, relax, take care of last minute personal needs, and then go back outside of the room to register. Boldly stroll into the examination room exuding extreme confidence to intimidate all those around you.
8. Bring *Zip-Loc®* bags of candy or other snacks to the examinations so that you can keep your blood sugar and energy reasonably constant throughout the day.
9. Bring aspirin, *Tylenol®*, *Motrin®*, *Excedrin®* (with caffeine), *Digel®*, and *Pepto-Bismol®* in a small container to the examinations. While not meant to endorse these products, this list covers a full range of over-the-counter drugs to control common, simple problems that might reduce your abilities to perform in an optimal fashion.
10. When you get to the examination room, fill out all required forms, and then await the start of the examinations.
11. Spend the next few minutes practicing "positive visualization" where you can see yourself answering the questions accurately, finishing the examinations on time, and then receiving your CIH, handsomely framing your certificate, and proudly boasting to all within the sound of your voice.
12. Do not waste time pondering a single question, thus leaving insufficient time to complete the examinations. Be mindful of your calculated average time allotted for each question.
13. If you finish the examinations early, carefully check every answer (especially your calculations for decimal points, units, reasonableness of answers, etc.).
14. Pack an umbrella and/or a raincoat. Sitting for examinations in soggy clothing and with cold, wet hair and shoes might cost a few percentage points off the final score.

15. Clean your glasses. Consider wearing comfortable ear plugs to reduce any noisy distractions. Pack a handkerchief or Kleenex® to handle any sneezing or crying. Eye drops aid weary eyes. To not distract others around you, refrain from sobbing loudly over problems you cannot answer. On the other hand, your wailing might reduce their vigilance.

EQUATIONS, CONSTANTS, CONVERSIONS, FORMULAE TO MEMORIZE

Know the empirical formulae of the common alcohols, aromatic hydrocarbons, alkanes, alkenes, ketones, phenols, ethers, and so on and their substituted products (e.g., trichloroethylene, 2-nitropropane, pentachlorophenol, bromobenzene, etc.).

Know the generic, alternative, and trivial names for the common solvents (e.g., methyl chloroform = 1,1,1-trichloroethane, 2-ethoxyethanol = ethylene glycol monoethyl ether, carbon "tet" = carbon tetrachloride, "tri" = trichloroethylene, "perc" = ?, MIBK = ?).

ATOMIC WEIGHTS OF MAJOR ELEMENTS

(rounded to the nearest whole number)

Na = 23	Cr = 52	Zn = 65	As = 75
Si = 32	Fe = 56	As = 75	Hg = 201[a]
Ca = 40	Cu = 64	Cd = 112	Pb = 207[a]
Cr = 52	Fe = 55	P = 31	Mn = 55[a]

[a] The intelectual robbers.

Memorize	H = 1	C = 12
	N = 14	O = 16
	S = 32	Cl = 35.5

From the above list of approximate atomic weights, know how to calculate molecular weights of common gases, metallic salts, oxides, solvents, and so on. For example, without consulting chemistry textbooks and references, calculate the molecular weights of sulfur dioxide, hydrogen cyanide, nitrogen dioxide, acetone, lead carbonate, toluene, carbon monoxide, chlorine, sulfuric acid, ozone, MEK, zinc oxide, limestone, benzene, caustic soda, arsenic trioxide, potash, rouge, and trichlorophenol.

This implies you have familiarity with chemical nomenclature including common and trivial names—important to the industrial hygienist because, for example, a worker might say that he's shoveling ash. You ask, "What kind of ash?" knowing there is coal-fired boiler bottom ash, air pollution control device fly ash, wood and oil ash, soda ash (Na_2CO_3), potash (K_2CO_3), and bicarb ash ($NaHCO_3$). He "knows" ash, but you must be certain. For example, if this is soda ash, you collect breathing zone air samples for total airborne particulates, perhaps respirable airborne particulates, and analyze your samples for total sodium. Then, divide the molecular weight of two sodiums into the molecular weight of soda ash (sodium carbonate) to get the correction factor for percent sodium that you then use to convert sodium to sodium carbonate.

Finally, to complete exposure assessment, carefully compare your results to the exposure standards and guidelines: total dust, respirable dust, total sodium carbonate, and respirable sodium carbonate. Moreover, since much soda ash is mined, and not synthesized today, it is prudent to analyze minerals for respirable quartz as well. Your work is cut out for you and your industrial hygiene analytical chemist particularly if the worker shoveled a mixture of soda ash and potash.

Molecular Weights of Common Gas and Vapor Air Contaminants

Acetic acid	60.05	Isoamyl acetate	130.18
Acetone	58.08	Isoamyl alcohol (*n, s*)	88.15
Ammonia	17.03	Isobutyl acetate	116.16
Benzene	78.11	Isobutyl alcohol	74.12
2-Butoxyethanol	118.17	Isopropyl alcohol	60.09
Butyl acetate (*n, s*)	116.16	MDI	250.25
Butyl alcohol (*n, s, t*)	74.12	Methanol	32.04
Chlorine	70.91	2-Methoxyethanol	76.09
Cyclohexane	84.16	2-Methoxyethanol acetate	118.13
Cyclohexanol	100.16	Methyl chloroform	133.42
Cyclohexanone	98.14	Methylene chloride	84.94
Diacetone alcohol	116.16	Methyl ethyl ketone	72.10
Diisobutyl ketone	142.24	Methyl isobutyl ketone	100.16
Dimethylformamide	73.09	Mineral spirits	$\cong 136$
Dioxane	88.10	Naphtha (VM & P)	$\cong 112$
2-Ethoxyethanol	90.12	Nitric oxide	30.01
2-Ethoxyethyl acetate	132.16	Nitrogen dioxide	46.01
Ethyl acetate	88.10	Ozone	48.00
Ethyl alcohol	46.07	Propyl acetate (*n*)	102.13
Ethyl benzene	106.16	Propyl alcohol (*n*)	60.09
Formaldehyde	30.03	Styrene	104.14
Gasoline ($\cong$ <1–3% benzene)	$\cong 78$–108	Sulfur dioxide	64.07
Hexane (*n*)	86.17	TDI (What's this?)	174.16
2-Hexanone (M*n*BK)	100.16	Toluene	92.13
Hydrogen bromide	80.92	Trichloroethylene	131.40
Hydrogen chloride	36.46	Triethylamine	101.19
Hydrogen cyanide	27.03	Vinyl acetate	86.09
Hydrogen fluoride	20.01	Vinyl chloride	62.50
Hydrogen sulfide	34.08	Vinyl toluene	118.18
Indene	116.15	Xylene (*o, m, p*)	106.16

Molecular weights above are, for example, grams/gram-mole or pounds/pound-mole. Water, for example, has a molecular weight of 18. That is, 18 g (or 18 mL) of water is 1 mole of water. Each mole of any substance contains 6.022045×10^{23} molecules or atoms. This huge number is Avogadro's.

Pandammonium trichloride = ?
2,5-Dimethyl chickenwire = ?
4,6′, 7-β-Triphenylawfulstuff = ?
Methyl ethyl death = ?

THE NOT-SO MYSTERIOUS MOLE

Moles (the chemical, not the underground kind; nor an FBI or Central Intelligence Agency counterspy; nor malignant skin blemish) confuse some people. Consider three balloons each containing 1 mol of a different gas and all at the same temperatures and pressures. How can they be so equivalent in most ways, but yet so different? As we see in the tables below, other than their molecular chemical properties and mass, each mole is the same as the others. If you know 1 mole, in a physicochemical sense, you know them all. A mole of anything has exactly Avogadro's number of molecules, and, arguably, one can have a mole each of acorns, chocolate bits, and rice—different in kind, but equal in numbers. Moles are not evil burrowing animals, but if we "Know thy Beast!" we will be better prepared to deal with these chemically helpful critters.

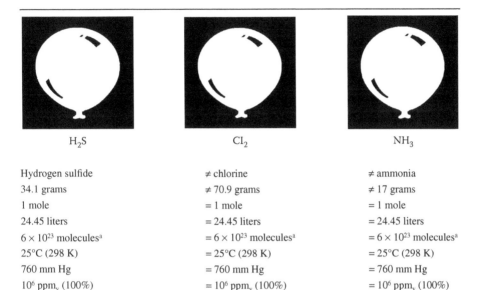

H_2S	Cl_2	NH_3
Hydrogen sulfide	≠ chlorine	≠ ammonia
34.1 grams	≠ 70.9 grams	≠ 17 grams
1 mole	= 1 mole	= 1 mole
24.45 liters	= 24.45 liters	= 24.45 liters
6×10^{23} molecules[a]	= 6×10^{23} molecules[a]	= 6×10^{23} molecules[a]
25°C (298 K)	= 25°C (298 K)	= 25°C (298 K)
760 mm Hg	= 760 mm Hg	= 760 mm Hg
10^6 ppm$_v$ (100%)	= 10^6 ppm$_v$ (100%)	= 10^6 ppm$_v$ (100%)

[a] Precisely, 1 mole is 6.022045×10^{23} molecules and is Avogadro's number—not his telephone number, which, if it were, and you dialed 100 numbers every second around the clock, it would take only 190,827,090,000,000 years to make the call. He probably would not be home, or his line is busy. Now, if he only had "Call Waiting!" After 3 billion years, only 0.0016% of numbers were dialed, the sun has become a "red giant," your telephone melts as the Earth burns to a charred cinder, and your line was disconnected because you did not pay your bill before you became burnt toast. Molecules, indeed, are incredibly tiny.

Indeed, 6.022045×10^{23} of anything is 1 mole of that thing. Imagine 1 mole of adult elephants!

Using 24.45 L of H_2S at 760 mm Hg and 25°C, as an example, if 34.1 grams = 10^6 ppm$_v$ (100%), then 34.1 micrograms = 1 ppm$_v$. Or, we could say:

$$1\,\text{ppm}_v\,H_2S = \frac{34.1\,\text{mcg}}{24.45\,\text{L}} = \frac{1.39\,\text{mcg}\,H_2S}{\text{liter}}$$

Expressed another way: 1 µL of H_2S gas diluted to 1 L with pure air at STP = 1 ppm$_v$ (1 part H_2S gas mixed with 999,999 parts air, both by volume).

Back to the balloons: if the temperature and/or the pressure are changed inside and/or outside the balloon, the volume of the balloon changes, but the mass of each gas inside remains the same—precisely 1 mole of each. However, the molecules may be closer together (less energy) or farther apart (more repelling, bouncing energy).

A gram–molecular volume of any gas or vapor occupies 22.4 liters at standard conditions of 0°C and 760 mm Hg pressure. This is the mass of 6.022045×1023 molecules of that gas or vapor. This assumes all gases and vapors behave as if they are "ideal"; for practical purposes at not too deviant temperature and gas pressure, this is the case. Note that the gram-molecular volume becomes 24.45 liters at 25°C and 760 mm Hg; that is, heat expands gas volume at constant pressure and constant mass. See Problem 495.

$$ppm_v = \frac{mg}{m^3} \times \frac{22.4}{mol.\,wt.} \times \frac{absolute\ temperature}{273.16\,K} \times \frac{760\,mm\,Hg}{pressure,\,mm\,Hg}$$

at 25°C and 760 mm Hg (1 gram-mole of a gas or vapor occupies 24.45 L at NTP):

$$ppm_v = \frac{(mcg/L) \times 24.45}{molecular\ weight}$$

$$mg/M^3 = \frac{ppm_v \times molecular\ weight}{24.45}$$

CONVERSION FACTORS AND CONSTANTS

There are several freeware programs on the Internet for conversions. The author finds www.onlineconversions.com very helpful, comprehensive, and robustly accurate.

$R = 0.0821$ L-atm/K-mole	ppm (volume/volume) $= x/10^6$	$1\% = 10,000$ ppm
mcg = µg = microgram	1 ft$^3 \cong 28.317$ L	µL/L = ppm$_v$
ppm$_v$ = micromoles of gas or vapor/mole of gas or vapor		1 mL = 0.001 L
$K = °C + 273.16$ K	$°C = 5/9(°F − 32)$	$°F = (1.8 \times °C) + 32$
1 kg = 2.2 lb	1 lb = 454 g 1" = 2.54 cm	1 m$^3 \cong 35.3$ ft^3
1 L $\cong 0.0353$ ft^3	1" = 25,400 microns	micron = 1 µm = 10^{-6} m
1 gallon = 3.785 L	1 fluid ounce $\cong 29.6$ mL	1 oz = 28.35 g
1 gallon H$_2$O = 8.345 lb (at 20°C)	1 gal/min = 0.134 ft^3/min	1 lb/hr = 10.9 kg/day
mg/L (liquids) = ppm	1 lb/day = 315 mg/min	1 m^3 = 10^6 cm^3 = 10^6 mL
volume of liquid × density (specific gravity) = mass of liquid		µ = micro-(1/10^6)
Area of circle = πr^2	volume of cylinder = πr^2 h	1 ton (2000 lb) = 907.185 kg
1 atmosphere = 760 mm Hg = 29.9" Hg = 14.7 lb/in^2		1 ton H$_2$O = 1.0183 m^3
Volume of sphere = 1.333 πr^3	1 m^3 = 1000 L	20 "drops" $\cong 1$ mL

Some might find the following standard metric descriptors helpful:

INTERNATIONAL SYSTEM OF UNITS (SI)

Quantity	Unit	Symbol
Length	Meter	m
Mass	Kilogram	kg
Time	Second	s
Electric current	Ampere	a
Thermodynamic temperature	Kelvin	K
Amount of substance	Mole	mol
Luminous intensity	Candela	cd

Questions also arise, now and then, about numerical prefixes. The following table puts these queries to rest:

Symbol	Prefix	Ratio	Symbol	Prefix	Ratio
T	Tetra-	10^{12}	c	Centi-	10^{-2}
G	Giga-	10^{9}	m	Milli-	10^{-3}
M	Mega-	10^{6}	μ	Micro-	10^{-6}
k	Kilo-	10^{3}	n	Nano-	10^{-9}
h	Hecto-	10^{2}	p	Pico-	10^{-12}
da[a]	Deca-	10	f	Femto-	10^{-15}
d[a]	Deci-	10^{-1}	a	Atto-	10^{-18}

[a] Avoid using these to prevent confusion. For example, 1 dL = 100 mL, but one decaliter = 10,000 mL—a 100-fold difference. I like milliliters. For reasons not clear to me, blood lead test results are traditionally reported as, for example, 9 micrograms/deciliter = 9 μg/dL.

IDEAL GAS (AND VAPOR) LAWS

Equations in boxes are worthy of committing to memory by those who practice industrial hygiene regularly.

> **General gas law: $PV = nRT$**

n = number of moles R = universal gas constant = 0.0821 L-atm/mole-K
i = initial gas condition f = final gas condition
P = gas pressure (atm) V = gas volume (L)

T = *absolute* temperature (in kelvin, K). Read: *ABSOLUTE* temperature! It is a common mistake to forget conversion of temperatures to their absolute, basal value—a state of zero energy.

$$\frac{P_i V_i}{T_i} = \frac{P_f V_f}{T_f}$$

or, rearranging

$$V_f = \frac{V_i \, T_f \, P_i}{T_i \, P_f}$$

Charles' law: The volume of a mass of gas is directly proportional to its temperature on the Kelvin scale when the pressure is held constant, or $V_i/T_i = V_f/T_f$.

Boyle's law: The volume of a mass of gas held at constant temperature is inversely proportional to the pressure under it is measured, or $P_i/V_i = P_f/V_f$.

Dalton's law of partial pressures: The total pressure of a mixture of gases is equal to the sum of partial pressures of the component gases and vapors.

Gay–Lussac's Law: At constant volume, the pressure of a mass of gas varies directly with the absolute temperature, or $P_i/T_i = P_f/T_f$.

Raoult's law: The partial vapor pressure of each component of a liquid solvent mixture equals its vapor pressure multiplied by its mole fraction in the mixture; that is, a solution's vapor pressure is proportional to the mole fraction of its component solvents.

$$V_o = \frac{298 \times V \times P}{760 \times T},$$

where V_o = standard air volume in liters, V = the indigenous volume of sampled air in liters, P = ambient pressure in mm of Hg. (Note: pressure is the actual barometric pressure adjusted to sea level), and T = the ambient temperature in kelvin, $K = 273 \text{ K} +$ ambient temperature, °C.)

To compare data on gases and vapors for direct-reading instruments with air quality standards, the meter reading in ppm must be converted into ppm at a normal temperature and pressure (NTP = 25°C and 760 mm Hg) by using the formula:

$$\text{ppm}_v \text{ (at NTP)} = \text{ppm}_{meter} \times \frac{P}{760 \, \text{mm Hg}} \times \frac{298 \, \text{K}}{T}$$

where
ppm_{meter} = meter reading in ppm_v
P = sampling site barometric pressure (in mm Hg)
T = sampling site air temperature (in kelvin, K)

Barometric pressure is obtained by checking a calibrated barometer or calling the local weather station, the airport, or the local National Oceanic and Atmospheric Administration office. Ask for the unadjusted barometric pressure. If these sources are unavailable, a good "rule-of-thumb" is for every 1000 feet of elevation, the barometric pressure decreases by approximately 1 inch of mercury (Hg).

DILUTION VENTILATION TO PURGE CONFINED SPACES AND ROOMS

Industrial hygienists are called on to predict concentration of an air contaminant remaining in a room, space, bin, or tank after the operation of an exhaust or air supply ventilation system. Dilution of air contaminants will follow first-order exponential decay kinetics with excellent mixing of dilution air with contaminated air. The key equation for these calculations is

$$C = kC_oe^{-[Q/V]t},$$

where

  C = concentration of contaminant remaining after operating the ventilation system for a specified time period, t, usually in minutes

  C_o = the original air contaminant concentration, usually in ppm_v or mg/m^3

  Q = ventilation rate [supply or exhaust, but not both (take larger of the two)– usually in cubic feet of air per minute, cfm]

  V = volume of the room or space, usually in cubic feet. If the room or confined space has contents, subtract their estimated volume from the space volume. If unknown, one may use 15% contents for average volume of rooms intended for normal occupancy. Storage bins, silos, tanks, and so forth usually have content volume level gauges.

  k, a ventilation mixing factor (with 1 = perfect, and 10 = extremely poor), is a subjective rating based on experience and judgment and/or applied as a safety factor.

VENTILATION PURGING TIME

This equation is used to calculate the time required to reduce concentration of an air contaminant with ventilation in a space of known volume. Depending on air mixing characteristics, application of safety factors requiring professional judgment is required.

$$t = \frac{-\ln\left[C/C_o\right]}{Q/V},$$

where

  C = concentration after time t (min),

  C_o = original concentration,

  Q = cfm, and

  V = volume (ft^3) of the space (room, tank, railroad boxcar, sealed silo, industrial plant, etc.).

This equation assumes perfect mixing of the dilution air with the contaminated air ($k = 1.0$). Otherwise, $k = >1$ to 10. In other words, if mixing is less than perfect (k, say, = 5), and 30 minutes are required for dilution if mixing is perfect, 150 minutes are required in this example.

Know the 10% rule and the 50% dilution ventilation rules:

2.3, 4.6, and 6.9 chamber, room, or tank volumes of clean air are needed, respectively, to dilute an air contaminant to 10%, 1%, and 0.1% of its initial concentration. This assumes perfect mixing of clean air with contaminated air and no further generation of air contaminant as the ventilation proceeds.

Under the same conditions, after a volume of air equal to 50% of room or a chamber volume has mixed with contaminated air, original concentration is reduced by nearly 40% (to 60.7% of the original concentration).

EXAMPLE

If $C_o = 1000$ ppm, $Q = 230$ cfm, and $V = 1000$ ft³, how long will it take to dilute vapor or gas concentration to 100 ppm$_v$? Assume perfect mixing of contaminated air in dilution air.

$$t = \frac{-\ln\left[100\,\text{ppm}_v/1000\,\text{ppm}_v\right]}{230\,\text{ft}^3/\text{minutes}/1000\,\text{ft}^3} = \frac{-\ln 0.1}{0.23} = \frac{-(-2.3)}{0.23} = 10\,\text{minutes}$$

$$\text{Time to dilute to 1 ppm}_v = t = \frac{-\ln\left[1\,\text{ppm}_v/1000\,\text{ppm}_v\right]}{230\,\text{cfm}/1000\,\text{ft}^3} = \frac{-(-6.9)}{0.23} = 30\,\text{minutes}$$

EXAMPLE

$C_o = 1000$ ppm$_v$
$V = 1000$ ft³ (e.g., 10' × 10' × 10')
$Q = 230$ cfm
$t = 10, 20, 30,$ and 40 minutes (or 2.3, 4.6, 6.9, 9.2 room volumes of clean dilution air)

After 10 minutes:

$$C = C_o\,e^{-[Q/V]t} = (1000\,\text{ppm}_v)\,e^{-\left[230\,\text{cfm}/1000\,\text{ft}^3\right]10\,\text{minutes}}$$
$$= (1000\,\text{ppm}_v)\,e^{-2.3} = 100\,\text{ppm}_v$$

After 20 minutes:

$$C = (1000\,\text{ppm}_v)\,e^{-\left[230\,\text{cfm}/1000\,\text{ft}^3\right]20\,\text{minutes}} = (1000\,\text{ppm}_v)\,e^{-4.6} = 10\,\text{ppm}_v$$

After 30 minutes: $C = 1$ ppm$_v$

After 40 minutes: $C = 0.1$ ppm$_v$

Note the concentrations are reduced to 10% of the original concentrations for every *2.3* chamber dilution volume.

$$C = C_{\rm o}\, e^{-q/V},$$

where
  C = resultant concentration, $C_{\rm o}$ = original concentration, q = withdrawn sample volume, and V = chamber or room volume.

EXAMPLE

Using 50% ventilation volume

$$C_{\rm o} = 1000\ {\rm ppm_v}$$

$$V = 1000\ {\rm ft}^3$$

$$q = 500\ {\rm ft}^3$$

$$C = (1000\,{\rm ppm_v})\, e^{-500\,{\rm ft}^3/1000\,{\rm ft}^3} = (1000\,{\rm ppm_v})$$

$$e^{-0.5} = (1000\,{\rm ppm_v})\,(0.6065) = 607\,{\rm ppm_v}$$

Using $\log_{10}$:

$$\frac{q}{V} = 2.3 \times \log\!\left[\frac{C_{\rm o}}{C}\right]$$

Use

$$\frac{C}{C_{\rm o}} = e^{-[Q/V]t}$$

to determine fraction remaining after operating the ventilation system for a specified time.

AIR CONTAMINANT HALF-LIFE

An air contaminant half-life, $C_{1/2}$ (the time required to dilute an air contaminant to 50% of its original concentration), is calculated by

$$C_{1/2} = 0.693\left[\frac{V}{Q}\right],$$

where
  V = volume of room, space, or container and
  Q = the uniform dilution ventilation rate.

For example, a 1000 ppm$_v$ gas or vapor concentration is reduced to 500 ppm$_v$ after about 14 minutes by 200 cfm in a room 20' × 20' × 10' (4000 ft³), to 250 ppm$_v$ after 28 minutes, and to 125 ppm$_v$ after $0.693 \times (4000\,\text{ft}^3/200\,\text{ft}^3/\text{minutes}) = 13.86$ minutes.

This applies only when there is no further generation of contaminant, air contaminant does not chemically decay, and there is excellent mixing of uncontaminated air with the polluted air.

DENSITY (SPECIFIC GRAVITY) CALCULATIONS

The term density denotes the ratio of a mass of a substance to that of an equal volume of a reference substance (usually water at a given temperature). Density is essentially identical to specific gravity of a substance, the terms being normally interchangeable. For gases and vapors, air is often used as the reference.

Water at 4°C is at its maximum density = 1.000 g/cm³ = 1.000 g/mL = 62.4 lb/ft³ = 1 kg/L (=0.997 g/cm³ at 25°C). Water is highly anomalous among liquids because it is denser at 4°C than at its freezing point of 0°C. Otherwise, ice sinks to the bottom of the lake! And life on Earth would be radically different than we know. Ice does not float because of entrained air bubbles—a common myth.

Density calculations are used to determine the mass of a volume of liquid. For example, the industrial hygienist might need to calculate the mass of a contaminant that evaporates from the spill of a specified volume of volatile liquid. To do this:

> **liquid volume × liquid's density = mass of liquid**

For example, 50 milliliters of toluene totally evaporated after being spilled on the floor:

$$50\,\text{mL of liquid phase toluene} \times \frac{0.87\,\text{g}}{\text{mL}} = 43.5\,\text{grams of vapor phase toluene}$$

Hydrocarbons that are not substituted with halogens (e.g., chlorine, bromine, fluorine) generally have densities less than water (i.e., <1 g/mL). That is, they float on water (e.g., oil, organic paint thinner, gasoline) if they do not appreciably mix or dissolve in the water. Example densities (in g/mL, at normal room temperature) are

Kerosene	0.82	Benzene	0.88	Ethanol	0.79
Acetone	0.79	Methanol	0.79	Toluene	0.87
n-Butanol	0.81	CH_2Cl_2	1.34	CCl_4	1.59

Acetone, methanol, ethanol, and butanol mix with water. Kerosene, toluene, and benzene do not and, because their densities are less than water, they float on water. Carbon tetrachloride and methylene chloride, with densities greater than water, sink below water because they do not appreciably dissolve in water. Adding halogen atoms to organic carbon molecules increases their mass and density.

VAPOR PRESSURE CALCULATIONS AND PROBLEMS

When the vapor pressure of a liquid or volatile solid is known, one can calculate the maximum, or saturation, vapor concentration that can exist in an enclosed space. Such calculations are very helpful when the industrial hygienist wishes to determine worst possible scenarios, and air-sampling instruments are not readily available. In the case of a closed vessel containing a volatile material, such as a chemical storage tank, the molecules will escape (i.e., evaporate, volatilize) into air or gas space above the liquid until eventually the enclosed atmosphere above the liquid can no longer hold any more vapor molecules at that pressure and temperature. The air (or other gas) is saturated at this point with molecules of the evaporating liquid (or certain solids sublime).

Be mindful that many solids have vapor pressures. Iodine crystals, naphthalene flakes, *p*-dichlorobenzene, organophosphate pesticides, DDT, and phthalates are some examples of solids that, sooner or later, evaporate. It should be obvious that liquids, in general, have higher vapor pressures than solids and that there is a tremendous range of vapor pressures between different chemicals. Again, many solids exhibit substantial vapor pressures at room temperature increasing, of course, if the temperature of the solid increases. Most do not regard motor oil as volatile. Spill a few drops on warm summer driveway and you will find none visible after a few days or less.

One should not be misled that a chemical with a lower vapor pressure, all other parameters being equal, is not necessarily less hazardous than another with a higher vapor pressure. A chemical with a low vapor pressure, when spread over a large area, can present a greater inhalation exposure hazard than a highly volatile material with a small surface area (see Problem 375).

To calculate maximum vapor concentration of a volatile material (at usually an assumed temperature of 25°C and 760 mm Hg) in an enclosed, unventilated space:

$$\frac{\text{vapor pressure at a temperature}}{\text{barometric pressure}} \times 10^6 = \text{ppm}_v \text{ of vapor}$$

Important: See Problem 677 for proper calculation of saturation concentrations of highly volatile chemicals or, in general, for those with vapor pressures exceeding 20 mm Hg at 20°C.

For example, the vapor pressure of 2-nitropropane is 13 mm Hg at 20°C. What is the saturation concentration of vapor at this temperature in a tank containing liquid 2-NP where the barometric pressure is 710 mm Hg? Assume sufficient liquid 2-nitropropane is in the tank to achieve saturation.

$$\frac{13 \text{ mm Hg}}{710 \text{ mm Hg}} \times 10^6 = 18{,}310 \text{ ppm}_v \ 2 - \text{NP}$$

Correct value = 17,981 ppm$_v$ per Problem 677.

This concentration of 2-nitropropane vapor (1.8%) is considerably greater than the ACGIH TLV® ceiling concentration of 10 ppm$_v$ (A3 carcinogen), but is slightly below LEL of 2.6%. Inhalation of this vapor concentration is a significant health hazard, and, since concentration is greater than 10–20% of the LEL, this also presents a potential fire and explosion hazard. The NIOSH IDLH for 2-NP is 100 ppm$_v$.

Some trivia:

1 ppm = 1 inch in 16 miles

1 ppb = 1 inch in 16,000 miles

1 ppb = 4 in^2 in a square mile

1 ppt = 1 second in 31,688 years

These help to explain units to lay people.

Dry air in a standard sewing thimble at sea level weighs about 3600 micrograms. A standard office staple 35,000 micrograms. Contrast this with the OSHA PEL of 50 mcg/m^3 for lead dust, fume, and mist aerosols (see Problem 245).

STANDARD REFERENCE MAN

Calculations in industrial hygiene and toxicology involve parameters of a "standard" or a reference man. Such an individual unlikely exists, but comparison to the standard reference can be helpful. For example, how many cubic meters of air are inhaled by a worker during an 8-hour workday? Obviously, to answer we must know a person's body mass, metabolic rate and time-weighted work effort, respiration rate, and so on. Knowing mg/m^3 or ppm$_v$ in breathing zone air, how many milligrams of the toxicant are inhaled during the air-sampling period can be calculated. The following parameters are helpful[*]:

Weight	70 kg
Height	175 cm
Lung weight	1000 g (900 g are blood and air)
Total lung capacity	5.6 L
Vital capacity	4.3 L
Functional residual capacity	2.2 L
Breathing rate	15/minutes
Tidal volume	1.45 L
Respiratory flow rate	43.5 L/minutes
Inspiratory period	2 seconds
Expiratory period	2 seconds

[*] *Reference Man*, International Commission on Radiological Protection, ICRP Publication 23, Pergamon Press, 1975.

ABIH CERTIFICATION EXAMINATION HOMEWORK

AIR CALCULATIONS AND VENTILATION

The following 14 problems are the most basic and fundamental to an understanding and comprehension of principles of industrial hygiene air sampling and ventilation. To help prepare for the American Board of Industrial Hygiene certification examinations in the comprehensive practice of industrial hygiene, solve the following problems. If one can understand and can solve these problems, he or she is in an excellent position to handle many related problems presented in the ABIH examinations.

The rules are

1. Do not look at the following problems until you are ready to solve them. Work alone. Do not solicit help.
2. Do not use books, notes, or other reference sources.
3. Use only pencils, a nonprogrammable calculator, a wrist watch, and note pad.
4. Solve as many problems as you can in one hour. (This is an average of 4.3 minutes per problem.)
5. Compare answers to correct answers in this book as given by the problem numbers in parentheses.

Note the many abbreviations given in the problems. The examiners assume one who is experienced in the professional practice of industrial hygiene knows and is familiar with these. Solutions to these are given in the parenthetical problem.

> **Good Luck!**

1. BZ air was sampled for total barley dust at 1.8 L/m for 5 hours, 40 minutes with a 37 mm MCE MF with respective pre-sampling and post-sampling weights of 33.19 and 38.94 mg. What was grain silo filler's 8-hour TWAE exposure to respirable dust if 85 mass-percent was nonrespirable? (304)
2. Air was sampled for HCl gas (mw = 36.45) in 15 mL of impinger solution at 0.84 L/m for 17 minutes, 20 seconds. HCl collection efficiency was 80%. A chemist analyzed 4.7 mcg Cl/mL in the sample and 0.3 mcg/mL in control blank impinger. What was the steel pickler's exposure in ppm_v? (305)
3. Determine an 8-hour TWAE of a scrap metal processor to Pb dust and fume with exposures of 3 hours, 15 minutes to 17 mcg Pb/m^3; 97 min to 565 mcg Pb/m^3; and 2 hours, 10 minutes to 46 mcg Pb/m^3. The worker wore an approved HEPA dust/fume/mist filter cartridge respirator for 5-½ hours. (306)
4. 172 grams of liquid phosgene splash onto a floor. What gas volume quickly results after evaporation at an air temperature of 23°C and an atmospheric pressure of 742 mm Hg? $COCl_2$ molecular weight = 99. Boiling point of phosgene = 47°F. (140)

5. A chemical plant operator had the following 8-hour TWAEs on Monday: 32 ppm$_v$ toluene, 19 ppm$_v$ xylene, and 148 ppm$_v$ MEK. Their respective TLVs are 50, 100, and 200 ppm$_v$. By what percent is the additive exposure limit exceeded? (308)

6. Air in an empty room (20' × 38' × 12') contains 600 ppm$_v$ cyclohexene vapor. How long will it take to dilute this to 6 ppm$_v$ with a 1550 cfm vane-axial exhaust fan? K factor = 3. (309)

7. 7.3μ L liquid styrene (MW = 104.2, density = 0.91 g/mL) are evaporated in a 21.6 L glass calibration bottle. What is the styrene vapor concentration in ppm$_v$? (310)

8. A rotameter was calibrated at 25°C and 760 mm Hg. What is the corrected air flow rate when rotameter indicates 2.0 L/m at 630 mm Hg and 33°C? (311)

9. What is the effective specific gravity of 13,000 ppm$_v$ of a gas in air when the gas has a specific gravity of 4.6? Will the mixture stratify with the denser gas at floor level? (312)

10. Analysis of an 866 L MF air sample detected 2667 mcg Zn. How much zinc oxide (ZnO) fume does this represent in the welder's breathing zone? (MW Zn and O = 65 and 16, respectively.) (313)

11. What is the air flow rate through an 8" diameter unflanged duct with a transport velocity of 2900 fpm? What capture velocity is expected 8" in front of the duct inlet? Without cross-drafts, what discharge velocity can be expected 20 ft from the exhaust outlet? What if the exhaust duct inlet has a wide flange? (314)

12. A 47' × 166' × 20' building is supplied with 7300 cfm. How many air changes occur per hour? How many minutes are required per air change? How many cubic feet are supplied per square foot of floor area (20' is the height of the ceiling)? Forty-seven people work in this single-story building. What is the outdoor air ventilation rate per person if 90% of the air is recirculated? (271)

13. An exhaust system operates at 19,400 cfm. A hood added to the system requires total system capacity of 23,700 cfm. By how much should fan speed be increased to handle the extra exhaust volume? (276)

14. In the preceding problem, what is the required increase in fan horsepower to handle the increased air volume? (277)

The great tragedy of Science—the slaying of a beautiful hypothesis by an ugly fact.

Thomas Huxley

1 725 Problems with Solutions (Industrial Hygiene, Ventilation, Toxicology, Chemical Risk Management, Control Methods)

1. Air was sampled in a carpenter's breathing zone with a 37 mm PVC (polyvinyl chloride) membrane filter for 7 hours and 37 minutes. The initial air flow rate was checked twice with a 1000 mL soap film calibrator at 49.7 and 50.1 seconds per liter. The post-sampling conditions were 53.7 and 53.3 seconds per liter. What volume of air was sampled in liters and cubic feet (at standard conditions of 25°C and 760 mm Hg)? Disregard water vapor content in the air.

7 hours and 37 minutes = 420 minutes + 37 minutes = 457 minutes

$$\frac{(49.7 + 50.1 + 53.7 + 53.3)\,\text{seconds}}{4} = 51.7\,\text{seconds}$$

$$\text{liters/minute} = \frac{60\,\text{seconds/liter}}{51.7\,\text{seconds/liter}} = 1.16\,\text{L/minute}$$

1.16 L/minute × 457 minutes = 530.12 liters

$$530.12\,\text{L} \times \frac{1\,\text{ft}^3}{28.317\,\text{L}} = 18.72\,\text{ft}^3$$

Answers: 530.12 L and 18.72 ft³.

The decrease in air flow rate over the air-sampling period was most likely due to the accumulation of wood dust and other particles on the filter surface. With many portable battery-powered air-sampling pumps, initial flow rate is artificially high by as much as 5–7%. This is overcome by allowing pumps to operate for at least 5 minutes before the initial and final calibrations.

2. A solvent degreaser operator's exposures to trichloroethylene (TCE) vapor on one work day were $2-\frac{1}{4}$ hours at 57 ppm, $3-\frac{1}{2}$ hours at 12 ppm, $1-\frac{3}{4}$ hours at 126 ppm, and 30 minutes at 261 ppm. What was her TWAE (time-weighted average exposure)? Did it exceed the Threshold Limit Value (TLV®)? The action level?

Haber's law is concentration $(C) \times$ time $(T) = CT$ and determines dose and effects.

8-hour TWAE worker dose calculation:

C		T		CT
57 ppm	$\times$	2.25 hours	=	128 ppm-hours
12 ppm	$\times$	3.5 hours	=	42 ppm-hours
126 ppm	$\times$	1.75 hours	=	221 ppm-hours
261 ppm	$\times$	0.5 hour	=	131 ppm-hours
		8.0 hour	=	522 ppm-hours

$$\frac{522\,\text{ppm} - \text{hours}}{8.0\,\text{hours}} = 65\,\text{ppm TWAE to TCE vapor}$$

The TLV is 10 ppm$_v$. TCE has a short-term exposure limit (STEL) of 25 ppm$_v$.

Answers: TWAE is 65 ppm$_v$. Yes, by 55 ppm$_v$. Yes, $13 \times$ the AL of 5 ppm$_v$. TCE is absorbed through healthy, intact skin and more so through breaches in one's skin. Careful attention to work practices is required to ensure that there is no skin contact by wearing appropriate gloves, aprons, and other gear coupled with prompt washing of skin and eyes when contact occurs, permits the industrial hygienist to focus on designing—or improving—local exhaust ventilation. Target organ adverse effects of TCE absorption are central nervous system impairment, cognitive impairments, and renal toxicity. A reversible cosmetic disorder known as "degreaser's flush" has been reported in TCE-exposed workers who concurrently or later consume ethyl alcohol. TCE is a suspected A2 human carcinogen. Therefore, all reasonable efforts should be taken to ensure that the worker's exposure to TCE is kept as low as possible.

3. In the previous problem, assume that the worker wore an organic vapor cartridge respirator with an overall efficiency of 90% (filtration + face-to-mask seal efficiency). What was the face piece penetration? What was the respirator protection factor? What was her true TWAE assuming equal protection at all vapor concentrations?

$$\text{respirator protection factor} = \frac{\text{ambient concentration}}{\text{concentration inside facepiece}}$$

65 ppm$_v \times 0.9 = 58.5$ ppm$_v$ (i.e., 90% of 65 ppm$_v$)

65 ppm$_v$ − 58.5 ppm$_v$ = 6.5 ppm$_v$ penetrated the respirator (10% penetration)

$$\frac{65\,\text{ppm}_v}{6.5\,\text{ppm}_v} = PF = 10$$

Answers: 10%. Protection factor = 10. TWAE $\cong$ 7 ppm$_v$.

4. What volume will 73 grams of dry ammonia gas occupy at 11°C and 720 mm Hg? How much air is needed to dilute the ammonia gas to 10 ppm$_v$?

$$PV = nRT$$

$$V = \frac{nRT}{P}$$

$$R = 0.0821\,\text{L-atm/mole-K}$$

Molecular weight NH_3 = 17 grams/gram-mole

$$n = \frac{73\,\text{g}}{17\,\text{g/mole}} = 4.29\,\text{moles}$$

$$T = 273\,\text{K} + 11°\text{C} = 284\,\text{K}$$

$$P = 720\,\text{mm Hg}/760\,\text{mm Hg} = 0.947\,\text{atmosphere}$$

$$V = \frac{(4.29\,\text{moles}\,NH_3)(0.0821)(284\,\text{K})}{0.947\,\text{atmosphere}} = 105.5\text{L}$$

$$10\,\text{ppm} = \frac{10}{10^6} = \frac{1}{10^5}$$

Answers: 105.5 liters. 105.5×10^5 liters of air dilutes 100% to 0.001%.

5. What is the cumulative error of several measurements if the day of air sampling was ±50% of the true daily exposure, analytical accuracy was ±10% of the true value, air sample timing was ±1% of the true value, and the air flow rate error was ±5% of the true value?

$$\text{cumulative error},\ E_c = \sqrt{(E_1)^2 + (E_2)^2 + \cdots (E_n)^2}$$

$$E_c = \pm\sqrt{50^2 + 10^2 + 1^2 + 5^2} = \pm\sqrt{2626} = \pm51.2\%$$

Answer: ±51.2% of the true value. This is referred to as the sampling analytical error (SAE) and is used to calculate the lower confidence limit (LCL) and upper confidence limit (UCL) for the air-sample test results (see Problem 430).

6. A paint sprayer had TWAEs to methyl ethyl ketone (MEK) of 68 ppm$_v$, toluene of 37 ppm$_v$, n-butyl alcohol of six ppm$_v$, and xylene of 23 ppm$_v$, all as vapor phase air contaminants and not as mist particles. Assume effects are toxicologically additive and that he did not wear a respirator. What was his equivalent exposure? American conference of governmental industrial hygienists (ACGIH) TLVs® are 200 ppm$_v$, 20 ppm$_v$, 50 ppm$_v$ (C), and 100 ppm$_v$, respectively.

$$\text{in ppm}_v : \frac{\text{exposure}}{\text{respective PEL or TLV}} + \cdots \frac{E_n}{PEL_n}, \quad \text{in TWAEs}$$

$$\frac{68}{200} + \frac{37}{20} + \frac{6}{50} + \frac{23}{100} = 2.54 \text{ (no units)}$$

Answer: 254% above the TLV for the mixture and 2.1 times the action level of 0.5 (unitless).

7. A polyurethane foam machine operator had TWAEs to vapors of toluene diisocyanate (TDI) of 0.003 ppm$_v$ and CH$_2$Cl$_2$ of 36 ppm$_v$. What was her equivalent exposure to this air contaminant mixture?

Answer: Generally, the additive mixture rule would not apply. Although both are irritants to respiratory tract and mucous membranes, TDI is a sensitizer, and CH$_2$Cl$_2$ primarily affects the central nervous system (CNS) and blood HgB (COHgb formation). CH$_2$Cl$_2$ is a potential liver and lung carcinogen. TDI is a recognized carcinogen in laboratory animals. Given extreme inhalation hazards both chemicals present, the author recommends controls to as low as reasonably achievable (ALARA).

8. What is specific gravity of a mixture of 10,000 ppm$_v$ Cl$_2$ gas in dry air? The specific gravity of Cl$_2$ gas is 2.5 (unitless, air = 1.000).

$$10,000 \text{ ppm}_v \text{ (volume/volume)} = 1\%$$

for air: $0.99 \times 1.0 = 0.990$

for Cl$_2$: $0.01 \times 2.5 = 0.025$

$$\overline{1.00} \qquad \overline{1.015}$$

Answer: The specific gravity (or relative density) = 1.015, or only 1.5% greater than air. In a practical sense with respect to designing work place ventilation, there is no great difference between air and a very high concentration of vapors or gases at, say, 1–5% in air (e.g., 2.5 times heavier than air). An exception, for example, are the so-called "paint kitchens" where large volumes of flammable and combustible organic solvents are processed. A spill of a flammable solvent releases a cloud of vapors slowly at first before mixing with air higher in the room. Supply and exhaust ventilation placed close to the floor sweeps and captures vapors before they reach the lower explosive

of flammable level. Regular ventilation and mechanical local exhaust ventilation are also installed to protect workers in such environments.

9. An industrial hygiene chemist analyzed 0.256 mg Zn on an air filter used for
a 96 L air BZ sample. What was the welder's exposure to zinc oxide fume
during the air-sampling period?

$$\text{Molecular weights: Zn} = 65, O = 16, ZnO = 81$$

$$\frac{65}{81} \times 100 = 80\% \text{ Zn in ZnO}$$

$$\frac{0.256\,mg\,Zn}{0.80} = 0.32\,mg\,ZnO = 320\,mcg\,ZnO\,fume$$

$$\frac{mcg}{L} = \frac{mg}{m^3}$$

$$\frac{320\,mcg\,ZnO}{96\,L} = 3.3\,mg\,ZnO/m^3$$

10. A one quart bottle of *n*-butanol broke upon falling to the floor entirely
evaporating in a $10' \times 40' \times 80'$ room with 10% room contents. What average vapor concentration exists assuming there is no ventilation? *n*-butanol
density = 0.81 g/mL.

$$(10' \times 40' \times 80') - 10\% = 32{,}000\,ft^3 - 3200\,ft^3 = 28{,}800\,ft^3$$

$$\frac{28{,}800\,ft^3}{35.3\,ft^3/m^3} = 816\,m^3 = \text{net volume of room}$$

one quart $= \frac{1}{4}$ (3785 mL/gal) = 946 mL × 0.81 g/mL = 766 g of liquid
n-butanol eventually evaporates into and mixes throughout the room air.
Remember that molecules like to move around following the second law of
thermodynamics and entropic disorder.

$$CH_3(CH_2)_2\,CH_2OH = C_4H_{10}O = (4 \times 12) + (10 \times 1) + 16 = \text{molecular}$$
weight = 74 grams/gram-mole

$$ppm = \frac{(766{,}000\,mg/816\,m^3) \times 24.45}{74} = 310\,ppm\,n\text{–butanol vapor}$$

"Phew, what a stench! Where's my mask?"

11. What can we conclude from the previous problem?
   a. Fire hazard, but no health hazard
   b. Health hazard, but no fire hazard
   c. Fire hazard and a health hazard

    d. No fire or health hazard
    e. Combustion risk, but not a flammability hazard
    f. None of the above

Answer: b. *n*-Butanol TLV and PEL (permissible exposure limit) = 50 ppm_v (C) and skin notation. LEL—UEL = 1.4% to 11.2%. 310 ppm_v = 0.031%. Therefore, below the LEL but greater than the PEL. 620% of PEL and 2.2% of LEL.

12. Salt mine air was sampled through a PVC filter that initially weighed 73.67 and 88.43 mg after sampling. The initial air flow rate was 2.18 L/minute. The final flow rate was 1.98 L/minute after 7 hours and 37 minutes. The analyzed filter had 5.19 mg sodium after blank correction. What were concentrations of salt dust and total dust in the air sample?

$$7 \text{ hours and } 37 \text{ minutes} = 420 \text{ minutes} + 37 \text{ minutes} = 457 \text{ minutes}$$

$$88.43 \text{ mg} - 73.67 \text{ mg} = 14.76 \text{ mg}$$

$$\text{L/min} = \frac{(2.18 + 1.98) \text{ L/min}}{2} = 2.08 \text{ L/min average}$$

$$457 \text{ minutes} \times 2.08 \text{ L/min} = 951 \text{ liters of air sampled}$$

$$\frac{14.76 \text{ mg}}{951 \text{ L}} = \frac{14.76 \text{ mg}}{0.951 \text{ m}^3} = 15.5 \text{ mg total dust/m}^3$$

$$\text{Molecular weights: Na} = 23; \text{Cl} = 35.5; \text{NaCl} = 58.5$$

$$58.5/23 = 2.54$$

$$5.19 \text{ mg Na} \times 2.54 = 13.2 \text{ mg NaCl}$$

$$\frac{13.2 \text{ mg NaCl}}{0.951 \text{ m}^3} = 13.9 \text{ mg NaCl/m}^3$$

difference = 15.5 mg/m³ – 13.9 mg/m³ = 1.6 mg/m³. Could this be diesel engine exhaust smoke from the underground mining equipment? Asbestos fibers from brake shoes? Tobacco smoke? Explosion dust?

Answers: 13.9 mg NaCl/m³. 15.5 mg of total dust, smoke, and fume/m³.

13. A closed 100,000 gallon storage tank in Houston contains 10,000 gallons of toluene. What is the equilibrium saturation vapor concentration in the tank at 20°C? The vapor pressure of toluene at 20°C is 22 mm Hg.
    Since Houston is close to sea level, assume the barometric pressure = 760 mm Hg.

$$\frac{22\,\text{mm Hg}}{760\,\text{mm Hg}} \times 10^6 = 28{,}947\,\text{ppm} = 2.89\%\,(\text{volume/volume})$$

Answer: $\cong 29{,}000\,\text{ppm}_v$ (2.9%). Remember $1\% = 10{,}000\,\text{ppm}_v$. That is, $100\% = $ one million parts per million. Saturation is ensured as long as liquid (or volatile solids) remain in the tank. That is, if only a few milliliters of toluene evaporated inside this tank, the saturation concentration will never be achieved. Moreover, if the tank walls are cooler than the air–vapor mixture, condensation will occur, and saturation might not occur because condensed toluene vapor drains back into its liquid phase.

14. In the preceding problem, is the tank atmosphere explosive? The LEL (lower explosive level) and UEL (upper explosive level) for toluene = 1.2% and 7.1%, respectively.

    Answer: Yes! BEWARE! 2.9% exceeds LEL, but is less than UEL. This is most hazardous because vapors are in the stoichiometric mid-range of explosion. The tank "carries its own match." Control all ignition sources. Consider use of inert gases [nitrogen, argon, carbon dioxide, helium, (steam)] to reduce O_2 concentration (e.g., to <6% O_2). Fully ventilate before entry, confined space entry practices, train workers, air testing, post and label, and so on.

15. In Problem 13, how many pounds of toluene are in the vapor phase?

$$22.4\,\text{L/gram-mole} \times \frac{273\,\text{K} + 20°\text{C}}{273\,\text{K}} = 24.04\,\text{L/gram-mole}$$

Toluene $= C_7H_8$

$$(7 \times 12) + (8 \times 1) = \text{molecular weight} = 92\,\text{grams/gram-mole}$$

$$\text{mg/m}^3 = \frac{\text{ppm} \times \text{mol. wt.}}{24.04} = \frac{28{,}947 \times 92}{24.04} = 110{,}779\,\text{mg/m}^3 = 110.8\,\text{grams/m}^3$$

$$100{,}000\,\text{gallon tank} - 10{,}000\,\text{gallons liquid } C_7H_8$$
$$= 90{,}000\,\text{gallons air-vapor space}$$

$$90{,}000\,\text{gallons} \times 3.785\,\text{L/gallon} = 340{,}650\,\text{L} = 340.65\,\text{m}^3$$

$$340.65\,\text{m}^3 \times 110.8\,\text{g/m}^3 = 37{,}744\,\text{grams of vapor phase toluene}$$

$$\frac{37{,}744\,\text{grams}}{454\,\text{grams/lb}} = 83.1\,\text{lb}$$

Answer: There are 83.1 pounds of vapor phase toluene in this tank.

16. In Problem 13, what is the toluene vapor concentration after 45 minutes operation of a 2000 cfm dilution blower? Assume good mixing of fresh dilution air or inert gas with the toluene-contaminated air, and there is negligible toluene evaporation as the dilution ventilation proceeds.

$$C = C_0 e^{-[Q/V]t} = \text{resultant air contaminant concentration}$$

$$90,000 \text{ gallons} \times 3.785 \text{ L/gal} = 340,650 \text{ L} = 12,037 \text{ ft}^3$$

$$C = (28,947 \text{ ppm}_v) \times e^{-[2000 \text{ cfm}/12,037 \text{ ft}^3] \times 45 \text{ minutes}}$$
$$= (28,947 \text{ ppm}_v) e^{-7.48} = 28,947 \text{ ppm}_v \times 0.00056 = 16.3 \text{ ppm}_v$$

Answer: 16.3 ppm$_v$ toluene vapor assuming static conditions and the negligible evaporation of toluene during dilution ventilation. The liquid toluene present would evaporate as ventilation commenced. Special dilution ventilation calculations can be used if the vaporization rate is known (see ACGIH's *Industrial Ventilation*). The use of air to initially ventilate could be very hazardous, especially if there is a fan that generates spark or heat of air friction. The better part of valor might be to reduce the oxygen concentration in the head space to less than 6% by volume with a nitrogen gas purge and then fully ventilate with air to reduce toluene vapor concentration to <TLV/PEL action level. Recent rodent teratology evidence reveals toluene is a reproductive health hazard. Accordingly, the TLV was reduced from 50 to 20 ppm$_v$.

17. Assume that the toluene solvent storage tank in Problem 13 is located in the Rocky Mountains where barometric pressure is 640 mm Hg. What is the saturation vapor concentration inside this tank assuming there is sufficient liquid toluene to saturate tank air?

$$\frac{22 \text{ mm Hg}}{640 \text{ mm Hg} + 22 \text{ mm Hg}} \times 10^6 = 33,233 \text{ ppm}_v = 3.32\% \text{ (volume/volume)}$$

Answer: 3.32%$_v$ toluene vapor exceeds the LEL, but this is below the UEL. However, the LEL–UEL limits will change with changes in the partial pressure of O_2, that is, altitude effects can alter the LEL–UEL range. Changes in temperature also affect the LEL and UEL range. Increased temperature and/or oxygen lower the LEL as well as increasing the UEL.

18. A solvent is 2% (v/v) benzene and 98% (v/v) toluene. What is the percent vapor phase concentration for each component? The vapor pressures of benzene and toluene are 75 and 22 mm Hg, respectively. The densities of benzene and toluene are 0.88 and 0.867 g/mL, respectively.

Use 100 mL of solvent mixture as the volume basis for Raoult's law calculations:

$$2\,\text{mL} \times 0.88\,\text{g/mL} = 1.76\,\text{g of benzene/100}\,\text{mL}$$

$$\text{Molecular weight of benzene} = 78.1$$

$$98\,\text{mL} \times 0.867\,\text{g/mL} = 85\,\text{g of toluene/100}\,\text{mL}$$

$$\text{Molecular weight toluene} = 92.1$$

$$\text{Partial vapor pressure from benzene} = \frac{1.76\,\text{g/(78.1}\,\text{g/mole)} \times 75\,\text{mm Hg}}{1.76\,\text{g/(78.1}\,\text{g/mole)} + 85\,\text{g/(92.1}\,\text{g/mole)}}$$
$$= 1.79\,\text{mm Hg}$$

$$\text{Partial vapor pressure from toluene} = \frac{85\,\text{g/(92.1}\,\text{g/mole)} \times 22\,\text{mm Hg}}{1.76\,\text{g/(78.1}\,\text{g/mole)} + 85\,\text{g/(92.1}\,\text{g/mole)}}$$
$$= 21.47\,\text{mm Hg}$$

Total saturation vapor pressure $= 1.79\,\text{mm Hg} + 21.47\,\text{mm Hg} = 23.26\,\text{mm Hg}$

$$\frac{1.79\,\text{mm Hg}}{23.26\,\text{mm Hg}} \times 100 = 7.7\%\,\text{benzene vapor}$$

$$\frac{21.47\,\text{mm Hg}}{23.26\,\text{mm Hg}} \times 100 = 92.3\%\,\text{toluene vapor}$$

Answers: Calculations made using Raoult's law demonstrate enrichment in the vapor phase by the more volatile component, that is, note how benzene enriched from 2% in the liquid phase to 7.7% (3.9 times) in the vapor phase. The following quotation from Harris and Arp (*Patty's Industrial Hygiene and Toxicology*, third edition) is noteworthy:

> Raoult's law should be used with caution in estimating emissions from partial evaporation of mixtures; not all mixtures behave as perfect solutions. Elkins, Comproni, and Pagnotto measured benzene vapor yielded by partial evaporation of mixtures of benzene with various aliphatic hydrocarbons, chlorinated hydrocarbons, and common esters, as well as partial evaporation of naphthas containing benzene. Most measurements for all four types of mixtures showed greater concentrations of benzene in air than were predicted by Raoult's law. Of five tests with naphtha-based rubber cements, one yielded measured values of benzene concentration in air in agreement with calculated values, the other four showed measured benzene concentrations in air to be 3–10 times greater than those calculated using Raoult's law.
>
> Substantial deviation from Raoult's law is not always the case, however, even with benzene. Runion compared measured and calculated concentrations in air of benzene in vapor mixtures yielded by evaporation from a number of motor gasolines and found excellent agreement.

19. A mine atmosphere averages 12 mg total dust/m³ of air. If this dust is 9% mass respirable and has 8% crystalline quartz in this respirable fraction, how long must one sample at 1.7 L/m if the analytical sensitivity is 50 mcg for α-quartz?

$$12 \text{ mg total dust/m}^3 = 12 \text{ mcg/L}$$

$$(12 \text{ mcg/L}) \times 0.09 \times 0.08 = 0.0864 \text{ mcg } \alpha\text{-SiO}_2/\text{L}$$

$$\frac{50 \text{ mcg quartz}}{0.0864 \text{ mcg quartz/L}} = 578 \text{ L}$$

$$\frac{578 \text{ L}}{1.7 \text{ L/m}} = 340 \text{ minutes}$$

Answer: 340 minutes = 5 hours and 40 minutes. Collect a full shift personal breathing zone air sample to determine the worker's 8-hour TWAE.

20. Two impingers are connected in series. Calculate collection efficiency of the first impinger if it contains 78.9 mcg, and the second impinger contains 6.3 mcg of the same air contaminant.

$$\% \text{ efficiency} = 100 \left[1 - \frac{C_2}{C_1} \right] = 100 \left[1 - \frac{6.3 \text{ mcg}}{78.9 \text{ mcg}} \right] = 100 \, (1 - 0.08) = 92\%$$

Answer: 92% of the total air contaminant is in the first impinger assuming both impingers collected 100%. A correction factor of 1.08 can be applied to the concentration analyzed in the first impinger (i.e., 92/(78.9 + 6.3) = 1.08). Of course, based on these data, a third (or more) impinger connected in series should arguably have diminishing concentrations of the analyte in each.

21. 6.7 ft³ of air at 53°F and 14.7 lb/in² are adiabatically compressed to 95 lb/in². What is the initial temperature of the air after compression? What is the final volume of the compressed air? 1.4 = the specific heat (the ratio of heat capacity at constant pressure to the heat capacity at constant volume, often expressed as k).

$$460 + 53°F = 513°R \text{ absolute temperature}$$

$$\text{Temperature} = 513° \left[\frac{95}{14.7} \right]^{(1.4-1)/1.4} = 875°R = 415°F$$

$$\text{volume} = 6.7 \text{ ft}^3 \left[\frac{14.7}{95} \right]^{1/1.4} = 6.7 \text{ ft}^3 \, (0.155)^{0.714} = 1.77 \text{ ft}^3$$

Answer: 415°F and 1.77 ft³. The value of k is a function of temperature and pressure. For air and several diatomic gases (e.g., N_2, O_2), k equals 1.4. Several hydrocarbons have k values typically between 1.1 and 1.2.

22. How many kilograms of ammonia are in a 3000 ft³ tank when the gauge pressure is 950 lb/in², and the ammonia temperature is 31°C?
Atomic weights of nitrogen and hydrogen are 14 and one, respectively.

$$NH_3 = 17 \text{ grams/gram-mole}$$

$$3000 \text{ ft}^3 \times \frac{28.32 \text{ L}}{\text{ft}^3} = 84{,}960 \text{ L}$$

$$P_{absolute} = \frac{950 \text{ lb/in}^2 + 14.7 \text{ lb/in}^2}{14.7 \text{ lb/in}^2 /\text{atmosphere}} = 65.6 \text{ atmospheres}$$

$$T = 31°C + 273 = 304 \text{ kelvin}$$

$$n = \frac{PV}{RT} = \frac{65.6 \text{ atm} \times 84{,}690 \text{ L}}{(0.0821 \text{ L-atm/mole-K}) \times 304 \text{ K}} = 223{,}579 \text{ gram-moles } NH_3$$

$$223{,}579 \text{ gram-moles} \times 17 \text{ grams/mole} = 3{,}800{,}843 \text{ grams}$$
$$= 3801 \text{ kg ammonia}$$

23. What is the volume if the gas in Problem 22 expanded to atmospheric pressure at a temperature of 20°C as might occur during a rapid tank rupture?
P, V, and T = pressure, volume, and absolute temperature of the gas, respectively
i and f = initial and final conditions, respectively

$$\frac{P_i V_i}{T_i} = \frac{P_f V_f}{T_f}$$

rearranging

$$V_f = \frac{P_i T_f V_i}{P_f T_i} = 3000 \text{ ft}^3 \times \frac{65.6 \text{ atm}}{1 \text{ atm}} \times \frac{293°K}{304°K} = 189{,}679 \text{ ft}^3$$

24. The dust concentration in a limestone mill is 41 mppcf. The density of $CaCO_3$ is 2.71 g/cc. If the calcite particles are uniformly spherical with a diameter of 1.42 microns, how much limestone dust is in the air of a 40,000 ft³ ball mill plant? How much dust is in every liter of mill air inhaled by ball mill operators?

$$\varnothing = 1.42 \text{ microns} = 0.000142 \text{ cm}$$

$$\text{radius} = 0.000071 \text{ cm}$$

$$\text{Volume of sphere} = \frac{4}{3}\pi r^3 = \frac{4}{3}\pi (0.000071 \text{ cm})^3$$

$$= 1.5 \times 10^{-12} \text{ cm}^3/\text{dust particle}$$

$$40{,}000 \text{ ft}^3 \times \frac{1.5 \times 10^{-12} \text{ cm}^3}{\text{particle}} \times \frac{41 \times 10^6 \text{ particles}}{\text{ft}^3} = 2.46 \text{ cm}^3 \times 2.71 \text{g/cm}^3$$

$$= 6.67 \text{ grams} = 6670 \text{ mg}$$

$$40{,}000 \text{ ft}^3 \times 28.32 \text{ L/ft}^3 = 1{,}132{,}800 \text{ liters}$$

$$6670 \text{ mg CaCO}_3/1{,}132{,}800 \text{ L} = 0.00589 \text{ mg/L} = 5.89 \text{ mcg/L}$$

Answers: 6670 milligrams. 5.89 milligrams of CaCO³ dust/liter of air = 5.89 mg/m³

25. Air was sampled with a midget impinger for 1 hour and 17 minutes at an average rate of 0.89 L/m. How much ozone gas was in the air if the chemist detected 3.6 mcg O_3 per mL, the impinger collection efficiency was 71%, and there were 13.5 mL of potassium iodide ozone collection solution?

$$0.89 \text{ L/min} \times (60 + 17) \text{ minutes} = 68.5 \text{ liters}$$

$$\text{Molecular weight } O_3 = 16 \times 3 = 48$$

$$(100/71) = \text{impinger inefficiency collection factor} = 1.408$$

$$3.6 \text{ mcg/mL} \times 13.5 \text{ mL} \times 1.408 = 68.4 \text{ mcg } O_3$$

$$\text{ppm} = \frac{(\text{mcg/L}) \times 24.45}{\text{molecular weight}} = \frac{(68.4 \text{ mcg/68.5 L}) \times 24.45}{48} = 0.51 \text{ ppm}$$

Answer: 0.51 ppm$_v$ ozone gas

26. Convert 136 micrograms of ethyl alcohol vapor per liter into ppm (volume/volume).

$$\text{Molecular weight CH}_3\text{CH}_2\text{OH} = 12 + 12 + 6 + 16 = 46$$

$$\text{ppm} = \frac{(136 \text{ mcg EtOH/L}) \times 24.45}{46} = 72.3 \text{ ppm}$$

Answer: 72.3 ppm$_v$ ethyl alcohol vapor.

27. An analyst counts 3.4 fibers/field on an aerosol filter. There are 27,900 fields/filter. What was the fiber concentration in f/cc³ if air sampling was 2 liters/minute for 89 minutes?

$$3.4 \text{ fibers/field} \times 27{,}900 \text{ fields} = 94{,}860 \text{ fibers}$$

$$2 \text{ Lpm} \times 89 \text{ minutes} = 178 \text{ liters} = 178{,}000 \text{ mL} = 178{,}000 \text{ cc}$$

$$(3.4 \text{ fibers/field} \times 27{,}900 \text{ fields})/178{,}000 = 0.53 \text{ fiber/cc of air}$$

28. What is the concentration of nitrogen in dry air in ppm_v?

$$\text{Air: } 78.09\% \text{ N}_2 + 20.95\% \text{ O}_2 + 0.93\% \text{ argon}$$
$$+ \cong 0.035\% \text{ CO}_2 + \text{trace gases } (\cong 79\% \text{ "inerts"} + \cong 21\% \text{ O}_2)$$

Answer: $780{,}900 \text{ ppm}_v$ N$_2$ (i.e., 100% air $= 10^6 \text{ ppm}_v$).

29. The PEL for a metal is 0.2 mg/m^3. A chemist can reliably detect 4 micrograms with good accuracy and precision. At air-sampling rate of 1.1 L/m, how long would an industrial hygienist have to sample air to detect 10% of the PEL?

$$\text{PEL} = 0.2 \text{ mg/m}^3 = 0.2 \text{ mcg/L}$$

$$10\% \text{ PEL} = 0.02 \text{ mcg/L}$$

$$\frac{4 \text{ mcg}}{0.02 \text{ mcg/L}} = \text{at least } 200 \text{ liters of air must be sampled}$$

$$\frac{200 \text{ L}}{1.1 \text{ L/min}} = \text{at least } 182 \text{ minutes}$$

Answer: >3 hours.

30. What gas concentration results when 5 mL of dry ammonia gas are injected by a gas syringe into a 13-gallon glass calibration carboy of air?
Method 1:

$$\frac{10^6 \times 5 \text{ mL}}{13 \text{ gallons} \times (3785 \text{ mL/gallon})} = 102 \text{ ppm}_v \text{ NH}_3$$

Method 2:

$$17 \text{ grams NH}_3/24.45 \text{ L} = 0.695 \text{ g/L} = 0.695 \text{ mg/mL}$$

$$0.695 \text{ mg/mL} \times 5 \text{ mL} = 3.475 \text{ mg}$$

$$13 \text{ gallons} \times 3.785 \text{ L/gallon} = 49.205 \text{ L} = 0.0492 \text{ m}^3$$

$$\frac{(3.475 \text{ mg}/0.0492 \text{ m}^3) \times 24.45}{17} = 102 \text{ ppm}_v \text{ NH}_3$$

Method 3:

$$\frac{5 \text{ mL}}{49,205 \text{ mL}} = \frac{x \, \text{ppm}_v}{10^6 \, \text{ppm}_v}$$

$$x = 102 \, \text{ppm}_v$$

Answer: The instrument calibration bottle contains $102 \, \text{ppm}_v$ NH_3 gas verified three ways.

31. What is the gas concentration after 75 mL of pure CO gas mixes with air containing 2 ppm CO in a 313 L instrument calibration tank? Assume negligible dilution loss as the CO gas is quickly injected into the tank.

$$\frac{10^6 \times 0.075 \text{ L}}{313 \text{ L} + 0.075 \text{ L}} = 240 \text{ ppm} = \left[\frac{75 \text{ mL}}{313,000 \text{ mL}}\right] \times 10^6$$

$$(240 + 2) \, \text{ppm}_v = 242 \, \text{ppm}_v$$

32. A chemist dropped a chlorine bottle releasing 2 pounds of gas into a laboratory with no ventilation. He immediately left and returned wearing a self contained breathing apparatus (SCBA) by the time the gas had mixed uniformly throughout the laboratory. The laboratory is $14' \times 20' \times 40'$. He turned on the exhaust hood with a uniform face velocity of 170 feet per minute. The hood face dimension is $40" \times 66"$. How long before Cl_2 gas concentration is reduced to $0.2 \, \text{ppm}_v$ (20% of $1 \, \text{ppm}_v$ STEL)? Assume ideal ventilation mixing.

$$\frac{40" \times 66"}{144 \text{ in}^2/\text{ft}^2} = 18.33 \text{ ft}^2 \times 170 \text{ fpm} = 3116 \text{ ft}^3/\text{minute}$$

$$\text{Molecular weight } Cl_2 = 71 \text{ grams/gram-mole}$$

$$14' \times 20' \times 40' = 11,200 \text{ ft}^3 = 317.3 \text{ m}^3$$

$$2 \text{ lb} = 908 \text{ grams}$$

$$\frac{908 \text{ g}}{317.3 \text{ m}^3} = 2.86 \text{ g/m}^3 = 2860 \text{ mg/m}^3$$

$$\text{ppm} = \frac{(2860 \text{ mg/m}^3) \times 24.45}{71} = 985 \text{ ppm} \, Cl_2$$

$$t = \frac{-\ln\left[C/C_0\right]}{Q/V} = \frac{-\ln\left[0.2 \text{ ppm}/985 \text{ ppm}\right]}{3116 \text{ cfm}/11,200 \text{ ft}^3} = \frac{-\ln 0.000203}{0.2782/\text{minutes}}$$

$$= \frac{-(-8.502)}{0.2782/\text{minutes}} = 30.6 \text{ minutes}$$

Answer: At least 31 minutes.

33. A measure of 20 kilograms of methyl chloroform (molecular weight = 133.4) evaporates into a $10' \times 25' \times 35'$ tank that has no ventilation. What is the equilibrium concentration in ppm_v? Is this a fire hazard? Could welding be permitted when this vapor level is reduced by ventilation to $50\,ppm_v$ (1/7 of the TLV)? Could a worker enter wearing a half-mask twin charcoal cartridge respirator without special ventilation and taking other precautions?

$$20\,kg = 2 \times 10^7\,mg$$

$$10' \times 25' \times 35' = 8750\,ft^3 = 247.8\,m^3$$

$$\frac{2 \times 10^7\,mg}{247.8\ m^3} = 80{,}710\ mg/m^3$$

$$ppm = \frac{(80{,}710\ mg/m^3) \times 24.45}{133.4} = 14{,}793\ ppm = 1.48\%\ (vol/vol)$$

Answers: $14{,}800\,ppm_v$. No, under normal conditions, this would not be an explosion hazard. However, at very high concentrations in air and, especially, with high oxygen levels, an explosion could occur with high-temperature ignition sources. $COCl_2$, HCl, and dichloroacetylene, and so on could be generated in the welding arc. No, this is the wrong respirator because the maximum use conditions are $1000\,ppm_v$ organic vapors for charcoal filters provided the concentration immediately dangerous to life and health (IDLH) is not less than $1000\,ppm_v$. In other words, since the IDLH concentration for methyl chloroform is $700\,ppm_v$, this is the maximum concentration permitted for use with organic vapor charcoal cartridge respirators, not $1000\,ppm_v$. This would be an acceptable respirator if the vapor levels did not exceed $700\,ppm_v$.

34. A volume of 10 mL of dry ammonia gas is injected into a 152 L tank of pure air. What is the NH_3 gas concentration? Can most people detect this ammonia gas concentration by smell or irritation?

$$Molecular\ weight\ NH_3 = 17$$

$$Density\ at\ NTP = 17\,g/24.45\,L = 0.7\,mg/mL$$

$$10\,mL\ NH_3 = 7\,g\ of\ NH_3$$

$$\frac{7\ g}{152\ L} = 0.046\ mg/L = 46\ mcg/L$$

$$ppm\,NH_3 = \frac{(46\ mcg/L) \times 24.45}{17} = 66.2\ ppm\,NH_3$$

Answers: Yes, everyone could be expected to respond to this gas concentration. Those exposed would cough, have profound nasal and respiratory irritation, teary eyes, and do whatever they could to escape the area.

35. If the apparent volume of sampled air is 570 L at 645 mm Hg and 33°C, what was the standard air volume that was sampled?

$$33°C + 273 = 306 \text{ kelvin}$$

$$\text{Liters} = \frac{298 \times V \times P}{760 \times T} = \frac{(298 \text{ K})(570 \text{ L})(645 \text{ mm Hg})}{(760 \text{ mm Hg})(306 \text{ K})} = 471 \text{ L}$$

36. The LEL for a solvent vapor is 1.7%. What is the vapor concentration in ppm_v if a calibrated CGI (computer-generated image) indicates a reading of 64% of the LEL?

$$1.7\% = 17,000 \, ppm_v$$

$$0.64 \, LEL = 0.64 \times 17,000 \, ppm_v = 10,880 \, ppm_v$$

37. An impinger contains 13.0 mL of a dilute alkali (e.g., 0.01 N NaOH). Each milliliter can neutralize 0.012 mg of HCl gas. Air was sampled at 0.86 L/m for 14.7 minutes when the alkali was neutralized as indicated by an abrupt color change in the solution. What was the average concentration of acid gas during the sampling period?

$$13 \, mL \times 0.012 \, mg/mL = 0.156 \, mg \, HCl$$

$$\text{Molecular weight HCl} = 1 + 35.5 = 36.5$$

$$0.86 \, L/m \times 14.7 \, \text{minutes} = 12.64 \, \text{liters}$$

$$ppm = \frac{(156 \text{ mcg}/12.64 \text{ L}) \times 24.45}{36.5} = 8.27 \text{ ppm HCl gas}$$

Answer: 8.3 ppm_v HCl, assuming 100% collection efficiency by the impinger.

38. 0.1 mL of a solvent mixture is evaporated in a 313 L calibration tank. The mixture is comprised (by volume) of 30% MEK, 30% toluene, and 40% methylene chloride with respective densities of 0.805, 0.870, and 1.335 g/mL. Calculate ppm_v of each vapor and the apparent molecular weight of the mixture. Molecular weights are 72, 92, and 85, respectively.

MEK $0.03 \, mL \times 0.805 \, g/mL = 0.02415 \, g$

toluene $0.03 \, mL \times 0.870 \, g/mL = 0.0261 \, g$

CH_2Cl_2 $\underline{0.04 \, mL} \times 1.335 \, g/mL = \underline{0.0534 \, g}$

            $0.10 \, mL$ $0.1037 \, g$

$$\text{ppm MEK} = \frac{(24{,}150 \text{ mcg/313 L}) \times 24.45}{72} = 26 \text{ ppm}$$

$$\text{ppm toluene} = \frac{(26{,}100 \text{ mcg/313 L}) \times 24.45}{92} = 22 \text{ ppm}$$

$$\text{ppm CH}_2\text{Cl}_2 = \frac{(53{,}400 \text{ mcg/313 L}) \times 24.45}{85} = 49 \text{ ppm}$$

$$26 \text{ ppm}_v + 22 \text{ ppm}_v + 49 \text{ ppm}_v = 97 \text{ ppm}_v \text{ total solvent vapors}$$

$$0.1037 \text{ g/313 L} = 331 \text{ mcg/L}$$

$$\text{apparent molecular weight} = \frac{(331 \text{ mcg/L}) \times 24.45}{97 \text{ ppm}} = 83.4$$

Answers: 26 ppm_v MEK, 22 ppm_v toluene, and 49 ppm_v CH_2Cl_2. The apparent molecular weight is 83.4 grams gram-mole^{-1}.

39. How many molecules of TDI are in every 2 liter inhalation if the air contains 0.001 ppm_v (one ppb_v) TDI vapor?

$$\frac{6.023 \times 10^{23} \text{ molecules TDI/gram-mole}}{24.45 \text{ L/gram-mole}} = 10^6 \text{ ppm} = 100\%$$

$$1 \text{ ppb} = \frac{10^{-9} \times 6.023 \times 10^{23} \text{ molecules TDI}}{24.45 \text{ L}} = \frac{2.46 \times 10^{13}}{\text{L}} \times 2 \text{ liters}$$

$$= 4.92 \times 10^{13} \text{ molecules}$$

Answer: $\cong 5 \times 10^{13}$ molecules of TDI in every 2-liter inhalation.

40. A rotameter calibrated at 25°C and 760 mm Hg indicates a rate of 1.45 L/min at 33°C and 690 mm Hg. What is the corrected standard air flow rate?

$$\text{L/min} = 1.45 \text{ L/min} \times \sqrt{\frac{690 \text{ mm Hg}}{760 \text{ mm Hg}}} \times \sqrt{\frac{25°\text{C} + 273 \text{ K}}{33°\text{C} + 273 \text{ K}}}$$

$$= 1.45 \times 0.9528 \times 0.9868 = 1.36 \text{ L/min}$$

Answer: 1.36 liters air per minute (i.e., the slightly colder, denser air at standard conditions has greater buoyancy for the rotameter float ball). The square root function is used only for rotameters and critical orifice air flow meters, but never for dry and wet gas meters. Refer to Problem 311.

41. Give two basic industrial hygiene air sampling "rules of thumb."

Answer:

1. Minimum collection volume (in m^3) = $\dfrac{\text{analytical sensitivity (mg)}}{0.1 \times \text{TLV (in mg/m}^3)}$

2. Minimum sampling time (in hours)

$$= \dfrac{\text{analytical sensitivity (mg)}}{(0.1 \times \text{TLV in mg/m}^3) \times (\text{rate in m}^3/\text{hour})}.$$

Generally, an integrated air sample should be taken to detect at least 10% of its respective TLV, PEL, STEL, NIOSH (National Institute for Occupational Safety and Health)-recommended exposure limit (REL), workplace environmental exposure limit, or ceiling limit. Sometimes, however, this is impossible because a 15-minute air sample may not be sufficient to collect enough air contaminant for analysis of a STEL concentration. For these, consult with an industrial hygiene chemist. Also consider a direct-reading instrument if such exists.

42. What is error of measurement if the true value was $13\,ppm_v$, and the amount found (the experimental value) was $16\,ppm_v$?

$$\% \text{ error} = \dfrac{EV - TV}{TV} \times 100 = \dfrac{16\,ppm - 13\,ppm}{13\,ppm} \times 100 = 23\%$$

Answer: +23%.

43. A bottle containing 600 grams of titanium tetrachloride breaks in a warehouse. The liquid quickly evaporates hydrolyzing in highly humid air according to the reaction:

$$TiCl_4 + 2H_2O \rightarrow TiO_2 \uparrow + 4\,HCl \uparrow$$

The warehouse ($20' \times 60' \times 185'$) is empty, unventilated, and unoccupied. Is it safe to enter without wearing respiratory protection? What concentration of HCl gas and TiO_2 fume could be present? Molecular weight of $TiO_2 = 47.09$ grams/gram-mole. Assume sufficient water vapor as atmospheric moisture to provide stoichiometric conversion.

$$20' \times 60' \times 185' = 222,000\,ft^3 = 6289\,m^3$$

$$\dfrac{600\,g\,TiCl_4}{189.73\,g/mole} = 3.16\,\text{moles }TiCl_4$$

Molecular weight

Therefore, 2×3.16 moles H_2O required   18

1×3.16 moles TiO_2 produced   79.9

4×3.16 moles HCl produced   36.5

$$1 \times 3.16 \text{ moles } TiO_2 \times 79.9 \text{ g/mole} = 252,484 \text{ mg } TiO_2$$

$$4 \times 3.16 \text{ moles HCl} \times 36.5 \text{ g/mole} = 461,360 \text{ mg HCl}$$

$$\frac{252,484 \text{ mg } TiO_2}{6289 \text{ m}^3} = 40.1 \text{ mg } TiO_2 \text{ fume/m}^3$$

$$\frac{(461,360 \text{ mg HCl}/6289 \text{ m}^3) \times 24.45}{36.5} = 49.1 \text{ ppm HCl gas}$$

Answer: 40 mg TiO_2/m³. 49 ppm$_v$ HCl. OSHA's (Occupational Safety & Health Administration) PEL for HCl = 5 ppm ceiling. No, STAY OUT! Ventilate. Otherwise, only enter with a SCBA or full-face acid gas cartridge respirator with HEPA pre-filters. A SCBA is preferred since the IDLH for HCl is 50 ppm$_v$, barely above the average concentration of HCl gas in the room.

44. How much liquid toluene is needed to make 100 ppm of vapor in a 10' × 14' × 20' chamber when the atmospheric pressure is 640 mm Hg and the air temperature is 21°C?

$$10' \times 14' \times 20' = 2800 \text{ ft}^3$$

mL

$$= \frac{100 \text{ ppm} \times (92 \text{ g/g-mol}) \times (2800 \text{ ft}^3/35.3 \text{ ft}^3/\text{M}^3) \times 273 \text{ K} \times 640 \text{ mm Hg}}{(0.867 \text{ g/mL}) \times (22.4 \text{ L/g-mol}) \times 294 \text{ K} \times 760 \text{ mm Hg} \times 10^3}$$

Answer: 29.4 milliliters.

45. What concentration of solvent vapor remains in a 1000 gallon tank (containing no liquid solvent) after 500 gallons of air: vapor mixture was removed and replaced with clean air? The initial solvent vapor concentration was 1000 ppm$_v$.

$$(2.3 \log 1000 \text{ ppm}) - (2.3 \log y \text{ ppm}) = \frac{500 \text{ gallons}}{1000 \text{ gallons}}$$

$$(2.3 \times 3) - (2.3 \log y \text{ ppm}) = 0.5$$

$$2.3 \log y \text{ ppm} = 6.4$$

$$\log y \text{ ppm} = \frac{6.4}{2.3} = 2.783$$

$$y = 606 \text{ ppm}_v$$

46. Calculate the vapor volume of 1 gram of water when boiled at 730 mm Hg.

$$PV = nRT$$

$$n = 1 \text{ gram}/18 \text{ grams/mole} = 0.0556 \text{ mole}$$

$$T = 100°\text{C} + 273 \text{ K} = 373 \text{ kelvin}$$

$$V = \frac{(0.0556 \text{ mole})(0.082)(373 \text{ K})}{730 \text{ mm Hg}/760 \text{ mm Hg}} = 1.7705 \text{ L}$$

Answer: A volume of 1771 milliliters of water vapor, that is, 1 mL of liquid water produces 1.77 liters of steam under these conditions of temperature and pressure.

47. A stack gas sample was collected using a dry gas meter calibrated at 32°F and 760 mm Hg. The water vapor pressure at this temperature (absolute humidity) is 0.08 lb/in². The wet and dry bulb temperatures of the stack gas were 96 and 111°F, respectively (58% relative humidity). Corresponding water vapor pressure is 0.78 lb/in². The indicated air volume was 930 ft³. The barometric pressure at the time of sampling was 740 mm Hg. What is the corrected dry gas volume?

Stack gas volumes are often calculated as if the gases are dry since the variable water vapor, especially at high temperatures, can account for a significant portion of the total gas volume. The variations due to pressure, temperature, and water vapor content are calculated by

$$V_1 = \frac{V_2(P_2 - W_2)(273 \text{ K} + T_1)}{(P_2 - W_1)(273 \text{ K} + T_2)}$$

where
V_2 = apparent gas volume at T_2 (°C)
V_1 = calculated gas volume at T_1 (°C)
W_1 and W_2 = mm H_2O vapor pressure at calculated and observed conditions, respectively
P_1 and P_2 = calculated and observed barometric pressures, respectively

$$(\text{lb/in}^2) \times 51.71 = \text{mm Hg}$$

$$0.02\,\text{lb/in}^2 = 1.03\,\text{mm Hg}$$

$$32°F = 0°C$$

$$0.78\,\text{lb/in}^2 = 40.33\,\text{mm Hg}$$

$$112°F = 44.4°C$$

$$V_1 = \frac{(930\ \text{ft}^3)(740\ \text{mm Hg} - 40.3\ \text{mm Hg})(273\ \text{K})}{(760\ \text{mm Hg} - 1\ \text{mm Hg})(273\ \text{K} + 44.4\ °\text{C})} = 737.4\ \text{ft}^3$$

Answer: 737.4 cubic feet of dry gas versus 930 ft³ of wet gas.

48. A volume of 1000 L of dry nitrogen gas at 1 atm pressure and 20°C are adiabatically compressed to 5% of the initial gas volume. What is the final temperature and pressure? For N_2 gas, $\alpha = 1.4$ (the specific heat for this diatomic gas. Refer to Problem 21).

$$P_1\,(V_1)^\alpha = P_1\,(V_2)^\alpha$$

$$(1\ \text{atmosphere})\,(V_1)^{1.4} = P_2\,(V_1/20)^{1.4}$$

$$P_2 = \frac{(1\ \text{atm})(V_1)^{1.4}}{(V_1/20)^{1.4}} = \frac{(1\ \text{atm})(1000\ \text{L})^{1.4}}{(50\ \text{L})^{1.4}} = 66.3\ \text{atmospheres}$$

$$T_1(V_1)^{\alpha-1} = T_2(V_2)^{\alpha-1} = (293\ \text{K})(1000\ \text{L})^{1.4-1} = T_2(50\ \text{L})^{1.4-1}$$

$$T_2 = \frac{293\ \text{K}(1000\ \text{L})^{1.4-1}}{(50\ \text{L})^{1.4-1}} = \frac{293\ \text{K}(1000\ \text{L})^{0.4}}{(50\ \text{L})^{0.4}} = 970.95\ \text{K}$$

Answers: 66.3 atmospheres and 971 kelvin.

49. An air sample filter contained 36 micrograms of chromium. If all chromium was from a lead chromate paint spray aerosol ("school bus yellow"), how much lead chromate was present?

$$PbCrO_4 \text{ molecular weight} = 207 + 52 + (4 \times 16) = 323$$

$$323/52 = 6.21$$

$$36\,\text{mcg Cr} \times 6.21 = 224\,\text{mcg PbCrO}_4$$

Answer: 224 micrograms $PbCrO_4$

50. Assume that a worker on an average inhales 15 L/min and the inhaled air contains 9 mcg cobalt/m³. If his absorption is 25%, how much cobalt would this worker accumulate every 8-hour work shift? Disregard excretion during the exposure period.

$$(9 \, mcg \, Co/m^3) \times 0.25 = 2.25 \, mcg \, Co/m^3 = 0.00225 \, mcg/L$$

$$\frac{0.00225 \, mcg \, Co}{L} \times \frac{15 \, L}{minute} \times \frac{60 \, minutes}{hour} \times \frac{8 \, hours}{shift} = 16.2 \, mcg \, Co/shift$$

Answer: A measure of 16 micrograms of cobalt are absorbed by this worker in an 8-hour work shift.

51. Air temperature is 42°C. Barometric pressure is 718 mm Hg. Air-sampling rate is 2.31 L/m. Sampling time is 17.5 minutes. Assume that the air is saturated with water vapor (61.5 mm Hg). What is the sampled volume of dry air at 25°C and 760 mm Hg?

$$17.5 \, minutes \times 2.31 \, L/m = 40.425 \, liters \, of \, wet \, air$$

$$40.425 \, L \times \frac{718 \, mm \, Hg - 61.5 \, mm \, Hg}{760 \, mm \, Hg} \times \frac{298 \, K}{273 \, K + 42°C} = 33 \, L$$

Answer: 33 liters of dry air.

52. What is the mercury vapor concentration in a dynamic vapor generation system if 100 mL/minute saturated at 18.8 mcg Hg/L are diluted with mercury-free air at 27 L/minute?

$$C_t = \frac{A \times C}{A + B} = \frac{\text{concentration of contaminant in dynamic generation}}{\text{calibration system}}$$

where

    A = contaminant flow rate
    B = clean air flow rate
    C = concentration of contaminant in A

$$100 \, mL/min = 0.1 \, L/min$$

$$C_t = \frac{(0.1 \, L/min) \times (18.8 \, mcg/L)}{(27 \, L/min) + (0.1 \, L/min)} = 0.069 \, mcg/L$$

Answer: 0.069 mg Hg vapor/m³.

53. How many particles result from crushing 1 cubic centimeter of quartz into 1 micron cubic particles?

$1\,cc = 1\,cm^3$

$1\,meter = 10^6\,microns$

$1\,cm = 10^4\,microns$

$(10^4\,microns)^3 = 10^{12}\,particles$

Answer: 10^{12} particles of respirable dust, that is, 1,000,000,000,000 particles.

54. What is the saturation concentration (in mg/m³) of mercury vapor at 147°F? Hg vapor pressure at this temperature is 0.0328 mm Hg. Assume that the air temperature is also 147°F or greater. Molecular weight of Hg is 200.6 grams/gram-mole.

$$\frac{0.0328\,mm\,Hg}{760\,mm\,Hg} \times 10^6 = 43.16\,ppm$$

$$147°F = 63.89°C$$

$$22.4\,L \times \frac{273\,K + 63.89\,°C}{273\,K} = 27.64\,L$$

$$\frac{mg}{m^3} = \frac{43.16\,ppm \times 200.6}{27.64\,L} = \frac{313\,mg}{m^3}$$

Answer: 313 mg mercury/m³ (OSHA's PEL = 0.05 mg/m³). The TLV for elemental mercury (all forms) is 0.025 mg/m³ with the SKIN notation.

55. How much mercury would have to evaporate to yield a vapor concentration of 0.1 mg/m³ in a chloralkali plant with a 2,250,000 ft³ interior volume? The density of liquid mercury is 13.6 g/mL.

$$0.1\,mg/m^3 = 2.83\,mcg\,Hg/ft^3$$

$$2.83\,mcg/ft^3 \times 2.25 \times 10^6\,ft^3 = 6.37 \times 10^6\,mcg$$

$$\frac{6.37\,g\,Hg}{13.6\,g/mL} = 0.47\,mL$$

Answers: 6.4 gram, 0.47 milliliter.

56. If 1 cc of air has a mass of 1.2 mg at 25°C and 760 mm Hg, what is the density of mercury vapor? Molecular weight of Hg is 200.6 g/g-mol.

$$1\,ppm\,Hg\,vapor = \frac{200.6 \times mg/L}{24,450\,mL/g\text{-}mol} = 0.0082\,mg/L = 8.2\,mcg/L$$

$$= 8.2\,g/1000\,cc = 0.0082\,g/cc$$

$$= 8.2\,mg/cc$$

$$8.2\,\text{mg/m}^3 = 1\,\text{cc/m}^3$$

$$1\,\text{cc Hg vapor} = 8.2\,\text{mg}$$

$$1\,\text{cc air} = 1.2\,\text{mg}$$

$$\frac{8.2\,\text{mg Hg/cc}}{1.2\,\text{mg air/cc}} = 6.83$$

Alternatively, one could use the ratio of their molecular weights, that is, the "apparent" molecular weight of air = 28.94, or 200.6/28.94 = 6.93—essentially identical to the above. See Problem 58.

Answer: 6.83 (air = 1.00, no units).

57. How many pounds of air are inhaled weekly by a person with a daily inhalation of 22.8 m³, the air inhalation volume of "standard man" (70 kg)?

$$\frac{22.8\,\text{m}^3}{\text{day}} \times \frac{7\,\text{days}}{\text{week}} \times \frac{35.3\,\text{ft}^3}{\text{m}^3} \times \frac{0.075\,\text{lb}}{\text{ft}^3} = \frac{423\,\text{lb}}{\text{week}}$$

Answer: About 400–450 pounds per week for a 154-pound man or woman who has an average daily metabolic rate. The inhaled volume, of course, increases as one's metabolic activity increases and as the person's body mass increases.

58. Calculate dry air's "apparent" molecular weight if the atomic weight of argon is 39.9 grams/gram-mole.

	% by volume		Molecular Weight		Proportion of Molecular Weight
O_2	21.0	×	32	=	6.72
N_2	78.1	×	28	=	21.866
Ar	0.9	×	39.9	=	0.355
	100.0				28.941

Answer: The apparent molecular weight of dry air is 28.94.

59. Worker breathing zone TWAE air contains 30 mcg Pb/m³ (PEL = 50 mcg/m³) and 0.8 mg H_2SO_4 mist/m³ (PEL = 1 mg/m³). Is the PEL for the mixture exceeded?

$$\frac{30\,\text{mcg/m}^3}{50\,\text{mcg/m}^3} = 0.6, \text{ or } 60\% \text{ of PEL}$$

30 mcg/m³ is OSHA's action level for Pb.

$$\frac{0.8 \text{ mg H}_2\text{SO}_4/\text{m}^3}{1 \text{ mg H}_2\text{SO}_4/\text{m}^3} = 0.8, \text{ or } 80\% \text{ of PEL}$$

Answer: No, toxic effects of these air contaminants are independent. However, since both exceed respective action levels, an aggressive industrial hygiene control program is needed.

60. By how much should workers' exposure limits be reduced if they are exposed for a 12-hour work shift?

$$\text{exposure reduction factor} = \frac{8}{h} \times \frac{24 - h}{16}$$

where
   h = hours exposed per work day

$$\frac{8}{12} \times \frac{24 - 12}{16} = 0.5$$

Answer: Reduce by at least 50%. This accounts, in part, for reduced time available each day to detoxify and reverse the daily adverse effects. This is the Brief and Scala model referenced in American Conference of Governmental Industrial Hygienists book of *threshold Limit Values®*.

61. A liquid contains (by weight) 50% heptane (TLV = 400 ppm$_v$ = 1600 mg/m³), 30% methyl chloroform (TLV = 350 ppm$_v$ = 1900 mg/m³), and 20% perchloroethylene (TLV = 50 ppm$_v$ = 335 mg/m³). Assume complete evaporation of each solvent in the mixture. What is the TLV for the vapor mixture?

n-heptane	1 mg/m³ $\cong$ 0.25 ppm$_v$
CH_3CCl_3	1 mg/m³ $\cong$ 0.18 ppm$_v$
"perc"	1 mg/m³ $\cong$ 0.15 ppm$_v$

$$\text{TLV of mixture} = \frac{1}{(0.5/1600) + (0.3/1900) + (0.2/335)} = 935 \text{ mg/M}^3$$

Of this mixture,

$$n\text{-heptane} \quad 935 \text{ mg/m}^3 (0.5) = \frac{468 \text{ mg}}{\text{m}^3} \times 0.25 = 117 \text{ ppm}_v$$

$$CH_3CCl_3 \quad 935 \text{ mg/m}^3 (0.3) = \frac{281 \text{ mg}}{\text{m}^3} \times 0.18 = 51 \text{ ppm}$$

$$\text{"perc" } 935 \text{ mg/m}^3 \, (0.2) = \frac{187 \text{ mg}}{\text{m}^3} \times 0.15 = 29 \text{ ppm}_v$$

Answer: $117 \text{ ppm}_v + 51 \text{ ppm}_v + 29 \text{ ppm}_v = 197 \text{ ppm}_v = 935 \text{ mg/m}^3$

62. A process evaporates 5.7 pounds of isopropyl alcohol of vapor per hour into a work area that has a general ventilation rate of 4500 cfm. What is average steady-state vapor concentration?

$$\text{ppm} = \frac{\text{ER} \times 24.45 \times 10^6}{\text{Q} \times \text{molecular} \times \text{weight}},$$

where
ER = evaporation (generation) rate (in grams/minute) and
Q = ventilation rate (in liters/minute)

$$CH_3CHOHCH_3 \text{ molecular weight} = C_3H_8O = 36 + 8 + 16 = 60$$

$$\frac{5.7 \text{ lb}}{\text{hour}} \times \frac{454 \text{ g}}{\text{lb}} \times \frac{\text{hour}}{60 \text{ minutes}} = \frac{43.13 \text{ grams}}{\text{minute}}$$

$$Q = \frac{4500 \text{ ft}^3}{\text{minute}} \times \frac{28.3 \text{ L}}{\text{ft}^3} = 127{,}350 \text{ L/m}$$

$$\text{ppm} = \frac{(43.13 \text{ grams/minute}) \times 24.45 \times 10^6}{(127{,}350 \text{ L/minute}) \times 60} = 138 \text{ ppm IPA}$$

Answer: 138 ppm_v isopropyl alcohol vapor at 43.13 grams per minute.

63. The dry bulb temperature in Miami Beach is 80°F at a barometric pressure of 760 mm Hg with relative humidity of 40%. Water vapor pressure at these conditions is 0.195 lb/in². A hurricane is approaching! What is the concentration of H_2O vapor in the air as the barometer decreases from 760 to 680 mm Hg?

$$(\text{lb/in}^2) \times 51.71 = \text{mm Hg}$$

$$(0.195 \text{ lb/in}^2) \times 51.71 = 10.08 \text{ mm Hg}$$

$$\frac{10.08 \text{ mm Hg}}{680 \text{ mm Hg}} \times 10^6 = 14{,}824 \text{ ppm } H_2O \text{ vapor}$$

$$\text{ppm} = \frac{\text{mg}}{\text{L}} \times \frac{22{,}400}{18} \times \frac{299.67 \text{ K}}{273 \text{ K}} \times \frac{760 \text{ mm Hg}}{680 \text{ mm Hg}},$$

or

$$\frac{mg\ H_2O}{L} = \frac{14{,}824\ ppm}{(22{,}400/18) \times (299.67\ K/273\ K) \times (760\ mm\ Hg/680\ mm\ Hg)}$$

$$= \frac{9.708\ mg}{L}$$

Answers: $14{,}824\ ppm_v$ ($\cong 1.48\%$) $\cong 9700\ mg\ H_2O/m^3$. Formerly, before the barometric pressure dropped, the water vapor concentration was

$$\frac{10.08\ mg\ Hg}{760\ mg\ Hg} \times 10^6 = 13{,}263\ ppm\ H_2O\ vapor$$

The increase in the water vapor concentration in the atmosphere as the barometric pressure decreases explains, in part, why heavy rains accompany hurricanes.

64. During an earthquake, two adjacent compressed gas lines in a $20' \times 125' \times 300'$ unventilated, closed room in a chemical plant simultaneously burst. One pipeline released 20 pounds of anhydrous ammonia, and the other released 4 pounds of anhydrous hydrogen chloride. Dense white fume and a pungent, highly irritating gas formed immediately. Assume complete reaction of both gases. What was the white fume? What was the remaining gas and its concentration? Should we boldly stroll into this room without respiratory protection? If not, what type of respirator should we use?

$$HCl + NH_3 \rightarrow NH_4Cl + heat$$

$$4\,lb\ HCl + 20\,lb\ NH_3 \rightarrow x\ lb\ NH_4Cl = 1816\,g\ HCl + 9080\,g$$
$$NH_3 \rightarrow y\ grams\ NH_4Cl$$

$$\frac{1816\,g\ HCl}{36.5\ g\ HCl/g\text{-mole}} = 49.75\ moles\ HCl$$

$$\frac{9080\ g\ NH_3}{17\ g\ NH_3/g\text{-mole}} = 534.1\ moles\ NH_3$$

Therefore, from the above equation, stoichiometrically, only $49.75\ moles$ of NH_4Cl can be formed (molecular weight $NH_4Cl = 53.5\,g/g\text{-mole}$):

$$y\ grams\ NH_4Cl = 49.75\ moles \times \frac{53.5\ grams}{mole\ NH_4Cl}$$

$$= y = 2662\ grams\ NH_4Cl = 5.86\ lb$$

$$20' \times 125' \times 300' = 750{,}000\ ft^3 = 21{,}247\ m^3$$

$$\frac{2262 \text{ g}}{22{,}247 \text{ m}^3} = \frac{0.125 \text{ g}}{\text{m}^3} = \frac{125 \text{ mg NH}_4\text{Cl fume}}{\text{m}^3}$$

20 lb NH_3 (an excess) + 4 lb HCl (totally reacts) $\rightarrow$ 5.86 lb NH_4Cl fume

24 lb total reactants − 5.86 lb product = 18.14 lb NH_3 gas remaining
= 8236 g NH_3

$$\text{ppm NH}_3 = \frac{(8{,}236{,}000 \text{ mg/21,247 m}^3) \times 24.45 \text{ L/gram-mole}}{17 \text{ grams NH}_3/\text{gram-mole}}$$

$$= 557 \text{ ppm NH}_3$$

Answer: NH_4Cl. 2660 grams, or 125 mg/m³. 557 ppm NH_3. No! SCBAs. The IDLH for ammonia is 300 ppm$_v$.

65. What is the 8-hour TWAE PEL for an insecticide mixture containing one part by weight "Parathion" (PEL = 0.1 mg/m³) and two parts "EPN" by weight (PEL = 0.5 mg/m³)?

$$\frac{C_1}{0.1 \text{ mg/m}^3} + \frac{C_2}{0.5 \text{ mg/m}^3} = \frac{C_{\text{mixture}}}{T_{\text{mixture}}} = \frac{C_m}{T_m}$$

$$C_2 = 2\,C_1$$

$$C_m = 3\,C_1$$

$$\frac{C_1}{0.1 \text{ mg/m}^3} + \frac{2\,C_1}{0.5 \text{ mg/m}^3} = \frac{3\,C_1}{T_m}$$

$$\frac{7 C_1}{0.5 \text{ mg/m}^3} = \frac{3 C_1}{T_m}$$

$$T_m = \frac{1.5}{7} = 0.21 \text{ mg/m}^3$$

Answer: The PEL for this insecticide mixture is 0.21 mg/m³.

66. Calculate the TLV for a mineral dust mixture of 40% X (TLV = 1 mg/m³) and 60% Y (TLV = 0.3 mg/m³). The adverse effects on respiratory health are assumed to be additive (pulmonary fibrosis).

$$\frac{C}{\text{TLV}} = \frac{0.4}{1} + \frac{0.6}{0.3}$$

$$\frac{1}{\text{TLV}} = 0.4 + 2.0 = 2.4$$

$$1 = 2.4 \times \text{the TLV}$$

$$\text{TLV} = 1/2.4 = 0.42 \, \text{mg/m}^3$$

Answer: $\text{TLV}_{X \text{ and } Y} = 0.42 \, \text{mg/m}^3$.

67. Workplace air contains $234 \, \text{ppm}_v$ acetone $(\text{TLV} = 750 \, \text{ppm}_v)$, $119 \, \text{ppm}$ *sec*-butyl acetate $(\text{TLV} = 200 \, \text{ppm}_v)$, $113 \, \text{ppm}$ MEK $(\text{TLV} = 200 \, \text{ppm}_v)$, and $49 \, \text{ppm}$ methyl chloroform $(\text{TLV} = 350 \, \text{ppm}_v)$. What is the concentration of the vapor mixture? Is the TLV exceeded if this was a breathing zone air sample?

$$234 \, \text{ppm}_v + 119 \, \text{ppm}_v + 133 \, \text{ppm}_v + 49 \, \text{ppm}_v = 535 \, \text{ppm}_v$$

$$\frac{234 \, \text{ppm}}{750 \, \text{ppm}} + \frac{119 \, \text{ppm}}{200 \, \text{ppm}} + \frac{133 \, \text{ppm}}{200 \, \text{ppm}} + \frac{49 \, \text{ppm}}{350 \, \text{ppm}} = 1.71$$

Answers: $535 \, \text{ppm}_v$. Yes, by 71%.

68. What is the partial pressure of O_2 in dry air at sea level?
    Oxygen: 20.95 volume % in air

$$0.2095 \times 760 \, \text{mm Hg} = 159.2 \, \text{mm Hg}$$

69. If, in Problem 68, the barometric pressure does not change, but the air is humidified to 100% relative humidity at 25°C, how will the air composition be altered? Water vapor pressure at 25°C is 23.8 mm Hg.

O_2	20.95% (760 mm Hg – 23.8 mm Hg)	= 154.23 mm Hg
N_2, and others	79.05% (760 mm Hg – 23.8 mm Hg)	= 581.97 mm Hg
H_2O		= 23.8 mm Hg
		760 mm Hg

Answers: $O_2 = 154.23$ mm Hg. N_2 + argon + and other trace gases = 581.97 mm Hg.

70. What is the percent oxygen in air at 12,000 feet altitude?

    Answer: Although partial pressure of O_2 decreases with altitude, the percent composition of air does not. The atmosphere remains $\cong 21\%$ O_2 at any altitude. A common myth is that the percentage oxygen in the atmosphere decreases with altitude.

71. If an air sample humidified to 50% at 25°C is taken from sea level to an altitude with a barometric pressure of 600 mm Hg and the same temperature, what will be the partial pressures of air gases and water vapor?

If 100% relative humidity (partial pressure due to water vapor) at 25°C = 23.8 mm Hg, then 50% relative humidity at 25°C = 11.9 mm Hg.

$$H_2O \text{ vapor} = \frac{600 \text{ mm Hg}}{760 \text{ mm Hg}} \times 11.9 \text{ mm Hg} = 9.39 \text{ mm Hg}$$

O_2	20.95% (600 mm Hg − 9.39 mm Hg) =	123.73 mm Hg
N_2, and others	79.05% (600 mm Hg − 9.39 mm Hg) =	466.88 mm Hg
		600.00 mm Hg

Answers: H_2O = 9.4 mm Hg. O_2 = 123.7 mm Hg. N_2, and others = 466.9 mm Hg.

72. An air filter is used at 1.36 L/m for 26 minutes to collect a monodisperse aerosol of uniformly spherical 1-micron particles with a density of 2.6. If the concentration of particles is 7.8 mppcf, how many particles are collected on the filter? How much mass is collected? What is the dust concentration in mg/m³?

$$\frac{1.36 \text{ L}}{\text{minute}} \times 26 \text{ minutes} = 35.36 \text{ L} = 1.25 \text{ ft}^3$$

$$7.8 \text{ mppcf} \times 1.25 \text{ ft}^3 = 9.75 \times 10^6 \text{ particles}$$

$$d = 1.0 \text{ micron} = 0.0001 \text{ cm}$$

$$r = 0.00005 \text{ cm}$$

$$V = \frac{4}{3} \pi r^3 = \frac{4}{3} \pi (0.00005 \text{ cm})^3 = \frac{5.236 \times 10^{-13} \text{ cm}^3}{\text{particle}}$$

$$\frac{5.236 \times 10^{-13} \text{ cm}^3}{\text{particle}} \times 9.75 \times 10^6 \text{ particles} = 5.11 \times 10^{-6} \text{ cm}^3$$

$$\frac{5.11 \times 10^{-6} \text{ cm}^3}{\text{total particles}} \times \frac{2600 \text{ mg}}{\text{cm}^3} = 0.0133 \text{ mg}$$

$$\frac{13.3 \text{ mcg}}{35.36 \text{ L}} = \frac{0.376 \text{ mg}}{M^3}$$

Answers: 9.8×10^6 particles, 0.376 mg/m³

73. Assuming a normal probability distribution of dust particles suspended in air, what is the standard geometric deviation if the geometric mean is 1.25 microns, and the 84.13% size is 2 microns?

$$\frac{2.0 \text{ microns}}{1.25 \text{ microns}} = 1.6 \text{ GSD}$$

Answer: The geometric standard deviation is 1.6.

74. 1.3 mL of nitrogen dioxide gas/minute are diluted into an air stream of 5.7 ft³ per minute. What is the NO_2 concentration in the mixed gas stream leading to the direct-reading instrument is being calibrated?

$$ppm = \frac{\text{volume of gas/minute}}{\text{volume of air/minute}} \times 10^6$$

$$= \frac{1.3 \text{ mL/minute}}{(5.7 \text{ ft}^3/\text{minute}) \times (28.3 \text{ L/ft}^3) \times (1000 \text{ mL/L})} \times 10^6$$

$$= 8.06 \text{ ppm } NO_2$$

75. An industrial hygienist calibrates his stopwatch by telephoning the "Time Lady." He starts the watch when she says "At the tone, the time is 2:13 and 20 seconds." He stops the watch after calling her again when she says "2:27 and 10 seconds." The watch's elapsed time is 13 minutes and 33 seconds. What is the percent error of his watch?

$$2{:}27{:}10 = 2{:}26{:}70 \qquad 2{:}26{:}70 - 2{:}13{:}20 = 0{:}13{:}50$$

$$\% \text{ error} = \frac{\text{experimental value} - \text{true value}}{\text{true value}} \times 100$$

$$= \frac{813 \text{ seconds} - 830 \text{ seconds}}{830 \text{ seconds}} \times 100 = -2.05\%$$

Answers: −2.05%. Watch correction factor = (830 seconds/813 seconds) = 1.021.

76. An aqueous solution of 250 ppm (w/v) lead nitrate is atomized into respirable mist droplets producing a total mist concentration of 360 mg/m³ (including water). What is the Pb concentration in air?
250 ppm w/v $Pb(NO_3)_2$ = 250 mg $Pb(NO_3)_2$/liter of solution

Molecular weight of $Pb(NO_3)_2$ = 207 + (2 × 14) + (6 × 16) = 331

$$\frac{207}{331} \times 100 = 62.5\% \text{ Pb}$$

$$\frac{250 \text{ mg } Pb(NO_3)_2}{L} \times 0.625 = \frac{156 \text{ mg Pb}}{\text{liter}}$$

$$156 \text{ ppm} = 1.56 \times 10^{-4} = 0.000156$$

$$\frac{360\,mg}{m^3} \times 0.000156 = \frac{0.056\,mg\,Pb}{m^3}$$

77. Total airborne particulates are emitted from an industrial power plant stack at a rate of 0.64 ton per day. What are milligram/minute and pound/minute emission rates?

$$\frac{0.64\,ton}{day} \times \frac{2000\,pounds}{ton} \times \frac{454\,grams}{pound} = \frac{581,120\,grams}{day}$$

$$\frac{581,120,000\,mg}{24\,hours} \times \frac{hour}{60\,minutes} = \frac{403,556\,milligrams}{minute}$$

Answer: 403,556 milligrams/minute = 0.89 pound/minute.

78. A gas mixture is 80% methane, 15% ethane, 4% propane, and 1% butane (by volume). Their respective LELs and UELs are 5%, 3.1%, 2.1%, 1.86%, and 15.0%, 12.45%, 9.5%, and 8.41%. What are the LEL and UEL (in air) for the gas mixture?
Calculations require application of Le Chatelier's law:

$$LEL = \frac{100}{(80/5) + (15/3.1) + (4/2.1) + (1/1.86)} = 4.30\%$$

$$UEL = \frac{100}{(80/15) + (15/12.5) + (4/9.5) + (1/8.41)} = 14.13\%$$

79. How much benzene must be evaporated inside a dry 20.3L Pyrex® bottle to obtain 50 ppm$_v$ of benzene vapor? The density of liquid benzene is 0.879 g/mL.

$$C_6H_6 \text{ molecular weight} = (6 \times 12) + 6 = 78$$

$$\frac{50}{10^6} \times \frac{20.3L}{(24.45\,L/gram\text{-}mole)} \times \frac{78}{(0.879\,g/mL)} = 0.00368\,mL$$

Answer: Inject 3.7 microliters of liquid analytical reagent grade benzene.

80. How much carbon monoxide gas must be added to a 29.8 L glass bottle to obtain a 35 ppm$_v$ gas mixture? Assume that the air balance is CO-free by passing ambient air containing 0.8 ppm$_v$ CO through a Hopcalite® filter.

$$CO \text{ molecular weight} = 12 + 16 = 28$$

$$\text{Density of CO gas (at NTP)} = \frac{28 \text{ g/gram-mole}}{24.45 \text{ L/gram-mole}} = \frac{1.145 \text{ g}}{L} = \frac{0.001145 \text{ g}}{mL}$$

$$\frac{35}{10^6} \times \frac{29.8 \text{ L}}{24.45 \text{ L/gram-mole}} \times \frac{28}{0.001145 \text{ g/mL}} = 1.043 \text{ mL CO}$$

$$\text{Alternative calculation method: } \frac{x}{29,800 \text{ mL}} = \frac{35}{10^6}$$

$$x = 1.043 \text{ mL CO gas}$$

Answer: 1.04 mL of 100% CO gas (or, e.g., 10.4 mL of 10% CO).

81. 680 tons of coal containing an average of 0.7 ppm mercury are burned daily in an electricity-generating power plant. Assume that 98% of all mercury compounds in the coal are released to the atmosphere. How much mercury is released every hour?

$$\frac{680 \text{ tons}}{\text{day}} \times \frac{2000 \text{ lb}}{\text{ton}} \times \frac{454 \text{ grams}}{\text{pound}} = 6.17 \times 10^8 \text{ grams of Hg/day}$$

$$\frac{6.17 \times 10^8 \text{ grams/day}}{24 \text{ hours/day}} = 2.57 \times 10^7 \text{ grams/hour}$$

$$0.98 \left[\frac{2.57 \times 10^7 \text{ grams}}{\text{hour}} \right] \times 0.7 \times 10^{-6} \text{ ppm Hg} = 17.6 \text{ g Hg/hour}$$

Answer: A weight of 17.6 grams of mercury emitted per hour (presumably as elemental Hg^o vapor and mercury oxides, sulfides, and sulfates).

82. How many kilograms of sulfur dioxide gas (SO_2) are produced when $6\text{-}\frac{1}{2}$ tons of coal containing 3.4% sulfur are completely burned (stoichiometrically oxidized)?

$$S + O_2 \rightarrow SO_2 \uparrow (2S + 3O_2 \rightarrow 2SO_3 \uparrow)$$

$$6.5 \text{ tons} \times 0.034 = 0.221 \text{ ton sulfur} = 200.5 \text{ kg sulfur}$$

$$\frac{200.5 \text{ kg sulfur}}{32 \text{ g sulfur/gram-mole}} \times \frac{1000 \text{ g}}{\text{kg}} = 6266 \text{ moles of sulfur}$$

Therefore, 6266 moles of sulfur dioxide (SO_2) are produced.

$$\text{Molecular weight of } SO_2 = 32 + (16 \times 2) = 64$$

$$(6266 \text{ moles } SO_2) (64 \text{ g } SO_2/\text{mole}) = 401,024 \text{ grams } SO_2 = 401 \text{ kg } SO_2$$

Answer: 401 kilograms SO_2, for example, 1 ton of sulfur $\Rightarrow$ 2 tons of SO_2 gas

83. The decay constant for a reactive gas in air is 4.9×10^{-2} molecular dissociations per minute. What is the half-life of this gas in seconds?

$$N_t = N_0 \, e^{-kt}$$

First-order exponential decay kinetics,

where
N_t = number of molecules remaining after time, t,
N_0 = original number of molecules, and
k = the molecular dissociation constant.

substitute $N_0/2$ for N_t and T for t:

$$\frac{N_0}{2} = N_0 \, e^{-kT}$$

solve for T: $0.5 = e^{-kT}$
taking the natural logs on both sides: $\ln 0.5 = -kT$
thus,

$$T = \frac{\ln 0.5}{-k} = \frac{\ln 0.5}{-(4.9 \times 10^{-2} \text{ dissociations/minute})} = 14.15 \text{ minutes}$$

Answer: 14.15 minutes $\times 60$ seconds/minute $= 849$ seconds, that is, 50% decay every 849 seconds.

84. What fraction of the gas in Problem 83 remains after 1 hour?

$$N_t = N_0 \, e^{-kt}$$

dividing both sides by N_0:

$$\frac{N_t}{N_0} = e^{-kt}$$

$$\frac{N_t}{N_0} = e^{-(0.049/\text{minute})(60 \text{ minutes})} = 5.286 \times 10^{-2} = 5.3\%$$

Answer: 5.3% of the unstable gas remains in the air after 1 hour.

85. What are the mean and the standard deviations for the following analytical results of the amounts of beryllium (in mcg) on 10×10 cm surface wipe samples:

6.3, 9.7, 9.4, 12.1, 8.5, and 7.7?

Answers: Mean (average) $= 9.0$ mcg Be/100 cm^2. Standard deviation $= 1.8$.

86. In question 85, what range most likely covers about 95% of the results?
$1.8 \times 2 = 3.6$
$9.0 - 3.6 = 5.4$
$9.0 + 3.6 = 12.6$

Answer: 5.4–12.6 mcg Be/100 cm²

87. What is the statistical correlation between the following corresponding "x" and "y" values: $x = 5, 13, 8, 10, 15, 20, 4, 16, 18,$ and 6; $y = 10, 30, 30, 40, 60, 50, 20, 60, 50,$ and 20?

Answer: $r = 0.866$

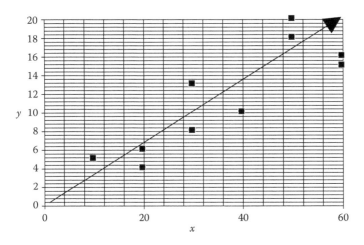

88. Workers inhale respirable dust that is 27% quartz and 11% cristobalite. What is the PEL for this fibrogenic dust mixture?

$$PEL = \frac{10 \text{ mg/m}^3}{2 + \% \text{ quartz} + 2 \, (\% \text{ cristobalite})} = \frac{10 \text{ mg/m}^3}{2 + 27 + (2 \times 11)} = 0.196 \text{ mg/m}^3$$

Answer: <0.2 mg/m³, assuming the balance of the inhaled dust mixture is biologically inert.

89. A 2 mL CS_2 extract of a small charcoal tube contained 3.7 mcg EO per mL. Air was sampled at 93 mL/min for 97 minutes at 77°F. What was the ethylene oxide concentration in air assuming 79% analytical recovery? The air sample was carried out in Boston.

$$EO = C_2H_4O$$

Molecular weight = 24 + 4 + 16 = 44 grams/gram-mole

$(93 \text{ mL/minute}) \times 97 \text{ minutes} = 9021 \text{ mL} = 9.021 \text{ L}$

$$2\,\text{mL} \times (3.7\,\text{mcg/mL}) = 7.4\,\text{mcg EO}$$

$$\frac{(7.4\,\text{mcg EO/9.021 L}) \times 24.45}{44} = 0.45\,\text{ppm}$$

$$\frac{100}{79} = 1.266$$

$$0.45\,\text{ppm}_v \times 1.266 = 0.6\,\text{ppm}_v$$

90. Determine cubic feet of vapor produced from evaporation of 1 gallon of methanol at 60°F in New Orleans. The density of methyl alcohol is 0.8 g/mL.
1 gallon = 3785 mL × 0.8 g/mL = 3028 grams = 6.67 pounds
CH_3OH molecular weight = 12 + 16 + 4 = 32.
1 pound of CH_3OH forms (359 ft³/32) vapor at 0°C = 11.2 ft³
At 60°F: °R = °F + 460 = 520°R
°R = 32°F + 460 = 492°R

$$\frac{\text{ft}^3}{\text{lb}} = \frac{(520°\text{R})\,(359\,\text{ft}^3)}{(492°\text{R})\,(\text{mol. wt.})} = \frac{379.4}{32} = \frac{11.86\,\text{ft}^3}{\text{lb}}$$

$$6.67\,\text{lb} \times \frac{11.86\,\text{ft}^3}{\text{lb}} = 79.1\,\text{ft}^3$$

Answer: 79.1 ft³ of vapor are produced by evaporating 1 gallon of CH_3OH at 60°F at sea level.

91. A 4-inch diameter Petri dish had 0.6 colony-forming units/cm² after the culture was incubated from a 27.4 ft³ air sample. How many viable organisms/m³ does this represent?

$$27.4\,\text{ft}^3 \times 28.3\,\text{L/ft}^3 = 775.4\,\text{L} = 0.775\,\text{m}^3$$

$$4''\text{ diameter} = 2''\text{ radius},\ r = 5.08\,\text{cm}$$

$$A = \pi r^2 = \pi(5.08\,\text{cm})^2 = 81.07\,\text{cm}^2$$

$$81.07\,\text{cm}^2 \times 0.6\,\text{CFU/cm}^2 = 48.6\,\text{CFU/Petri dish}$$

$$\frac{48.6\,\text{CFU}}{0.775\,\text{m}^3} = \frac{63\,\text{CFU}}{\text{m}^3}$$

Answer: 63 colony-forming units per cubic meter of air.

92. 0.062 mL of ethyl acetate is evaporated/min into a 34 liter/minute air stream. What is the EA vapor concentration in ppm at 77°F and 760 mm Hg? Molecular weight of EA = 88. Density of liquid EA = 0.9 g/mL.

$$0.062 \, \text{mL} \times (0.9 \, \text{g/mL}) = 0.0558 \, \text{gram}$$

$$\frac{55.8 \, \text{mg}}{34 \, \text{L}} = \frac{1.64 \, \text{mg}}{\text{L}} = \frac{1640 \, \text{mg}}{\text{m}^3}$$

$$\text{ppm}_v = \frac{(1640 \, \text{mg/m}^3) \times 24.45}{88} = 456 \, \text{ppm}_v \text{ ethyl acetate vapor}$$

93. A direct-reading instrument to measure CO gas in air is calibrated at sea level and 25°C. What is the corrected reading for the instrument if it indicates 47 ppm$_v$ CO at 7500 feet elevation and 29°C?

"Rule of thumb": For every 1000 feet of elevation above sea level, the barometric pressure decreases by about 1 inch of mercury. Therefore, for 7500 feet, the equivalent pressure = 7.5" Hg × (25.4 mm/inch) = 190.5 mm Hg.

$$760 \, \text{mm Hg} - 190.5 \, \text{mm Hg} = 569.5 \, \text{mm Hg}$$

$$\text{ppm at NTP} = \text{meter reading} \times \frac{P}{760} \times \frac{298 \, \text{K}}{T} = 47 \, \text{ppm} \times \frac{P}{760} \times \frac{298 \, \text{K}}{T}$$

$$= 47 \, \text{ppm} \times \frac{569.5 \, \text{mm Hg}}{760 \, \text{mm Hg}} \times \frac{298 \, \text{K}}{302 \, \text{K}} = 34.8 \, \text{ppm}$$

Answer: Instrument correction factor $= \dfrac{34.8 \, \text{ppm CO}}{47 \, \text{ppm CO}} = 0.74.$

94. The concentration of a ketone vapor in inhaled air is 48 ppm$_v$. It is 19 ppm$_v$ in the end-exhaled air of a worker. What is the average body retention of this ketone?

$$\% \text{ retention} = \frac{C_i - C_e}{C_i} \times 100 = \frac{48 \, \text{ppm} - 19 \, \text{ppm}}{48 \, \text{ppm}} \times 100 = 60\%$$

Answer: "Retention" is a function of catabolism-excretion, detoxification, and storage.

95. Two airborne dust samples obtained in a worker's breathing zone had the following results for combined work place daily exposure to respirable silica:

Air Sample	Duration, Minutes	Liters	Respirable Weight, mg	mg/m³	Silica
Morning	161	274	0.961	3.51	6.9% quartz
					1.8% cristobalite
					0.0% tridymite
Afternoon	247	420	0.530	1.26	7.3% quartz
					1.9% cristobalite
					0.0% tridymite
Total	408	694	1.491		

Calculate the percent quartz, cristobalite, and tridymite in the respirable particulate fractions. Calculate the PEL for the mixture and the employee's exposure. Adjust his exposure to an 8-hour time-weighted average. Assume that the remainder of the worker's exposure was in an area with no airborne silica dust. Also assume that this clean-shaven worker used an organic vapor cartridge respirator.

$$\text{quartz: } 6.9\% \times \frac{0.961\,\text{mg}}{1.491\,\text{mg}} + 7.3\% \times \frac{0.530\,\text{mg}}{1.491\,\text{mg}} = 7.0\%$$

$$\text{cristobalite: } 1.8\% \times \frac{0.961\,\text{mg}}{1.491\,\text{mg}} + 1.9\% \times \frac{0.961\,\text{mg}}{1.491\,\text{mg}} = 2.4\%$$

$$\text{PEL} = \frac{10\,\text{mg/m}^3}{2 + \%\text{ quartz} + 2\,(\%\text{ cristobalite}) + 2\,(\%\text{ tridymite})}$$

$$= \frac{10\,\text{mg/m}^3}{2 + 7 + 2\,(2.4) + 2\,(0)} = \frac{0.72\,\text{mg}}{\text{m}^3}$$

$$\text{actual exposure} = \frac{(0.961\,\text{mg/m}^3) + (0.530\,\text{mg/m}^3)}{0.694\,\text{m}^3} = 2.15\,\text{mg/m}^3$$

$$\text{TWAE} = \frac{2.15\,\text{mg}}{\text{m}^3} \times \frac{408\,\text{minutes}}{480\,\text{minutes}} = 1.83\,\text{mg/m}^3$$

Answers: quartz = 7.0%; cristobalite = 2.4%. PEL = 0.72 mg/m³. Exposure period = 2.15 mg/m³. Adjusted 8-hour TWAE = 1.83 mg/m³. Exposures are calculated without regard to the use of a respirator. Besides, organic vapor cartridges make this a totally unacceptable respirator for silica dust exposures.

96. A precision rotameter was calibrated at 1.7 L/m at sea level and 70°F. Air will be sampled for respirable silica dust at 90°F and at a barometric pressure of 633 mm Hg with a cyclone dust sampler. What corrected rotameter reading must be used with this cyclone? What is the rotameter correction factor, C_f?

$$90°F = 32.2°C$$

$$32.2°C + 273\,K = 305.2\,\text{kelvin}$$

$$\frac{1.7\,L}{min} \times \frac{633\,mm\,Hg}{760\,mm\,Hg} \times \frac{298\,K}{305.2\,K} = \frac{1.38\,L}{min}$$

$$C_f = \frac{1.7\,L/m}{1.38\,L/m} = 1.23$$

Answer: In other words, sample at 1.7 L/min × 1.23 = 2.09 L/min to obtain the desired air-sampling rate of 1.7 L/min.

97. A clothing dry cleaning plant purchases two 55-gallon drums of perchloro-ethylene every month. Losses to the environment are entirely due to evaporation. If 75% of the total loss is through a vent exhausting 200 cfm, what is average "perc" vapor concentration in the vent if the plant operates 6 days/week and 9 hours daily. The density of liquid "perc" is 1.62 g/mL.

    110 gallons/month × 3.785 L/gallon = 416.4 liters "perc" evaporated every month.

$$\frac{6\,days}{week} \times \frac{9\,hours}{day} \times \frac{60\,minutes}{hour} \times \frac{4.5\,weeks}{month} \times \frac{200\,ft^3}{minute} = \frac{2.916 \times 10^6\,ft^3}{month}$$

$$\frac{416.4\,L\,\text{``perc''}/month}{2{,}916{,}000\,ft^3/month} = \frac{416{,}400\,mL\,\text{``perc''}}{82{,}581{,}120\,L\,\text{of air and vapor}}$$

$$= \frac{0.05\,mL}{L} = \frac{5\,mL}{m^3}$$

$$5\,mL \times (1.62\,g/mL) = 8.1\,\text{grams of ``perc''}\ 8.1\,g/m^3 = 8100\,mg/m^3$$

Molecular weight of "perc" = (4 × 35.5) + (2 × 12) = 166

$$ppm = \frac{(8100\,mg/m^3) \times 24.45}{166} = 1193\,ppm$$

$$1193\,ppm\,\text{``perc''} \times 0.75 = 895\,ppm$$

Answer: ≅ 900 ppm "perc" vapor

98. How much water vapor is in every 1000 ft³ of building air if the relative humidity is 50% at an air temperature of 80°F? The vapor pressure of H_2O at 80°F = 26.2 mm Hg. Assume this building is located in Norfolk, Virginia.

$$80°F = 26.67°C$$

$$26.67°C + 273\,K = 299.76\,K \text{ (Kelvin)}$$

$$\frac{22.4\,L}{\text{gram-mole}} \times \frac{299.76\,K}{273\,K} = \frac{24.59\,L}{\text{gram-mole}}$$

$$\text{ppm}\,H_2O\,\text{vapor} = 0.5 \times \frac{26.2\,mm\,Hg}{760\,mm\,Hg} \times 10^6 = 17,237\,\text{ppm}$$

$$\frac{mg}{m^3} = \frac{\text{ppm} \times \text{mol. wt.}}{24.59\,L/\text{gram-mol}} = \frac{17,237 \times 18}{24.59} = 12,612\,mg/m^3$$

$$\frac{12.61\,\text{grams}}{35.3\,ft^3} = \frac{0.357\,\text{gram}}{ft^3}$$

$$\frac{357\,g}{10^3\,ft^3} = \frac{0.79\,lb}{10^3\,ft^3}$$

Answer: 0.79 pound of water vapor per 1000 ft³ of air.

99. Determine the amount (in grams) of CO_2 produced by completely burning 25 L of ethane in air at standard temperature and pressure according to the stoichiometric combustion equation:

$$2C_2H_6 + 7O_2 \rightarrow 4CO_2\uparrow + 6H_2O\uparrow$$

$$C_2H_6 \text{ molecular weight} = (2 \times 12) + 6 = 30$$

$$30\,\text{grams/gram-mole}$$

$$22.4\,L = 30\,g$$

$$25\,L = x\,\text{grams}$$

$$\frac{22.4\,L}{25\,L} = \frac{30\,\text{grams}}{x\,\text{grams}} \quad \frac{33.48\,g}{30\,g/g\text{-mole}} = 1.116\,\text{moles of } C_2H_6$$

$$\frac{2}{7} = \frac{1}{3.5}$$

$$1.116\,\text{moles} \times 2 = 2.232\,\text{moles of } CO_2 \text{ produced}$$

$$2.232\,\text{moles of } CO_2 \times (44\,\text{grams/gram-mole}) = 98.2\,\text{grams}$$

Answer: 98.2 grams of CO_2 are produced.

100. If the concentration of a hydrocarbon vapor in inhaled air is $70 \, mg/m^3$, exposure is 8 hours, respiratory ventilation rate is $1.2 \, m^3/hour$, and the average retention is 73%, what is the biologically absorbed dose?

$C \, (mg/m^3) \times T \, (hour) \times V \, (m^3/hour) \times R \, (\%) = $ absorbed dose in milligrams

$$= \frac{70 \, mg}{m^3} \times 8 \, hour \times \frac{1.2 \, m^3}{hour} \times 0.73 = 490 \, milligrams$$

101. Workers in chemical plants such as petroleum refineries often work in 6-week schedules of three 12-hour days for 3 weeks followed by four 12-hour work days for 3 weeks. What is the PEL and TLV reduction for such workers?

Note that the weekly average exposures of 36 and 48 hours are slightly different than a normal work week schedule of 40 hours (weekly average is 42 hours).

$$\text{TLV reduction factor} = \frac{8 \, hours}{12 \, hours} \times \frac{24 \, hours - 12 \, hours}{16 \, hours} = 0.5$$

Answer: Note that the reduction factor of 0.5 applies to 12-hour work days regardless if the exposures are 3, 4, or 5 or more days per week.

102. Hydrogen chloride has a TLV® (C) of 5 ppm. What is the modified TLV for HCl gas for a work schedule of 12 hours per day for a 3-day work week?

Answer: Since the basis of the TLV is prevention of acute respiratory irritation, lowering of the limit might not be justifiable. The concept of reducing 8-hour limits applies to systemically toxic agents and not usually to air contaminants with acute effects and with C, or ceiling, limits.

103. Determine the mass of vapor expelled when 1000 gallons of benzene are added to a 5000 gallon storage tank if the temperatures of the tank space and benzene are 20°C. Assume the vessel wall is warmer than the liquid benzene, the tank is in Baltimore, and the tank contained 2500 gallons of benzene for the past 3 weeks. The vapor pressure of benzene at $20°C = 75 \, mm \, Hg$.

The American Petroleum Institute provides a formula addressing the release of solvent vapors from tanks during filling. This is used to estimate vapor emissions into work room or ambient air by displacement of the saturated vapors of volatile hydrocarbons.

The mass of vapor expelled by displacement when a volatile liquid is transferred into a tank using the API formula is

$$M = 1.37 \, V \, S \, P_v \left[\frac{M \, W}{T} \right],$$

where

M = mass of vapor expelled, lb
V = volume of liquid transferred to tank, ft³
S = fraction of vapor saturation of expelled air
P_v = vapor pressure of liquid, atm
MW = molecular weight of solvent or hydrocarbon
T = temperature of tank vapor space, °R

For splash filling of a tank initially free of vapors, or for the refilling of tanks from which the same liquid was just withdrawn, the value of S is normally 1.0.

$$V = 1000 \text{ gallons} = 3785 \text{ liters} = 133.7 \text{ ft}^3$$

$$P_v = 75 \text{ mm Hg}$$

$$75 \text{ mm Hg}/760 \text{ mm Hg} = 0.09868 \text{ atmosphere}$$

$$C_6H_6 = \text{molecular weight} = (6 \times 12) + 6 = 78$$

$$20°C = 68°F$$

$$°R = °F + 459.69 = 68°F + 459.69 = 527.69°F$$

$$M = (1.37)\,(133.7 \text{ ft}^3)\,(1.0)\,(0.09868 \text{ atm}) \left[\frac{78}{527.69} \right] = 2.67 \text{ lb}$$

Alternate calculation method:
1000 gallons benzene vapor = 3785 liters of benzene vapor

$$\text{ppm} = \frac{75 \text{ mm Hg}}{760 \text{ mm Hg}} \times 10^6 = 98,684 \text{ ppm}$$

$$\frac{\text{mg}}{\text{M}^3} = \frac{\text{ppm} \times \text{mol. wt.}}{24.04} = \frac{98,684 \text{ ppm} \times 78}{24.04} = \frac{98,684 \times 78}{24.04}$$
$$= 320,190 \text{ mg/m}^3$$

$$(320.19 \text{ g/m}^3) \times 3.785 \text{ m}^3 = 1211.9 \text{ g} = 2.67 \text{ lb}$$

Answer: 2.67 pounds of benzene vapor are expelled to the atmosphere.

104. Workers are exposed to systemically toxic air contaminants 12 hours/day and 6 days/week. By how much should the exposure limits be reduced to

help protect their health? Apply Brief and Scala method for air toxicants with chronic effects.

$$\frac{40}{h} \times \frac{168 - h}{128},$$

where

h = hours worked/week
40 = hours in a normal work week
168 = total hours in a week
128 = hours available in a normal work week to detoxify, excrete toxicants, and so on.

$$\frac{40}{72} \times \frac{168 - 72}{128} = 0.42$$

Answer: Reduce their exposure limits (including action levels) by at least 42%.

105. Using the OSHA model, what is the equivalent PEL for cumulative air toxicants such as lead and mercury (PELs = 0.05 mg/m³) if exposures are, for example, 50 hours per week?

$$\text{equivalent PEL} = \frac{40 \text{ hours}}{\text{exposure hours/week}} = \text{weekly adjustment}$$

for example,

$$\frac{40 \text{ hours}}{50 \text{ hours}} \times 0.05 \text{ mg/m}^3 = 0.04 \text{ mg/m}^3$$

Answer: In other words, in this case, if the exposure week is increased from the normal 40–50 hours (20%), reduce the exposure limit by 20%.

106. How many molecules of mercury vapor exist per cm³ in cold traps of mercury diffusion pumps maintained at minus 120°C? The vapor pressure of mercury at this temperature is 10^{-16} mm Hg.

$$\frac{\text{moles of Hg}}{\text{cm}^3} = n = \frac{PV}{RT} = \frac{10^{-16} \text{ atmosphere} \times 10^{-3} \text{ L}}{0.082 \text{ L} - \text{atm}^{-1} \text{ mole}^{-1} \times 153°\text{K}} = 1 \times 10^{-23} \text{mole Hg}$$

10^{-23} mole Hg × 6.023 × 10^{23} molecules/mole = 6 molecules of Hg/cm³.

Answer: Six mercury molecules per cubic centimeter. Contrast this to 1.5×10^{11} mercury molecules in each cm³ of air at the PEL of 50 mcg Hg/m³.

107. Determine the weight of solvent vapor emitted per day from a paint baking oven drying 40 gallons/day as a blend of four gallons of un-thinned enamel to 1 gallon of thinner. The enamel weighs 9.2 lb/gallon of which 51% is volatile. The thinner weighs 7 lb/gallon. Assume no particulates form and all volatiles exist in the vapor phase.

This is a blend of 4 gallons un-thinned enamel plus 1 gallon of thinner. Thus, $(40/5) \times 4 = 32$ gallons of un-thinned enamel are used daily + $(40/5) \times 1 = 8$ gallons of thinner are used daily.

Volatiles in the un-thinned enamel are

$$32 \text{ gallons} \times \frac{9.2 \text{ lb}}{\text{gallon}} \times (1 - 0.49) = 150 \text{ lb}$$

Volatiles in the thinner are: (8 gallons) (7 lb/gallon) = 56 lb

$$150 \text{ lb} + 56 \text{ lb} = 206 \text{ lb/day}$$

Answer: 206 pounds of solvent vapor are emitted to an air pollution control device (or to the atmosphere) daily.

108. The average solvent vapor concentration in a plant is 130 ppm$_v$ and cannot be further reduced by improved work practices. It is necessary to reduce this to at least 15 ppm$_v$ to protect the health of the exposed workers. By how much must the dilution air be increased?

If the solvent vapor contaminant is generated at a steady rate:

$$\frac{Q}{Q_o} = \frac{C}{C_o}$$

$$\frac{C_o}{C} = \frac{15 \text{ ppm}_v}{130 \text{ ppm}_v}$$

Therefore, $Q/Q_o = 8.7$, that is,
The supplied air volume must be increased at least 8.7 times.

Answers: Supply at least 8.7 times more clean dilution air with excellent air mixing. This is uneconomical in frigid climates. Carefully study each source of vapor release to see if mechanical local exhaust ventilation can be applied after a diligent search to control solvent vapor leaks.

109. A $100' \times 50' \times 12'$ workshop is provided with 2 cfm of outside air per ft^2 of floor area. A gaseous contaminant is evolved in this shop at a rate of 6.6 cfm. An air sample taken when the exhaust fan is off reveals a concentration of 580 ppm$_v$. How long must this fan operate before the air concentration reaches 100 ppm$_v$? Assume excellent mixing of the dilution air with the contaminated air.

$$100' \times 50' \times (2\,\text{cfm/ft}^2) = 10,000\,\text{cfm}$$

$$t,\,\text{minutes} = \frac{2.303 \times \text{ft}^3}{Q} \times \log \frac{G - (Q \times C_a)}{G - (Q \times C_b)}$$

$$t = \frac{2.303 \times 60,000\,\text{ft}^3}{10,000\,\text{cfm}} \times \log \frac{6.6\,\text{cfm} - (10,000\,\text{cfm} \times (100/10^6))}{6.6\,\text{cfm} - (10,000\,\text{cfm} \times (580/10^6))} = 13.82$$

$$\times \log \left[\frac{6.6 - 1}{6.6 - 5.8} \right] = 13.82 \times \log 7 = 13.82 \times 0.845 = 11.7\,\text{minutes}$$

Answer: At least 12 minutes if there is good mixing—longer if there is poor mixing of fresh air with pockets of contaminated air.

110. Ambient air has been reported to contain from 0.01 to 0.02 microgram of mercury per cubic meter. How many molecules of mercury are in each liter of air inhaled at this concentration?

$$\frac{6.023 \times 10^{23}\,\text{molecules Hg/gram-mole}}{200.59\,\text{g Hg/gram-mole}} = \frac{3 \times 10^{21}\,\text{molecules}}{\text{gram}}$$

$$= \frac{3 \times 10^{15}\,\text{molecules}}{\text{microgram}}$$

$$0.01\,\text{mcg/m}^3 = 0.00001\,\text{mcg/L}$$

$$\frac{10^{-5}\,\text{mcg}}{L} \times \frac{3 \times 10^{15}\,\text{molecules}}{\text{mcg}} = \frac{3 \times 10^{10}\,\text{molecules Hg}}{L}$$

Answer: 3×10^{10} molecules of mercury per liter of air.

111. Seven mL of 100% CO gas was added by a gas syringe to a plastic bag into which 127 liters of CO-free air had been metered. What is the resultant CO gas concentration?

$$\text{ppm CO} = 10^6 \times \frac{0.007\,\text{L}}{127\,\text{L} + 0.007\,\text{L}} = 55.1\,\text{parts CO/}10^6\,\text{parts of air}$$

$$\text{Alternative calculation method:}\quad \frac{7\,\text{mL}}{127,007\,\text{mL}} \times \frac{x}{10^6}$$

Answer: $x = 55\,\text{ppm}_v$ CO

112. An incinerator produces 0.032 lb of phosgene gas per hour for every 100 lb of pentachlorophenol processed per hour at extreme temperatures with methane and excess air. If the total gas flow rate leaving the incinerator is

4600 cfm at NTP, what is the average phosgene gas concentration in this exhaust gas?

$$C_6OHCl_5 + CH_4 + \text{excess air } (O_2) \rightarrow CO_2, CO, Cl_2, HCl, H_2O, \text{etc.} + COCl_2$$

$$4600 \, \text{cfm} = 276,000 \, \text{ft}^3/\text{hour} = 7.81 \times 10^6 \, \text{L/hour}$$

$$\frac{0.032 \, \text{lb/hour}}{276,000 \, \text{cfh}} = \frac{14.53 \times 10^6 \, \text{mcg}}{7.81 \times 10^6 \, \text{L}} = 1.86 \, \text{mcg/L}$$

$$\text{Molecular weight } COCl_2 = 12 + 16 + 2 \, (31.5) = 99$$

$$\text{ppm} = \frac{(1.89 \, \text{mcg/L}) \times 24.45}{99} = 0.46 \, \text{ppm} = 460 \, \text{ppb}_v$$

Answer: 0.46 ppm$_v$ $COCl_2$ gas. TLV is 0.1 ppmv to help prevent upper respiratory tract irritation, pulmonary edema, and emphysema.

113. A worker was exposed to a 6 mg/m³ dust cloud of radioactive mercuric cyanide [Hg^{203}(CN)$_2$] for 30 minutes. Assuming no other exposures that day, what were his TWAEs to mercury and cyanide? Hg^{203} is a β-emitter with a $T_{1/2}$ of 47 days. What are the three health hazards? What is the greatest health hazard? What is the least acute? Assume that the dust is entirely respirable with all particles less than 2 microns.

$$\text{Molecular weight } Hg(CN)_2 = 203 + (12 \times 2) + (14 \times 2) = 255$$

$$(203/255) \times 100 = 79.6\% \, Hg$$

$$100\% - 79.6\% = 20.4\% \, CN^-$$

$$(6 \, \text{mg/m}^3) \times 0.796 = 4.78 \, \text{mg} \, Hg^{203}/\text{m}^3$$

$$(6 \, \text{mg/m}^3) \times 0.204 = 1.22 \, \text{mg} \, CN^-/\text{m}^3$$

$$\frac{(4.78 \, \text{mg} \, Hg/M^3) \times 0.5 \, \text{hour} + 0}{8 \, \text{hours}} = 0.30 \, \text{mg} \, Hg/\text{m}^3$$

$$(\text{TLV inorganic } Hg = 0.05 \, \text{mg} \, Hg/\text{m}^3)$$

$$\frac{(1.22 \, \text{mg} \, CN^-/\text{m}^3) + 0}{8 \, \text{hours}} = 0.08 \, \text{mg} \, CN^-/\text{m}^3$$

$$\text{TLV } CN^- = 5 \, \text{mg/m}^3$$

Answer: 0.30 mg Hg/m³ and 0.08 mg CN⁻/m³. Acute inorganic mercury and cyanide poisoning and internal ionizing radiation (pulmonary deposition and systemic absorption of an internal β-emitter). The ionizing radiation hazard more than likely outweighs the mercury and cyanide hazards because of delayed, chronic effects (cancer?). Although the mercury exposure exceeded the TLV by 6 times, it was a single occurrence, and the long term, chronic effects from this are probably negligible.

114. 13.7 liters of air at 19°C and 741 mm Hg were sampled with a midget impinger containing 14.4 mL of collection solution with 100% absorption efficiency for SO_2. The SO_2 concentration was analyzed at 11.6 micrograms/mL. Calculate the ppm_v of SO_2 in this air sample.

Total SO_2 sampled = (11.6 mcg/mL) × 14.4 mL = 167 mcg

The volume of 1 micromole SO_2 gas at 19°C and 741 mm Hg =

$$1.0 \text{ micromole } SO_2 = \frac{224 \text{ microliters}}{\text{micromole}} \times \frac{760 \text{ mm}}{741 \text{ mm}} \times \frac{292 \text{ K}}{273 \text{ K}}$$

$$= 24.6 \text{ microliters } SO_2$$

$$SO_2 \text{ concentration in ppm} = \frac{167 \text{ mcg } SO_2}{13.7 \text{ L air}} \times \frac{\text{micromole } SO_2}{64 \text{ mcg } SO_2}$$

$$\times \frac{24.57 \text{ microliters } SO_2}{\text{micromole } SO_2} = \frac{4.68 \text{ microliters } SO_2}{\text{liter of air}}$$

Answer: 4.7 ppm_v SO_2

115. One drop (0.05 mL) of TDI evaporated in an unventilated, closed telephone booth (remember those?). What is the vapor concentration of TDI if interior volume of the booth is 1.5 m³? Is atmosphere hazardous to health? TDI molecular weight is 174. Density of liquid TDI is 1.22 g/mL.

0.05 mL × 1.22 g/mL = 0.061 gram = 61 milligrams of TDI

$$ppm = \frac{(mg/M^3) \times 24.45}{\text{molecular weight}} = \frac{(61 mg/1.5 M^3) \times 24.45}{174}$$

$$= 5.7 \, ppm_v = 5700 \, ppb_v$$

$(5700 \, ppb_v/5 \, ppb_v \text{ PEL}) = 1140 \text{ times} > \text{PEL}. \, LC_{Lo} = 500 \, ppb_v$

Answers: 5.7 ppm, or 1140 times the PEL and 11.5 times the LC_{Lo}. Clark Kent (a.k.a. "Superman") would not survive this exposure. TDI has two isomers (2,4-TDI and 2,6-TDI). Both are equally toxic with respect to their ability to sensitize the lungs.

116. How long must one sample at 2 L/min with a 25 mm diameter cellulose ester membrane filter to evaluate asbestos fibers at 0.04 fiber/cubic centimeter of air? The filter area is 385 mm². A minimum fiber count is taken as 100 per mm².

$$\text{Sampling time (minutes)} = \frac{385\,\text{mm}^2 \times (100\,\text{fibers/mm}^2)}{2.0\,\text{L/minute} \times 0.04\,\text{fiber/cc} \times 1000\,\text{cc/L}}$$

$$= 481.25\,\text{minutes}$$

Answer: 481 minutes (= 8 hours), that is, this is a full work shift breathing zone air sample to determine asbestos exposure at 20% or greater of the OSHA PEL (0.04 f/cc).

117. Air contaminants in a 50,000 gallon tank are 80 ppm$_v$ CO, 3 ppm$_v$ HCN, and 8 ppm$_v$ H$_2$S. Three pipe fitters have work to do in the tank. Would you:
  a. Permit entry to not exceed 30 minutes every 8 h?
  b. Permit entry to not exceed 60 minutes every 8 h?
  c. Permit entry to not exceed more than 25% of the work shift?
  d. Permit entry only if the pipe fitters are outside for 1 hour for every hour that they are inside the tank?
  e. Ventilate the tank with a 500 cfm blower for 20 minutes?
  f. Permit entry only if the pipe fitters have escape masks?
  g. Not permit entry?
  h. Duck and cover?
  i. Contact a board-certified industrial hygienist?

Answer: g. First, it would be highly unusual to have three chemical asphyxiant gases exist in the same confined space. Second, since it is likely that the pipe fitters could breach the integrity of the pipes and tanks containing these gases, much higher gas concentrations might occur in their work area. Because of this, entry must not be permitted until further evaluation is done. All elements of a comprehensive confined space entry procedure are required to protect the health and safety of these pipe fitters (refer to OSHA 29 *CFR* 1910.1025).

118. Determine the volume and mass conversion of ozone gas at 25°C and 760 mm Hg (NTP).

$$\text{Molecular weight of O}_3 = 3 \times 16 = 48\,\text{g/g-mole}$$

$$T = 273\,\text{K} + 25°\text{C} = 298\,\text{K}$$

$$R = 0.0821\,\text{L-atm/mole-K}$$

$$\text{ppm} = \frac{W}{V} \times \frac{RT}{PM} \times 10^6 = \frac{W\,(\text{g}) \times 10^6\,(\text{mcg/g})}{V\,(\text{L}) \times 10^{-3}\,\text{m}^3/\text{L}}$$

$$\frac{(0.0821 \text{L} - \text{atm/mole-K}) \times 298 \text{K}}{1 \text{ atmosphere} \times (48 \text{ g/mole}) \times 10^6} = 0.00000051 \text{ ppm}_v$$

$$1 \text{ ppm} = 1960 \text{ mcg/m}^3$$

$$1 \text{ mcg/m}^3 = 0.51 \times 10^{-0.3} \text{ ppm}_v = 0.51 \text{ ppb}_v$$

Answer: $1 \text{ ppm}_v = 1960 \text{ mcg/m}^3$. $1 \text{ mcg/m}^3 = 5.1 \times 10^{-4} \text{ ppm}_v$.

119. Calculate the dust emission from a lumber scrap incinerator stack with the following stack sampling conditions:

V_m = volume of gas by meter = 105 ft^3

T_m = temperature at meter = $83°\text{F} + 460 = 543°\text{A}$

P_b = barometric pressure = $27.8" \text{ Hg}$

P_m = average suction at meter = $2.5" \text{ Hg}$

V_w = condensed water = 138 cc

W_t = filtered dust = 23 grams

V_o = Pitot traverse exhaust volume = $37,200 \text{ cfm}$

Volume of total gas samples (at meter conditions): convert H_2O condensate to water vapor volume at meter conditions:

$$V_v = 0.00267 \times \frac{V_w \times T_m}{P_b \times P_m} = 0.00267 \times \frac{138 \text{ cm}^3 \times 543°\text{A}}{27.8" - 2.5"} = 7.92 \text{ ft}^3$$

Total gas sampled = $V_v + V_m = 7.92 \text{ ft}^3 + 105 \text{ ft}^3 = 112.92 \text{ ft}^3$

Moisture content of the gas sampled (by volume) = M_m = volume of the moisture remaining in the metered gas at meter conditions in ft^3

Vapor pressure of H_2O at $83°\text{F} = 1.138" \text{ Hg}$

$$M_m = \frac{V_p \times V_m}{P_b - P_m} = \frac{1.138" \text{ Hg} \times 105 \text{ ft}^3}{27.8" \text{ Hg} - 2.5" \text{ Hg}} = 4.72 \text{ ft}^3$$

$$\% \text{ moisture} = \frac{V_v + M_m}{V_v + V_m} \times 100 = \frac{7.92 \text{ ft}^3 + 4.72 \text{ ft}^3}{7.92 \text{ ft}^3 + 105 \text{ ft}^3} \times 100 = 11.2\%$$

Convert total sampled gas volume to stack conditions:

$$V_t = (V_m + V_v) \times \frac{P_b - P_m}{P_s} \times \frac{T_s}{T_m}$$

$$= (105\,\text{ft}^3 + 7.92\,\text{ft}^3) \times \frac{27.8'' \text{ Hg} - 2.5'' \text{ Hg}}{27.8'' \text{ Hg}} \times \frac{700°\text{A}}{543°\text{A}} = 132.4\,\text{ft}^3$$

Dust concentration is generally referred to as grains per cubic foot and pounds per hour. To express concentration in grains/ft³, divide filtered dust weight (W_t) by total sampled gas at stack conditions (V_t), and convert grams to grains:

grains/ft³ = (W_t/V_t) × 15.43 = (23/132.4) × 15.43 = 2.68 grains/ft³

pounds/hour = (grains/ft³) × V_o × (60/70)

$$= \frac{2.68\,\text{grains}}{\text{ft}^3} \times 37,200\,\text{cfm} \times \frac{60}{760} = \frac{854\,\text{lb}}{\text{hr}}$$

Answer: 854 pounds of dust are emitted from this stack every hour.

120. 6.1 mL of TCE are evaporated in a sealed container with dimensions of 1.65 × 1.65 × 2.35 meters. Chamber air and wall temperature are 25.5°C. The barometric reading is 730 mm Hg. The density of TCE is 1.46 g/mL. What is the TCE vapor concentration in the container?

Molecular weight of CHCl=CCl₂ = (2 × 12) + 1 + (3 × 35.5) = 131.5

6.1 mL × 1.46 g/mL = 8.906 grams of liquid TCE

$$\text{molar volume} = 24.45\,\text{L} \times \frac{760\,\text{mm Hg}}{730\,\text{mm Hg}} \times \frac{298.5\,\text{K}}{298\,\text{K}} = 25.5\,\text{liters}$$

$$\text{ppm} = \frac{10^6 \times ((8.906\,\text{grams}/131.5\,\text{grams})/\text{gram-mole})}{1000 \times (1.65 \times 1.65 \times 2.35)\,\text{meters}/25.5\,\text{L}} = 269.9$$

Answer: 270 ppm$_v$ TCE vapor.

121. A drying process using acetone is done on an open bench in the center of a 20' × 20' × 10' room. The room has two to three air changes every hour. Between 7 and 10 gallons of acetone evaporate every 8 hours as determined by the department's solvent purchase records. What are the hazards? Risks? The vapor volume of acetone is 44 ft³/gallon. Calculate

average concentration of acetone vapor in the room assuming good mixing of outside air with the vapor-contaminated air.

Estimate the hazards and the risks by assuming the worst-case situation, that is, the evaporation of 10 gallons every 8 h and an air exchange rate of two room volumes every hour.

$$20' \times 20' \times 10' = 4000\,ft^3$$

$$4000\,ft^3 \times \frac{2\,air\,changes}{hour} \times 8\,hours = \frac{64,000\,ft^3}{8\,hours}$$

Total 100% acetone vapor volume = (44 ft³/gallon) × 10 gallons = 440 ft³

Total volume of air required to dilute to the TLV and PEL

$$= \frac{volume\,of\,100\%\,solvent\,vapor \times 10^6}{TLV\,(in\,ppm)} = \frac{440\,ft^3 \times 10^6}{750\,ppm} = 58,670\,ft^3$$

$$\frac{58,670\,ft^3 \times 750\,ppm}{64,000\,ft^3} = 688\,ppm$$

Answer: Nearly 700_v ppm acetone vapor. Use local exhaust ventilation! Note: Use the lowest air exchange rate and highest acetone consumption rate in calculations. Apply generous safety factor, for example, 10, to ensure vapor levels are substantially below the action level and as low as technically and economically feasible. Give careful consideration to improved work practices, alternative drying techniques, using mechanical local exhaust ventilation, use of respirators, and so on.

122. The major component of natural gas is methane (CH_4) which burns in excess air (i.e., excess oxygen) according to the gas-phase reaction:

$$CH_4 + 2O_2 + 2\,(3.78)\,N_2 \rightarrow CO_2 + 2H_2O + 7.56N_2$$

which is 1 mole of methane + 9.56 "moles" of air yields 1 mole of CO_2 + 2 moles of H_2O + 7.56 moles of N_2.

Determine the weight and volume of air required to burn 1000 ft³ of CH_4 if the gas and air are at 70°F and 15 pounds per square inch absolute pressure.

According to the reaction, 9.56 "moles" of air are required for each mole of fuel gas. Since the molar volume of both gases is the same at the same temperature and pressure, (9.56) (1000 ft³) = 9560 ft³ of air are needed.

The molar volume at 70°F and 15 psia is

$$358\,ft^3 \times \frac{70°F + 460°F}{32°F + 460°F} \times \frac{14.7\,lb/in^2}{15\,lb/in^2} = 378\,ft^3$$

The molar volume is $358\,ft^3$ at 32°F and 14.7 lb per square inch absolute, that is, this is the volume of a pound-mole of any ideal gas or vapor at these conditions. Therefore,

$$\text{weight of air required} = \frac{9560\,ft^3}{378\,ft^3/\text{pound-mole}}$$

$$= 25.3\text{ "moles" air.}$$

29 = the "apparent" molecular weight of air

(25.3 moles) ("29 lbs/mole" of air) = 734 pounds of air

Answer: $9560\,ft^3$ of air $\times 0.075\,lb/ft^3 = 717$ pounds of air precisely

123. Calculate the gas density from an Orsat analysis of the gas with moisture content of 20% (from the wet and dry bulb temperatures): $CO_2 = 10.5\%$, $CO = 6.2\%$, $O_2 = 3.0\%$, $N_2 = 80.3\%$. Orsat analyses are on a dry gas basis.

$$
\begin{aligned}
H_2O &= 0.20 \times 1.00 \ \times 18 = \ 3.60 \\
CO_2 &= 0.80 \times 0.105 \times 44 = \ 3.70 \\
CO &= 0.80 \times 0.062 \times 28 = \ 1.39 \\
O_2 &= 0.80 \times 0.03 \ \times 32 = \ 0.77 \\
N_2 &= 0.80 \times 0.803 \times 28 = \underline{17.99} \\
& \qquad\qquad\qquad\qquad\quad 27.45
\end{aligned}
$$

Gas density = $(27.45/28.966) = 0.947$ (28.966 is the "apparent" molecular weight of air).

Answer: 0.947 (air = 1.000).

124. A correction factor for excess air is often required in combustion processes. The flue (exhaust) gases are analyzed for CO_2, O_2, and CO with an Orsat apparatus. Nitrogen gas is determined by difference. A flue gas analyzed by an Orsat device contained 10.1% CO_2, 11.1% O_2, and 0.8% CO. Based on allowable 50% excess air, correct the flue gas dust loading. The flue gas contained 0.493 lb of dust per 1000 lb of gas.

$$\text{Ratio of actual:theoretical air} = \frac{N_2}{N_2 - 3.782\,(O_2 - 1/2\,CO)}$$

$$= \frac{N_2}{N_2 - 3.782\,(O_2 - 1/2\,CO)} = 0.209$$

% total air = 100% + percent excess air

Ratio of actual to theoretical air = percent total air/100

$$\frac{(2.09)(0.493\,\text{lb dust}/1000\,\text{lb gas})}{150/100} = 0.687\,\text{lb dust}/1000\,\text{lb gas}$$

Answer: 0.687 pound of dust per 1000 pounds of flue gas.

125. An industrial hygienist would like to estimate a welder's actual exposure to welding fume during "arc time" discounting interval periods when the welding arc fume generation does not occur. He obtains a 3-hour and 36-minute air sample at an average air flow rate of 2.13 L/min. The difference in the filter's weight after air sampling was 6.77 milligrams. If seven measurements of the actual electric arc time on this production welder were 21, 19, 14, 13, 26, 20, and 15 seconds of arc time per minute, what was the approximate average welding fume concentration during the actual welding process?

$$\frac{21 + 19 + 14 + 13 + 26 + 20 + 15}{7} = \frac{18.3\,\text{seconds}}{60\,\text{seconds}} \times 100 = 30.5\%\ \text{"arc time"}$$

$$3\,\text{hours and}\ 36\,\text{minutes} = 216\,\text{minutes}$$

$$216\,\text{minutes} \times (2.13\,\text{liters/minute}) = 460\,\text{liters}$$

$$\frac{6.77\,\text{milligrams}}{0.460\,\text{m}^3} = \frac{14.71\,\text{mg total fumes}}{\text{m}^3}$$

$$\frac{1}{0.305} \times \frac{14.71\,\text{mg}}{\text{m}^3} = \frac{48.33\,\text{mg fume}}{\text{m}^3}$$

Answers: 30.5% actual arc time. TWAE level during sample period was 14.71 mg/m^3. The average concentration during the welding arc generation was 48.33 mg/m^3 which, of course, must not be construed as the welder's actual TWAE.

126. Relative evaporation rate of toluene has been reported by the American Alliance of Insurers (*Handbook of Organic Solvents*) to be 4.5 times slower than diethyl ether. If 25 mL of "ether" completely evaporates from a flat surface in 21 seconds, how long before 1 gallon of toluene evaporates at similar conditions of liquid and surface temperatures, air temperatures, exposed surface area, and air flow rate over the liquid surface?

"ether": 25 mL/25 second = average of 1.19 mL evaporates/second

Toluene: $(1.19 \, \text{mL/second})/4.5 = \text{average of } 0.264 \, \text{mL evaporates/second}$

$$1 \, \text{gallon} \times \frac{3785 \, \text{mL}}{\text{gallon}} \times \frac{\text{second}}{0.264 \, \text{mL}} = 14{,}337 \, \text{seconds} = 240 \, \text{minutes} = 4.0 \, \text{hours}$$

Answer: 4 h (very approximate).

127. A 43-mm long NO_2 gas permeation tube was used to calibrate a direct reading air-sampling instrument. What is the outlet concentration of NO_2 in
parts per million if the flow rate of NO and NO_2-free nitrogen over the tube
is 43 mL/minute, and the diluent pure air flow rate is 11.6 L/min? System
temperature is maintained at 30°C, and permeation rate, PR, for NO_2 gas at
this temperature is 1200 nanograms/minute-centimeter. The permeation K
value (diverse density in L/g) for NO_2 at 30°C is 0.541.

$$43 \, \text{mm} = 4.3 \, \text{cm}$$

$$\frac{1200 \, \text{ng}}{\text{minute-cm}} \times 4.3 \, \text{cm} = \frac{5160 \, \text{nanograms}}{\text{minute}}$$

$$\text{ppm} = \frac{PR \times K}{A + B},$$

where
   PR = generation rate of permeation tube, micrograms/minute
   K = generation rate constant supplied by the manufacturer
   A = flow rate of diluent air, L/min
   B = flow rate of diluent nitrogen, L/min

$$\text{ppm} \, NO_2 = \frac{(5.16 \, \text{mcg/min}) \times 0.541}{(11.6 \, \text{L/min}) + (0.043 \, \text{L/min})} = 0.239$$

Answer: $0.24 \, \text{ppm}_v \, NO_2 = 240 \, \text{ppb}_v \, NO_2$

128. A large activated charcoal air-sampling tube has a desorption efficiency of
68% for an alcohol vapor. Weight of this alcohol reported by an industrial
hygiene chemist is 4.23 milligrams for a 119 minute air sample obtained at
an average air flow rate of 770 mL/min. What is concentration in ppm if
molecular weight of the alcohol is 74? Assume air-sampling temperature
was 77°F, and barometric pressure was 760 mm Hg.

$$\frac{\text{corrected milligrams}}{\text{sample}} = \frac{\text{detected weight, milligrams}}{\text{desorption efficiency}} = \frac{4.23 \, \text{mg}}{0.68} = 6.22 \, \text{mg}$$

Air volume sampled = 199 minutes × 0.77 L/min = 91.63 liters

$$\text{ppm} = \frac{(6220 \text{ mcg}/91.63 \text{ L}) \times 24.45}{74} = 22.43 \text{ ppm R} - \text{OH}$$

Answer: 22.4 ppm$_v$ of alcohol vapor. MeOH, especially, is very poorly collected on, and desorbed, from charcoal. Silica gel or other adsorbents must be used.

129. A critical air-sampling orifice was calibrated at an elevation of 2000 feet at 65°F and at 1.40 L/min. The orifice will be used at 9000 feet at 80°F. What is the air flow meter correction?

2000 feet = 13.7 psia = 27.82″ Hg

65°F = 18.33°C = 291.3 kelvin

9000 feet = 10.5 psia = 21.39″ Hg

80°F = 26.67°C = 299.7 kelvin

$$Lpm_{actual} = Lpm_{indicated} \times \sqrt{\frac{CP \times AT}{AP \times CT}},$$

where
  P and T = pressure and temperature in absolute units
  C = calibration conditions
  A = actual sampling and critical orifice use conditions

$$Lpm_{actual} = 1.4 \, Lpm \times \sqrt{\frac{13.7 \text{ psia} \times 299.7 \text{ K}}{10.5 \text{ psia} \times 291.3 \text{ K}}} = 1.62 \text{ L/min}$$

Answer: 1.62 liters per minute

$$\text{Correction factor} = \frac{1.62 \text{ lpm}}{1.40 \text{ lpm}} = 1.157.$$

See Problem 311 for an explanation of the square root function for orifice air flow meters such as rotameters and critical orifices. Do not use Charles' and Boyle's laws for correction of temperature and pressure for orifice air flow meters.

130. One milligram of typical mineral dust is equivalent to about 30–50 million dust particles as determined by standard impinger counting techniques. In a respirable dust cyclone breathing zone the air sample obtained at 1.67 L/min for 450 minutes, the filter weight gain was 0.35 milligrams. What is the weight of the respirable dust per cubic meter? How many particles does this represent? What is this in mppcf?

$$(1.67\,\text{L/min}) \times 450\,\text{minutes} = 751.5\,\text{liters}$$

$$\frac{350\,\text{micrograms}}{751.5\,\text{liters}} \times \frac{30\,\text{to}\,50 \times 10^6\,\text{particles}}{\text{mg dust}} = \frac{14\,\text{to}\,23 \times 10^6\,\text{particles}}{m^3} \times \frac{m^3}{35.3\,\text{ft}^3}$$

$$= 0.40\,\text{to}\,0.65\,\text{mppcf}$$

$$350\,\text{mcg}/751.5\,\text{L} = 0.466\,\text{mcg/L}$$

Answers: $0.466\,\text{mg/m}^3$. 14–23 million. 0.40–0.65 mppcf.

131. NIOSH reports a coefficient of analysis variation of 0.06 for 1,1-dichloroethane (DCE). The coefficient of variation (CV) for rotameter measurements is 0.05 when air sampling is carried out with activated charcoal tubes. What is the total sampling and analysis CV?

$$CV = \sqrt{(0.06)^2 + (0.05)^2} = \sqrt{0.0061} = 0.078$$

Answer: The combined sampling and analysis CV is 0.078.

132. What volume of air is needed to dilute the vapors from 1 gallon of varnish maker and painter's naphtha below 20% of LEL? LEL of VM & P naphtha is 0.9%. Cubic feet of VM & P naphtha per gallon at 70°F = 22.4 ft³. When 7 gallons solvent are evaporated from parts every hour in a drying oven, what rate of ventilation (in cfm) is required to keep VM & P naphtha vapor concentration below 20% of the LEL?

Dilution volume (ft³) required for each gallon of solvent

$$= \frac{(100 - \text{LEL})\,(\text{cubic feet of vapor per gallon})}{20\%\,\text{LEL}}$$

$$= \frac{(100 - 0.9)\,(22.4\,\text{ft}^3/\text{gallon})}{0.20 \times 0.9\%} = \frac{12{,}332\,\text{cubic feet of air}}{\text{gallon of VM \& P naphtha}}$$

$$\frac{7\,\text{gallons}}{\text{hour}} \times \frac{12{,}332\,\text{ft}^3}{\text{gallon}} \times \frac{\text{hour}}{60\,\text{minutes}}$$

$$= 1439\,\text{ft}^3 \text{ of air (at NTP, i.e., oven supply air) per minute}$$

Answers: 12,332 ft³/gallon. 1439 cfm are needed with good, uniform mixing of the dilution air with the vapors and gases in the oven.

133. A 170-cubic inch gasoline engine is running in an enclosed garage at 850 rpm. What volume of exhaust gases is produced every hour? If exhaust gases contain 0.76% CO by volume (7600 ppm$_v$), how many cubic feet of CO gas are generated per hour?

$$\frac{\text{engine displacement (in}^3) \times \text{engine rpm (60 minutes/hour)}}{2* \times (1728 \, \text{in}^3/\text{ft}^3)}$$

= exhaust volume in ft^3/hour (not corrected for temperature of the hot gas)

(*Note: The denominator of 2 allows for 50% volume of pistons in the engine's cylinders, that is, some up, some down.)

$$\frac{170 \, \text{in}^3 \times 850 \, \text{rpm} \times (60 \, \text{minutes/hour})}{2 \times (1728 \, \text{in}^3/\text{ft}^3)} = 2509 \, \text{ft}^3/\text{hour}$$

(2509 ft^3 total exhaust gases/hour) $\times$ 0.0076 = 19.1 ft^3 of CO/hour

Answers: 2509 ft^3 of total exhaust gases per hour. This engine generates 19.1 ft^3 of CO gas per hour.

134. If the garage in Problem 133 has natural ventilation air exchange rate of 0.5/ hour, what is the CO concentration 5 minutes after starting the engine? This garage is 4000 ft^3.

   Contamination of air in enclosed spaces is calculated from the rate of generation of the air pollutant and ventilation of the space [assumes contaminant generation is steady and the ventilation provides uniform air mixing (increase and decay kinetics)].

$$C = \frac{100 \, \text{K} \, (1 - e^{-Rt})}{RV},$$

where

C = % (volume/volume) of the gas or vapor in the space after time, t
R = air changes of the space/hour
t = time, hours
V = volume of the space, ft^3
K = contaminant generation rate, ft^3/hour

$$C = \frac{100 \left[19.1 \, \text{ft}^3/\text{hour} \right] (1 - e^{-(0.5)(0.083)})}{(0.5)(4000 \, \text{ft}^3)} = \frac{(1910 \, \text{ft}^3) \, (1 - e^{-0.0415})}{2000 \, \text{ft}^3}$$

$$= \frac{(1910)(1 - 0.959)}{2000} = 0.0392\% \, \text{CO} = 392 \, \text{ppm CO}$$

Answer: 392 ppm$_v$ CO. Gas concentration will increase and eventually will plateau as the engine continues to operate and the generation rate is balanced by ventilation loss. 392 ppm$_v$ CO does not exceed the NIOSH IDLH level of 1500 ppm$_v$, but exceeds OSHA's 200 ppm$_v$ ceiling concentration. Evacuate, stop the engine stat, ventilate, and test the air for CO before allowing reentry.

135. A calibrated length of stain CO detector tube indicates the breathing zone concentration of 100 ppm$_v$ ± 25% at a barometric pressure of 625 mm Hg. What is the CO gas concentration corrected to sea level?

$$100 \text{ ppm} \pm 25\% \times \frac{760 \text{ mm Hg}}{625 \text{ mm Hg}} = 122 \text{ ppm} \pm 25\% = 90\text{–}153 \text{ ppm CO}$$

Answer: 90 to 153 ppm$_v$ CO gas.

136. What is resultant density of air that contains 1% (vol/vol) CO gas? Density of 100% CO gas = 0.97, and 100% air = 1.00 (no units).

$$1\% \text{ CO (v/v)} = 10{,}000 \text{ ppm}_v \text{ CO in } 990{,}000 \text{ ppm}_v \text{ air.}$$

$$\text{Air} \quad 0.99 \times 1.00 = 0.9900$$

$$\text{CO} \quad 0.01 \times 0.97 = 0.0097$$

$$\text{Resultant density} = 0.9900 + 0.0097 = 0.9997$$

Answer: Density of a 10,000 ppm$_v$ concentration of CO gas in air is 0.9997 (essentially equal to that of air). CO gas stratification will never occur. However, warm exhaust gases tend to ascend, and this mixture would tend to rise if the CO source was organic fuel combustion.

137. A breathing zone air sample was obtained from a paint sprayer for 443 minutes at an average rate of 0.83 liter/minute using a large charcoal tube preceded by a 37 mm PVC membrane filter cassette. The difference in filter weight after sampling was 6.37 milligrams. The charcoal tube contained 2.97 mg n-butyl alcohol, 14.66 mg toluene, 48.49 mg mineral spirits, and 7.01 mg xylene. Benzene was not detected. What was the painter's 8-hour TWAE to these vapor air contaminants? Approximate molecular weights = 74.1, 92.1, 99, and 106.2, respectively. What was painter's exposure to airborne particles? Assume an 8-hour exposure.

$$443 \text{ minutes} \times (0.83 \text{ L/minute}) = 367.7 \text{ liters}$$

$$\text{ppm}_v = \frac{(\text{micrograms/liter}) \times 24.45}{\text{molecular weight}}$$

$$n\text{-butanol:} \frac{(2970/367.7) \times 24.45}{74.1} = 2.7\,\text{ppm}_v$$

$$\text{Toluene:} \frac{(14{,}660/367.7) \times 24.45}{92.1} = 10.6\,\text{ppm}_v$$

$$\text{Mineral spirits:} \frac{(48{,}490/367.7) \times 24.45}{99} = 32.6\,\text{ppm}_v$$

$$\text{Xylene:} \frac{(7010/367.7) \times 24.45}{106.2} = 4.4\,\text{ppm}_v$$

$$6370\ \text{micrograms}/367.7\ \text{liters} = 17.32\,\text{mg/m}^3$$

$$\frac{480\ \text{minutes/8 hour work shift}}{443\ \text{minutes}}$$

$$= 1.084\ \text{multiplier factor for 480 minutes exposure}$$

Answers: 2.9 ppm$_v$ n-butanol, 11.5 ppm$_v$ toluene, 35.3 ppm$_v$ mineral spirits, and 4.8 ppm$_v$ xylene. 18.77 mg/m^3 total airborne particulates. Particulate exposure exceeds the PEL. See Problem 138 for solvent vapor TWAEs.

138. What is the additive exposure to solvent vapors in Problem 137? Does it exceed the PEL for the mixture? PELs are 50 (C), 100, 100, and 100 ppm$_v$, respectively. How many parts of vapor per million parts of air are in the mixture?

$$2.9\,\text{ppm}_v + 11.5\,\text{ppm}_v + 35.3\,\text{ppm}_v + 4.8\,\text{ppm}_v = 54.5\,\text{ppm}_v$$

$$\frac{2.9}{50} + \frac{11.5}{100} + \frac{35.3}{100} + \frac{4.8}{100} = 0.574\ (57.4\%\ \text{of the PEL for the mixture})$$

There is compliance with the additive mixture PEL, however, the action level is exceeded. It is, therefore, probable that the painter's exposure exceeds the PEL on other days. Even though solvent vapor exposures may not routinely exceed the PEL, the inhalation of airborne paint solids is excessive. A good organic vapor and paint spray mist respirator or, better, an air-line respirator is needed until the ventilation, work practices, and other industrial hygiene controls are substantially improved.

"Nuisance" particulates (not otherwise classified, i.e., there are no compounds of lead, chromates, isocyanates, epoxy resins, etc. in the paint) have an 8-hour TWAE PEL of 10 mg/m^3. The author loathes the term "nuisance" because, in his experience, others often discount adverse effects of these particles on respiratory function.

188% of "nuisance" particulate limit assuming paint mist contains no hazardous components. What can be done to reduce exposures?—improve ventilation, use solvents of lower toxicity, modify work practices, reduce paint spray gun pressures, and so on.

139. A high-volume air sampler ran for 8 hours and 7 minutes at an average air flow rate of 47.3 ft³/minute. How many cubic meters of air were sampled?

$$480 \text{ minutes} + 7 \text{ minutes} = 487 \text{ minutes}$$

$$487 \text{ minutes} \times (47.3 \text{ ft}^3/\text{minute}) = 23{,}035.1 \text{ ft}^3$$

$$\frac{23{,}035 \text{ ft}^3}{35.315 \text{ ft}^3/\text{m}^3} = 652.3 \text{ m}^3$$

140. 172 grams of liquid phosgene splash on a floor. What gas volume quickly results after evaporation at an air temperature of 24°C and an atmospheric pressure of 742 mm Hg? The boiling point of phosgene is 447°F.

$$COCl_2 \text{ molecular weight} = 12 + 16 + (2 \times 35.5) = 99$$

$$PV = nRT$$

$$n = \frac{172 \text{ grams}}{99 \text{ grams/gram-mole}} = 1.737 \text{ gram-moles}$$

$$T = 23°C + 273 = 296 \text{ kewin}$$

$$V = \frac{nRT}{P} = \frac{(1.737 \text{ gram-moles})(0.0821 \text{ L-atm/K-mole})(296 \text{ K})}{(742 \text{ mm Hg}/760 \text{ mm Hg}) = 0.976 \text{ atmosphere}} = 43.183$$

Answer: 43.2 × 10⁸ liters of air would be needed to dilute this gas to 10% of the TLV of 0.1 ppm (dilution to 10 ppb).

141. A detection limit of 40 micrograms/m³ was reported for 1,2-dichloroethane. How long must one sample at 100 mL/minute with a small charcoal tube to detect this concentration? The analytical sensitivity for DCE is 1 microgram.

$$\text{Sampling time, minutes} = \frac{1 \text{ mcg}}{(40 \text{ mcg}/10^6 \text{ mL}) \times (100 \text{ mL/minute})}$$

$$= 250 \text{ minutes}$$

Answer: Sample for 250 minutes (4 hour and 10 minutes).

142. An electroplater was exposed to HCl gas at four operations: parts dipping (1.6 ppm$_v$ for 2.5 hours), parts draining (2.9 ppm$_v$ for 3.25 hours), acid replenishment (8.8 ppm$_v$ for 15 minutes), and cleanup (0.7 ppm$_v$ for 0.5 hours). Assuming that he had negligible exposure for the balance of his 8-hour work shift, what was his TWAE to hydrogen chloride gas?

$$C \times T = \text{concentration} \times \text{exposure time} = CT = \text{dose (Haber's law)}$$

Dipping	1.6 ppm$_v$ × 2.5 hours	= 4.00 ppm$_v$-hours
Draining	2.9 ppm$_v$ × 3.25 hours	= 9.43 ppm$_v$-hours
Replenishment	8.8 ppm$_v$ × 0.25 hour	= 2.20 ppm$_v$-hours
Cleanup	0.7 ppm$_v$ × 0.5 hour	= 0.35 ppm$_v$-hour
Other	0.0 ppm$_v$ × 1.5 hours	= 0.00 ppm$_v$-hours
	8.0 hours	15.98 ppm$_v$-hours

Total dose = (15.98 ppm$_v$ hours/8 hours) = 2.0 ppm$_v$ TWAE to HCl gas.

Exposure when replenishing acid exceeds 5 ppm$_v$ OSHA STEL. A full-face acid-gas respirator must be worn until ventilation is improved or better methods of replenishing are adopted.

143. A 4-$\frac{1}{2}$ pound chunk of calcium phosphide fell into a vat of water and released phosphine gas according to the following reaction. How much phosphine gas was generated disregarding water solubility and assuming stoichiometric conversion?

$$Ca_3P_2 + 6H_2O \rightarrow 3Ca(OH)_2 + 2PH_3\uparrow$$

$$\text{Molecular weight of } Ca_3P_2 = 182.19 \text{ grams/gram-mole}$$

$$\text{Molecular weight of } PH_3 \text{ gas} = 34.00$$

$$4.5 \text{ lb } Ca_3P_2 \times 454 \text{ grams/lb} = 2043 \text{ grams } Ca_3P_2$$

$$\frac{2043 \text{ grams } Ca_3P_2}{182.19 \text{ grams/mole}} = 11.21 \text{ moles of } Ca_3P_2$$

$$11.21 \text{ moles of } Ca_3P_2 \rightarrow 22.42 \text{ moles of } PH_3 \uparrow$$

$$22.42 \text{ moles} \times 34.00 \text{ grams/mole} = 762.28 \text{ g } PH_3$$

Answer: 762 grams of phosphine gas were released to the air.

144. In Problem 143, what would be the phosphine gas concentration in ppm_v if this occurred in an unventilated room with dimensions of $20' \times 50' \times 50'$? Assume a uniform mixing of the PH_3 gas with the room air.

$$20' \times 50' \times 50' = 50,000\,ft^3 = 1416\,m^3$$

$$ppm = \frac{(762,280\ mg/1416\ m^3) \times 24.45}{34} = 387\ ppm\ PH_3$$

Answer: $387\,ppm_v$ PH_3! TLV and PEL of phosphine gas are only $0.3\,ppm_v$ with a STEL of $1\,ppm_v$. SCBAs, training, ventilation, and so on are necessary.

145. ASHRAE recommends dust collectors, not air filters, to clean exhaust air when the dust concentrations exceed four grains of dust for every thousand cubic feet of exhaust air. What is this dust concentration in mg/m^3?

$$\frac{4\ grains}{1000\ ft^3} \times \frac{35.315\ ft^3}{m^3} \times \frac{0.06480\ gram}{grain} \times \frac{1000\ mg}{gram} = \frac{9.15\ mg}{m^3}$$

Answer: $9.15\,$milligrams of total particulates per cubic meter of air.

146. A vertical hazardous liquid waste incinerator has a gas flow rate of 6900 cfm at 70°F. The interior dimensions of the incinerator are 9 feet square and 37 feet high. The design operating temperature of the incinerator is 2200°F. What is the residence time for a molecule of vapor in this incinerator? What is the actual flow rate of gas? What incinerator dimensions are required for 3-second residence time? This residence time is sufficiently long to fully oxidize the supplied waste.

$$70°F = 530°R$$

$$2200°F + 460 = 2660°R$$

$$\text{Applying Charles' law: } \frac{V_i}{V_f} = \frac{T_i}{T_f}:$$

$$V_f = \frac{T_f \times V_i}{T_i} = \frac{6900\ cfm \times 2660°R}{530°R} = 34,630\ cfm$$

$$Q = AV$$

$$V = Q/A = \frac{34,630\ cfm}{9' \times 9'} = 428\ fpm$$

$$\frac{428\ fpm}{60\ seconds/minute} = 7.1\ feet/second$$

$$\frac{37 \text{ feet}}{7.1 \text{ feet/second}} = 5.2 \text{ seconds average residence time/molecule}$$

$$\text{Incinerator volume} = 9' \times 9' \times 37' = 2997 \text{ ft}^3$$

$$\text{Incinerator volume for a 3 second residence time} = \frac{2997 \text{ ft}^3}{5.2 \text{ seconds}} = \frac{x \text{ ft}^3}{3 \text{ seconds}}$$

$x = 1729 \text{ ft}^3$. The length to cross-sectional area ratios must remain identical to ensure necessary transit velocity and molecule residence time in the combustion space.

Answers: 5.2 seconds. 34,630 cfm. 1727 ft³. Perhaps a smaller incinerator could be used, say, 7' × 7' × 37' that would provide a 3.1 second residence time.

147. Industrial waste liquid chlorobenzene is fed into a large vertical hazardous waste incinerator at an atomizer nozzle feed rate of 1670 pounds/hour. What are the stack gas combustion products assuming 100% oxidation with 70% excess air?

$$C_6H_5Cl + 7O_2 + \left[\frac{0.79}{0.21} \times 7N_2\right] \rightarrow HCl\uparrow + 6CO_2\uparrow + 2H_2O\uparrow$$
$$+ \left[\frac{0.79}{0.21} \times 7N_2\right]$$

at 70% excess air:

$$C_6H_5Cl + (1.7 \times 7)O_2 + \left[1.7 \times \frac{0.79}{0.21} \times N_2\right] \rightarrow 6CO_2\uparrow + 2H_2O\uparrow + HCl\uparrow$$
$$+ (0.6)7O_2 + 7\left[1.6 \times \frac{0.79}{0.21}\right]N_2$$

molecular weights (grams/gram mole^{-1}): chlorobenzene = 112.5, oxygen = 32, nitrogen = 28, HCl = 36.5, carbon dioxide = 44.
   One pound-mole of chlorobenzene requires 7 pound-moles of O_2

$$\frac{1670 \text{ pounds/hour}}{112.5 \text{ lb/lb-mole}} = \frac{7 \times 1670 \text{ pound-moles } O_2}{112.5}$$

$$\frac{1670}{112.5} = 14.84$$

$$CO_2: \frac{1670}{112.5} \times 6 = 89.07$$

$$H_2O: \frac{1670}{112.5} \times 2 = 29.69, \text{ and so on.}$$

Incinerator combustion products:

CO_2	6 lb-mole	$\times 44 \times 14.84$	= 3918 lb
H_2O	2 lb-mole	$\times 18 \times 14.84$	= 534 lb
HCl	1 lb-mole	$\times 36.5 \times 14.84$	= 542 lb
O_2	0.6 lb-mole	$\times 7 \times 32 \times 14.84$	= 1995 lb
N_2	$\dfrac{1.6 \times 0.79}{0.21}$ lb-mole	$\times 7 \times 28 \times 14.84$	= 17,507 lb
	Total		24,496 lb

3918 lb/24,496 lb	=	0.160
534 lb/24,496 lb	=	0.022
542 lb/24,496 lb	=	0.022
1995 lb/24,496 lb	=	0.081
17,507 lb/24,496 lb	=	0.715
		1.000

Answers: 24,500 pounds/hour. 16% CO_2, 2.2% H_2O, 2.2% HCl, 8.1% O_2, and 71.5% nitrogen-argon.

148. A solvent paint stripper is 30% by volume methylene chloride and 70% by volume methanol. What is the volume percent composition of the vapor at normal room temperature? The vapor pressures of CH_2Cl_2 and MeOH are 350 and 92 mm Hg, respectively. Their respective densities are 1.33 and 0.79 g/mL, and their molecular weights are 84.9 and 32.1 grams/gram mole^{-1}, respectively.

Using 100 mL of the solvent mixture as a basis for Raoult's law calculations:

$$30 \text{ mL} \times 1.33 \text{ g/mL} = 39.9 \text{ g } CH_2Cl_2$$

$$70 \text{ mL} \times 0.79 \text{ g/mL} = 55.3 \text{ g MeOH}$$

Partial pressure CH_2Cl_2

$$= \frac{(39.9 \text{ g}/84.9 \text{ g/mole}) \times 350 \text{ mm Hg}}{(39.9 \text{ g}/84.9 \text{ g/mole}) + (55.3 \text{ g}/32.1 \text{ g/mole})} = 75 \text{ mm Hg}$$

Partial pressure MeOH

$$= \frac{(55.3 \text{ g}/32.1 \text{ g/mole}) \times 92 \text{ mm Hg}}{(39.9 \text{ g}/84.9 \text{ g/mole}) + (55.3 \text{ g}/32.1 \text{ g/mole})} = 72.3 \text{ mm Hg}$$

75 mm Hg + 72.3 mm Hg = 147.3 mm Hg total vapor pressure for both solvents

$$\frac{75 \text{ mm Hg}}{147.3 \text{ mm Hg}} \times 100 = 50.9\% \text{ methylene chloride vapor}$$

$$\frac{72.3 \text{ mm Hg}}{147.3 \text{ mm Hg}} \times 100 = 49.1\% \text{ methyl alcohol vapor}$$

Answers: 50.9% CH_2Cl_2 and 49.1% MeOH in the vapor phase. Refer to Problem 18 for a discussion of the application and deviations from Raoult's law.

149. A rotameter calibrated in cubic feet per hour was used in an air-sampling train operating for 2 hours and 39 minutes. What volume of air was sampled in liters if the average rotameter reading was 4.6?

$$\frac{4.6 \text{ ft}^3}{\text{hour}} \times \frac{\text{hour}}{60 \text{ minutes}} \times 159 \text{ minutes} \times \frac{28.32 \text{ L}}{\text{ft}^3} = 345 \text{ L}$$

Answer: 345 liters of air.

150. A solvent drum filling operator has an average exposure to 83 ppm_v methylene chloride for 3-$\frac{1}{2}$ hours, 32 ppm_v isopropyl alcohol for 1-$\frac{1}{2}$ hours, and 17 ppm_v toluene for 3 hours. If their respective TLVs are 50, 400, and 20 ppm_v, are solvent vapor exposure controls warranted?

$$\frac{83 \text{ ppm} \times 3.5 \text{ hours}}{50 \text{ ppm} \times 8 \text{ hours}} \times 100 = 72.6\% \text{ of the TLV for } CH_2Cl_2$$

$$\frac{12 \text{ ppm} \times 1.5 \text{ hours}}{100 \text{ ppm} \times 8 \text{ hours}} \times 100 = 1.5\% \text{ of the TLV for IPA}$$

$$\frac{17 \text{ ppm} \times 3 \text{ hours}}{20 \text{ ppm} \times 8 \text{ hours}} \times 100 = 31.9\% \text{ of the TLV for toluene}$$

72.6% + 1.5% + 31.9% = 106% of the additive TLV for this work shift exposure.

Answer: Yes, industrial hygiene controls are required because combined exposure exceeds the action level based on additive toxicological effects.

Remember: *Permissible exposure limits are worst acceptable conditions, and must not be regarded as goals, objectives, and end points.* A major industrial hygiene premise is the control of every exposure to as low as possible using best available technology. Control of methylene chloride vapors is especially important since it is a possible human carcinogen with multiple toxic effects, and contributes the highest percent distributional exposure to the mixture of vapors.

151. A railroad hopper car filler has an 8-hour TWAE to mixed grain dust (barley, oats, wheat) of 3.4 mg/m³ (TLV = 4 mg/m³). He has a simultaneous exposure to respirable silica (α-quartz) released from the cascading dry grain dust of 0.036 mg SiO$_2$/m³ (TLV = 0.1 mg respirable quartz/m³). Are industrial hygiene controls needed? If so, what would you prescribe?

$$\frac{3.4 \text{ mg/m}^3}{4 \text{ mg/m}^3} + \frac{0.036 \text{ mg/m}^3}{0.1 \text{ mg/m}^3} = 0.85 + 0.36 = 1.21$$

$$21\% > \text{TLV}_{\text{mixture}}$$

Answer: Industrial hygiene controls (powered air-purifying respirator?, working in filtered air booth?, PFTs, training, etc.) are necessary because grain dust exposure exceeds the action level, and the combined exposure to both dusts is 2.4 times the action level. Since adverse pulmonary effects (fibrosis) are common to both dusts, good control of inhalation exposures assumes great importance.

152. A calibrated combustible gas indicator was used to measure explosive vapors in a gasoline tank. A gas/vapor-sampling probe attached to the CGI was lowered into the tank. The needle immediately "pegged" above 100% of LEL and then quickly dropped to less than 2% of the LEL. Assuming a nonabsorptive sampling gas/vapor-sampling probe, which of the following is most likely?
    a. The meter correctly indicates less than 2% of the LEL.
    b. A CGI Wheatstone bridge electronic circuit is quickly "poisoned" by organic lead vapors emitted from the gasoline and then reads low.
    c. There is less than 4% oxygen in the tank's vapor space.
    d. Since gasoline is a mixture of numerous hydrocarbons, an inaccurate reading results if the instrument was not calibrated with the actual gasoline vapor being tested.
    e. The vapor concentration is not less than 2% of the LEL; it is actually above the UEL.

Answer: e. The concentration of gasoline vapors is so rich that no combustion can occur in the instrument. This is a very dangerous situation because unsuspecting persons may assume a safe atmosphere with respect to explosion. As vapors are diluted down into the LEL–UEL range with air, an explosion will occur if there is an ignition source. Dilute the explosive vapors with nitrogen to or argon <20% LEL, then use air to <TLV or <PEL. Verify concentrations of oxygen, flammable vapors, and toxics with recently calibrated combustible gas indicator, oxygen level meter, and atmospheric toxicant instruments.

153. A paint sprayer has 8-hour TWAEs to toluene at 19 ppm$_v$ (TLV is 20 ppm$_v$), n-butyl alcohol at 16 ppm$_v$ ("C" is 50 ppm$_v$/skin), and noise at 84 dBA. What are the major industrial hygiene issues?

    Answer: In addition to the additive narcotic effects of these two solvent vapors, reports in the occupational medical literature have indicated loss of hearing in people with chronic work place exposures to these solvents. Therefore, effects of the inhaled solvent vapors are presumed additive to physical effects of noise on the painter's hearing. All elements of a hearing conservation program must be applied along with careful practices to reduce vapor and skin and eye contact exposures. Also, toluene is a reproductive health toxicant that may be additive to reported effects of noise (especially high frequency) on the embryo-fetus.

154. A maintenance welder is exposed to 3.6 mg iron oxides/m^3 and 27.4 mg of total dust per m^3 in a corn flour mill. What must you recommend to protect his health from these dust and fume exposures?

    Answer: Perish forbid! First, get him to stop welding before he blows the mill and everybody who works there to smithereens! After getting his attention, start a fire safety and welding safety training program including "hot work" permits, grain dust explosion prevention practices, ventilation engineering, respiratory protection, supervision, plant housekeeping, and so on. He and others must never use compressed air wands to remove settled dust from surfaces. Seal all leaks. Use vacuum suction to control fugitive dust leaks.

155. How much air is exhausted from a drying oven operating at 320°F if 1360 cubic feet of air at 65°F are supplied per minute to the oven?

$$65°F + 460°F = 525° \text{ absolute (Rankine)}$$

$$320°F + 460°F = 780° \text{ absolute (Rankine)}$$

$$\frac{780°R}{525°R} \times \frac{1360 \text{ ft}^3}{\text{minute}} = \frac{2020.6 \text{ ft}^3}{\text{minute}}$$

    Answer: 2021 cfm at 320°F, although the mass of hot air remains the same.

156. A gray iron foundry cupola attendant leaves work after 6 hours going home with an intense headache, dizziness, and nausea. This 26-year old man, feeling better after an hour, decides to strip paint from an old chair in his basement. His wife finds him an hour later collapsed on their basement floor. He died from cardiac arrest after ventricular fibrillation. His medical history includes chronic hemocytic anemia. What is a reasonable forensic conclusion?

   a. He should not have gone to work that morning.

   b. There was an oxygen-deficient atmosphere near the cupola.

   c. Silica dust and formaldehyde gas in the foundry act in an additive manner with methylene chloride vapors that volatilize from paint strippers.

   d. He should have worn a good dust respirator while at work so his lungs would be capable of detoxifying any solvent vapors he inhaled while working in his basement.

   e. The carbon monoxide he inhaled at work was additive to the *in vivo* conversion of inhaled methylene chloride vapor to carbon monoxide while at home.

   f. Anemia was a significant risk factor.

Answer: e. Cupola workers have potential exposures to high concentrations of CO gas. His symptoms are consistent with COHgb concentration exceeding 10–20% or more. His symptoms occurred late in work shift indicating he went home with a significant body burden of COHgb. Methylene chloride vapor, in addition to being a potential sensitizer of the myocardium, is present in most paint strippers and metabolizes partly *in vivo* to carbon monoxide. His anemia was likely contributive.

157. Air was sampled at an average flow rate of 2.14 L/m through a 37-mm membrane filter for 7 hours, 47 minutes. The sample was obtained to measure welder's TWAE to aluminum fume as oxide, Al_2O_3. An industrial hygiene chemist analyzed 7.93 millligrams of aluminum on the filter after filter blank correction. What was the concentration of aluminum oxide fume? The atomic weight of aluminum is 27, and the molecular weight of aluminum oxide is 102.

$$\frac{1 \text{ aluminum oxide}}{2 \text{ aluminum}} = \frac{102}{2 \times 27} = 1.89 = \text{the Al to } Al_2O_3 \text{ molar conversion factor}$$

$$7 \text{ hours, } 47 \text{ minutes} = 467 \text{ minutes} = (467/480) \times 100$$
$$= 97.3\% \text{ of an 8-hour work shift}$$

$$467 \text{ minutes} \times 2.14 \text{ L/m} = 999.4 \text{ liters}$$

$$7.93 \text{ mg Al/m}^3 \times 1.89 = 14.99 \text{ mg } Al_2O_3/m^3 \text{ TWAE}$$

$$14.99 \text{ mg/m}^3 \times 467 \text{ minutes} = 7000.33 \text{ mg/m}^3\text{-minutes}$$

$$0 \text{ mg/m}^3 \times 13 \text{ minutes} = 0 \text{ mg/m}^3\text{-minutes}$$

$$\frac{7000.33\,\text{mg/m}^3 - \text{minutes}}{480\,\text{minutes}} = 14.58\,\text{mg/m}^3 \text{ TWAE}$$

(assuming that the balance of his/her work shift is free of Al_2O_3 fume exposure).

This result is anomalous because the TLV for aluminum oxide is $10\,\text{mg/m}^3$ (as aluminum). This exposure to the fume meets the TLV when calculated as Al, but exceeds the TLV and PEL of $10\,\text{mg/m}^3$ for "nuisance" particulates. (The author abhors the term "nuisance.") Clearly, however, exposure to welding fume requires better control. Since welding of aluminum is often done by MIG technique, ozone gas can be generated in large amounts. Note that there is a specific TLV of $5\,\text{mg/m}^3$ for aluminum welding fume that resolves apparent anomaly. Al_2O_3 dust has a justifiably higher TLV than that for Al_2O_3 fume.

158. Convert 1.78 mg HF gas per cubic meter of air to ppm. Molecular weight of HF is 20.

$$\text{ppm}_v = \frac{(\text{mg/m}^3) \times 24.45}{\text{molecular weight}} = \frac{(1.78\ \text{mg/m}^3) \times 24.45}{20} = 2.18\,\text{ppm}_v \text{ HF gas}$$

HF gas ceiling concentration is $5\,\text{ppm}_v$.

159. An industrial hygiene chemist detected less than 5 micrograms of toluene in a charcoal tube used for collecting a 9.76 liter air sample. What was the toluene vapor concentration?

$$\frac{\leq 5\ \text{mcg toluene}}{9.76\ \text{L}} \leq 0.512\ \text{mcg/L}$$

Molecular weight of toluene = 92 grams gram-mole^{-1}

$$\frac{(\leq 0.512\ \text{mcg/L}) \times 24.45}{92} \leq 0.136\ \text{ppm}$$

Answer: Less than $0.14\,\text{ppm}_v$ toluene ($\leq 140\,\text{ppb}_v$).

160. What was the average flow rate of an air-sampling pump if pre-sampling and post-sampling elapsed times through a 1000 mL soap bubble apparatus were 83.5 and 84.9 seconds, respectively?

$$\frac{83.5\ \text{seconds} + 84.9\ \text{seconds}}{2} = 84.2\ \text{seconds average}$$

$$\frac{60 \text{ seconds/minute}}{84.2 \text{ seconds/liter}} = 0.713 \text{ L/m}$$

Answer: Average air flow rate $= 0.713$ liter per minute $= 713 \text{ mL/min}$.

161. What is the density of methyl chloroform vapor? The density of air at NTP $=$ 1.2 milligrams per cubic centimeter $= 1.2 \text{ mg/mL}$.

CH_3CCl_3 molecular weight $= (2 \times 12) + (3 \times 35.5) + (3 \times 1) = 133.5 \text{ grams/gram-mole}$

$$1 \text{ ppm } CH_3CCl_3 = \frac{\text{molecular weight} \times (\text{mcg/L})}{24.45} = \frac{133.5}{24.45} = \frac{5.46 \text{ mg}}{m^3} \approx \frac{1 \text{ mL}}{m^3}$$

$$\frac{5.46 \text{ mg MC/mL}}{1.2 \text{ mg air/mL}} = 4.55$$

Answer: The density of 100% methyl chloroform vapor is 4.55 (air $= 1.00$).

162. Convert 1 ppm_v hydrogen selenide gas to mg/m^3. The molecular weight of H_2Se is $81.0 \text{ grams/gram-mole}$.

$$1 \text{ ppm } H_2Se = \frac{(\text{mcg/L}) \times 24.45}{81} = \frac{3.31 \text{ mcg}}{L} = \frac{3.31 \text{ mg}}{m^3}$$

Answer: $1 \text{ ppm } H_2Se = 3.31 \text{ mg } H_2Se/m^3$

163. Knowing the concentration of CO gas in inhaled air and a few other important parameters permits one to estimate the percent unsaturated hemoglobin in a CO-exposed person. An equation for this is

$$\% \text{COHb} = (2.76 \times e^{h/7000}) + (0.0107 \ a \ C^{0.9} \ t^{0.75}),$$

where
  $h =$ height above sea level (ft), $a =$ activity ($3 =$ rest, $5 =$ light activity, $8 =$ light work, and $11 =$ heavy work), $C = \text{ppm}_v$ inhaled CO, and $t =$ the exposure duration (hours).

What is the predicted COHb concentration in a man performing light work at 8000 feet for 6 hours at 50 ppm_v CO?

$$\% \text{COHb} = (2.76 \times e^{h/7000}) + 0.0107 \ (8 \times 50^{0.9} \times 6^{0.75})$$
$$= (2.76 \times e^{1.142}) + 0.0107 \ (8 \times 33.8 \times 3.83) = 19.74\%$$

Answer: About 20% COHgb; that is, a worker who has an intense head-ache, weakness, and perhaps nausea, vomiting, and dimness of vision, tin-nitus, and so on and possibly a myocardial infarction if she/he has a history of ischemic heart disease, ventricular fibrillations, myocardial infarction, or other compromised cardiac capacity.

164. "Standard man" in toxicology weighs 70 kilograms (= 154.35 pounds). He inhales 22,800 liters of air per day: 9600 L at 8 hours of light work, 9600 L at 8 hours nonoccupational, and 3600 L resting and sleeping. What weight of dry air does he inhale yearly? Dry air weighs $1.2\,kg/m^3$ at 21°C at sea level. Assume this "standard man" lives in Seattle.

$$22,800\,L = 22.8\,m^3$$

$$\frac{22.8\,m^3}{day} \times \frac{365.25\,days}{year} \times \frac{1.2\,kg}{m^3} = \frac{9993\,kg}{year}$$

$$\frac{9993\,kg}{year} \times \frac{2.205\,lb}{kg} = \frac{22,035\,lb}{year}.$$

Answer: About 22,000 pounds (11 tons!) per year. Does this mean exhalation of 11 tons of bad breath per year? Should there be a halitosis TLV of 1 ppb_v?

165. What factors are of minor significance in stack sampling?
    a. Air/gas density, condensed water, dust thimble weight
    b. Stack gas flow rate, isokinetic sampling, temperature
    c. Suction pressure at meter, condensed water, gas flow rate
    d. Molecular weight of air, friction losses, season of the year
    e. Flow rate through sampling train, barometric pressure, sample duration
    f. Pitot correction factor, gas:dust ratio, presence of reactive gases, temperature

    Answer: d.

166. An air pollution chemist wants to determine the emission rate of chlorine gas from a vent in a chemical synthesis reactor. Which of the following should be of minor, negligible significance to her?
    a. Concentration of chlorine in the gas stream
    b. Isokinetic sampling
    c. Density of air or gas stream containing the Cl_2 gas
    d. Process operation or cycle time
    e. Volumetric total air or gas emission rates
    f. Moisture content of the air or gas stream

    Answer: b. Isokinetic stack sampling is not required when only gases or vapors are determined. Isokinetic sampling is mandatory for airborne particulates in ducts, stacks, and exhaust vents (e.g., dust, fume, smoke, mist, fibers, aerosols, spray, etc.). All other parameters are important in her analysis.

167. A 10-inch internal diameter stack discharges air containing 23 ppm$_v$ chlorine gas at 670 feet per minute. What is the chlorine gas mass emission rate?

$$\text{Stack area} = \pi \frac{(5 \text{ in})^2}{(144 \text{ in}^2/\text{ft}^2)} = 0.5454 \text{ ft}^2$$

$$Q = AV = 0.5454 \text{ ft}^2 \times 670 \text{ fpm} = 365.4 \text{ cfm}$$

$$\frac{\text{mg}}{\text{m}^3} = \frac{\text{ppm} \times \text{molecular weight}}{24.45} = \frac{23 \times 71}{24.45} = \frac{66.79 \text{ mg}}{\text{m}^3}$$

$$\frac{66.79 \text{ mg}}{\text{m}^3} \times \frac{\text{m}^3}{35.3 \text{ ft}^3} \times \frac{365.4 \text{ ft}^3}{\text{minute}} = \frac{691.4 \text{ mg Cl}_2}{\text{minute}}$$

Answer: 691 milligrams of chlorine are released per minute.

168. The LEL for methyl ethyl ketone at normal room temperature is 1.7%. What is the MEK vapor level expressed in grams/m^3?
Molecular weight of CH_3-(C=O)-CH_2CH_3 = (4 carbon × 12) + (1 oxygen × 16) + (8 hydrogen × 1) = 72 grams/gram-mol.

$$1.7\% = 17,000 \text{ ppm (vol/vol)}$$

$$\frac{\text{mg}}{\text{m}^3} = \frac{\text{ppm} \times \text{molecular weight}}{24.45} = \frac{17,000 \times 72}{24.45} = \frac{50,061.14 \text{ mg 2-butanone}}{\text{m}^3}$$

Answer: MEK LEL is 50 grams of vapor per cubic meter of air.

169. The LEL for MEK (2-butanone) is 1.7% (vol/vol) at standard room temperatures. A vapor concentration of 1.4% (vol/vol) in air explodes. What might be concluded?
   a. The LEL of 1.7% must be wrong
   b. The 1.4% vapor-in-air mixture must have been hotter than the standard room temperature
   c. The data provided are insufficient to conclude anything
   d. The ignition source for the 1.4% vapor must have been hotter than normal
   e. The oxygen concentration for the 1.7% LEL determination was higher than existed in the explosion of the 1.4% concentration

Answer: b. Raising the temperature lowers the LEL, for example, the LEL for MEK is 1.4% for this air:solvent vapor mixture at 200°F. Intuitively, we understand this because the hotter air and vapor mixture has more energy among the molecules. Therefore, less energy is required to ignite the explosive mixture.

170. An air pollution control baghouse collects 93 pounds of dust for every 8-hour operating period. The ventilation duct leading to the baghouse plenum has an 18-inch internal diameter with average duct velocity of 3250 feet per minute. What is the average airborne dust concentration in the influent duct in mg/m³?

$$18'' \text{ diameter} = 1.5' \text{ diameter}$$

$$\text{duct radius} = 0.75 \text{ foot}$$

$$Q = AV = \pi(0.75' \times 0.75')^2 \times 3250 \,\text{fpm} = 5743 \,\text{cfm}$$

$$\frac{93 \text{ pounds}}{480 \text{ minutes}} = \frac{0.194 \text{ pound}}{\text{minute}}$$

$$\frac{(0.194 \text{ lb/minute})}{(5743 \text{ ft}^3/\text{minute})} = \frac{0.0000338 \text{ pound}}{\text{ft}^3}$$

$$\frac{0.0000338 \text{ lb}}{\text{ft}^3} \times \frac{454 \text{ g}}{\text{lb}} \times \frac{1000 \text{ mg}}{\text{gram}} \times \frac{35.3 \text{ ft}^3}{\text{m}^3} = \frac{542 \text{ mg}}{\text{m}^3}$$

171. A carbon adsorption unit air pollution control device operating at 6600 cfm has an inlet vapor concentration of 360 ppm$_v$ and outlet concentration of 25 ppm$_v$. What is the annual savings if the unit operates 100 hours per week and 50 weeks per year? The cost of this solvent is $4.83/gallon. The solvent's molecular weight is 114. The solvent's density is 0.83 g/mL. Assume a 97% recovery of the solvent vapor adsorbed on the charcoal.

360 ppm$_v$ vapor at inlet − 25 ppm$_v$ at outlet = 335 ppm$_v$ (a 92.5% collection efficiency)

$$\frac{\text{mg}}{\text{m}^3} = \frac{\text{ppm} \times \text{molecular weight}}{24.45} = \frac{335 \times 114}{24.45} = \frac{1562 \text{ mg}}{\text{m}^3}$$

$$\frac{1562 \text{ mg}}{\text{m}^3} \times \frac{\text{m}^3}{35.3 \text{ ft}^3} \times \frac{\text{gram}}{1000 \text{ mg}} \times \frac{\text{pound}}{454 \text{ grams}} = \frac{0.0000975 \text{ lb}}{\text{ft}^3}$$

$$\frac{0.0000975 \text{ lb}}{ft^3} \times \frac{6600 \text{ ft}^3}{\text{minute}} \times \frac{60 \text{ minutes}}{\text{hour}} \times \frac{5000 \text{ hours}}{\text{year}} \times 0.97$$

$$= \frac{187,259 \text{ lb solvent recovered}}{\text{year}}$$

$$\frac{187,259 \text{ lb}}{\text{year}} \times \frac{454 \text{ grams}}{\text{lb}} \times \frac{\text{mL}}{0.83 \text{ g}} \times \frac{\text{gallon}}{3785 \text{ mL}} \times \frac{\$4.83}{\text{gallon}} = \frac{\$130,708}{\text{year}}$$

Answer: An annual savings of $130,708 minus maintenance costs, labor costs, taxes, operating costs, and capital equipment depreciation. Other opportunities for savings include alternative technologies, improved vapor collection efficiency, improved liquid solvent recovery efficiency, and using a less costly solvent.

172. A 210 ft³ compressed gas cylinder containing 10% by volume CO in air crashes to the floor in a 10' × 30' × 60' room that has no ventilation. There is no protective cover over the gas valve. The valve snaps off, and the CO gas mixture is quickly released. The de-pressurizing, rapidly whirling cylinder provides uniform gas mixing in the room. What is the final CO gas concentration after completely mixing with the room air? Disregard the small increase in the overall atmospheric pressure in the room from the released gas.

$$210 \text{ ft}^3 \times 0.10 = 21 \text{ cubic feet of CO}$$

$$10' \times 30' \times 60' = 18,000 \text{ cubic feet room volume}$$

$$\frac{21 \text{ ft}^3 \text{ of CO}}{18,000 \text{ ft}^3} \times 10^6 = 1167 \text{ ppm CO}$$

Answer: 1167 parts of CO gas per million parts of air. It should be noted that this answer is slightly incorrect because an amount of CO equal to the cylinder volume will remain in the cylinder when it comes to atmospheric pressure. For example, subtract 3 cubic feet from 210 ft³ if this is the cylinder's interior volume. See Problem 327.

173. What volume of propane gas is required to yield a concentration of 1000 ppm$_v$ in a gas calibration chamber with an internal volume of 765 liters?

$$\text{ppm} = \frac{\text{volume of gas}}{\text{volume of air}} \times 10^6$$

$$\text{volume of gas} = \frac{1000 \times 765 \text{ L}}{10^6} = 0.765 \text{ L}$$

Answer: 765 milliliters of propane gas.

174. The average dust concentration in a baghouse inlet plenum is $348\,mg/m^3$. The outlet dust concentration is $4.7\,mg/m^3$. What is the average collection efficiency of the baghouse filters for this dust?

Percent collection efficiency

$$= 100\left[1 - \frac{C_{out}}{C_{in}}\right] = 100\left[1 - \frac{4.7\ mg/m^3}{348\ mg/m^3}\right] = 100\,(1 - 0.0135) = 98.65\%$$

Note that this is the average *mass* collection efficiency. Since particle weights are proportional to the cube of their diameters, the mass collection efficiency does not equal particle collection efficiency. Put in another way, depending upon the type of dust, for example, 80% of the particles by weight might only comprise 2% of the particles by count. Recirculating the exhaust air from a dust bag collector with a collection "efficiency" of 95%, for example, could be highly hazardous to workers. Always ask manufacturers of air pollution control devices how they determined the collection efficiency of their products.

175. One gram-mole of SO_2 occupies _____ liters at _____ °C and _____ mm Hg and contains _____ molecules.
    a. 22.4, 25, 760, 6×10^{23}
    b. 24.45, 0, 760, 6×10^{21}
    c. 24.45, 25, 745, 6×10^{23}
    d. 22.4, 0, 760, 2.023×10^{23}
    e. 24.45, 25, 760, 6×10^{23}

    Answer: e.

176. One milliliter of methyl chloroform becomes what vapor volume when evaporated at a pressure of 720 mm Hg and a temperature of 25°C?

$$Density = 1.34\,g/mL$$

$$PV = nRT$$

$$\frac{1.34\ gram}{133.4\ g/mole} = 0.010\ mole$$

$$V = \frac{n\,R\,T}{P} \frac{(0.010\ mole)(0.0821\ atm\text{-}liter/mole\text{-}K)(298\ K)}{(720\ mm\ Hg/760\ mm\ Hg) = 0.947\ atmosphere} = 0.258\ liter$$

Answer: 258 milliliters of 100% methyl chloroform vapor.

177. _____ refers to the number of molecules per cubic centimeter of an ideal, perfect gas at 0°C (2.6782×1019).
     a. Avogadro's number
     b. Loschmidt's number
     c. Fanning's friction factor
     d. Boyle's number
     e. Bunny's number (313-123-4567)
     f. Dalton's number

Answer: b.

178. A $20' \times 100' \times 120'$ building has a ventilation system providing 6000 cfm of the outside air. The outdoor design condition is 20°F at 60% relative humidity (nine grains of moisture per pound). If we desire to maintain 50% relative humidity at 75°F (66 grains of moisture per pound of air) at 13.78 cubic feet per pound of air in the air-conditioned space, how much liquid water must be added to the air stream?

$$\text{The formula for humidification load} = \frac{(\text{CFH})(G)}{(V)(7000)}$$

where
     CFH = cubic feet of air per hour (= cfm $\times$ 60 minutes/hour)
     G = grains of moisture per pound of the inside air minus grains per pound outside air
     V = specific volume of inside air in cubic feet per pound of air
     7000 = conversion factor, grains of moisture per pound

$$\frac{(6000 \times 60)(66 - 9)}{(13.78)(7000)} = 212.7 \text{ pounds of water per hour}$$

$$\frac{212.7 \text{ pounds H}_2\text{O/hour}}{8.33 \text{ pounds/gallon}} = \frac{25.5 \text{ gallons to be evaporated}}{\text{hour}}$$

Answer: About 25–26 gallons of water per hour. If there were no ventilation system, an air infiltration rate of two air changes/hour could be assumed for a building with tight construction in a cold climate. Instead of the (6000 $\times$ 60) for CFH, we would use 2 (100' $\times$ 120' $\times$ 20'). The humidification load then becomes 285 lb/hour.

179. One milligram of quartz dust in the particle size range reported for typical foundry atmospheres contains 200–300 million particles. What does one hypothetical silica dust particle weigh?

$$\frac{250 \times 10^6 \text{ particles}}{1000 \text{ micrograms}} = \frac{250,000 \text{ particles}}{\text{microgram}} = \frac{250 \text{ particles}}{\text{nanogram}} = \frac{0.25 \text{ particle}}{\text{picogram}}$$

Answer: This average particle weighs four picograms (4×10^{-12} gram).

180. The behavior of airborne particles is best described by which following physical laws and factors:
    a. Stokes' law, Archimedes' principle, Newtonian gravity, and Brownian motion
    b. Frank's friction factor, Cunningham's factor, gravity, and Newtonian gravity
    c. Brownian motion, Avogadro's law, gravity, and ion flux
    d. Newtonian gravity, Stokes' law, Drinker's coefficient, McWelland's aerodynamic slip factors
    e. Gravity, Stoke's law, Cunningham's factor, and Brownian motion

Answer: e.

181. A granite quarry worker's exposures to dust were 192 mppcf for 3-$\frac{3}{4}$ hours during drilling, 1260 mppcf for 15 minutes blowing out holes, 12 mppcf for 1-$\frac{1}{4}$ hours changing drills, 9.8 mppcf for 2-$\frac{1}{4}$ hours watching drills, and lowest at 8.1 mppcf for 1/2 hour at broaching. What was TWAE to airborne dust?

Activity	mppcf	×	hours	=	mppcf-hours
Drilling	19	×	3.75	=	720
Blowing out holes	1260	×	0.25	=	315
Changing drills	12	×	1.25	=	15
Watching drills	9.8	×	2.25	=	22
Broaching	8.1	×	0.5	=	4
			8.0	=	1076

1076 mppcf-hours/8.0 h = 134.5 mppcfw.

Answer: His 8-hour TWAE = 134.5 mppcf. Dusty! How about prescribing a PAPR (Physician Assistants' Prescribing Reference), wet methods of dust suppression, mechanical local exhaust ventilation systems, and improvements in his work practices? Very importantly, what was his exposure to α- and β-quartz?

182. A drying oven vents MEK vapor into a work space. If solvent consumption rate is steady at one pint every 4 minutes, and the concentration of MEK vapor leaving the oven is 200 ppm, what is the ventilation rate of the oven? The vapor volume for MEK = 4.5 cubic feet of vapor per pint evaporated.

$$\text{Concentration} = \frac{\text{contaminant released (cfm)}}{\text{dilution air required (cfm)}}$$

$$\frac{4.5 \text{ ft}^3}{\text{pint}} \times \frac{1 \text{ pint}}{4 \text{ minutes}} = \frac{1.13 \text{ ft}^3 \text{ released}}{\text{minute}}$$

$$\text{Dilution air required} = \frac{\text{contaminant released (cfm)}}{\text{concentration}} = \frac{1.13 \text{ cfm}}{200/10^6} = 5650 \text{ cfm}$$

Answer: 5650 cubic feet of air per minute—based on the oven's inlet, not the oven's outlet air temperature. See Problem 188.

183. What is the maximum ground-level concentration of gas from a stack designed for emergency venting of chlorine gas? The stack is 110 feet high. The maximum emission rate is 5 cubic feet of Cl_2 per second. The effective stack height is 120 feet due to the gas injection velocity into the ambient air. Assume wind velocity at the top of the stack is 10 miles per hour.

The Bosanquet–Pearson equation can be used to estimate maximum ground-level concentration of a vent or stack-emitted air pollutant:

$$C_{max} = \frac{2.15 \times Q \times 10^5}{V \times H^2} \times \frac{p}{q},$$

where

$p/q = x, y,$ and z plane diffusion parameters, often collectively taken as 0.63
$H =$ effective stack height = the sum of the physical stack height plus plume height resulting from the discharge velocity and the thermal buoyancy of the gas, in feet
$C_{max} =$ maximum ground level concentration of the air pollutant, in ppm_v
$Q =$ emission rate of gas contaminant at ambient temperatures, in ft³/second
$V =$ wind velocity, in feet/second

$$\frac{10 \text{ miles}}{\text{hour}} = \frac{5280 \text{ feet}}{\text{mile}} = \frac{\text{hour}}{60 \text{ minute}} \times \frac{\text{minute}}{60 \text{ seconds}} = \frac{14.7 \text{ feet}}{\text{second}}$$

$$C_{max} = \frac{2.15 \times 5 \text{ ft}^3/\text{second} \times 10^5}{14.7 \text{ feet/second} \times (120 \text{ feet})^2} \times 0.63 = 3.2 \text{ ppm } Cl_2$$

Answer: 3.2 ppm_v chlorine gas at ground level. The approximate distance from the stack to the C_{max} point is 10H or, in this case, 10 × 120' = 1200 feet. Check the residential set-back distances. Consider an alkaline scrubber (higher stack), injecting air into base of stack to promote dilution, wind vane, and community emergency response plans.

184. 97 air sample results obtained at an industrial process were linearly distributed when plotted on log-normal graph paper. The graph gave a 50% concentration as 1.03 ppm_v and an 84% concentration as 2.41 ppm_v. What is geometric standard deviation for these air sample results?

$$GSD = \frac{84\% \text{ value}}{50\% \text{ value}} = \frac{2.41 \text{ ppm}}{1.03 \text{ ppm}} = 2.34$$

Answer: Geometric standard deviation is 2.34 (no units).

185. Air containing hydrogen sulfide gas was bubbled at 1 liter/minute for 12 minutes through an impinger containing 10 mL of 0.001 N iodine solution before the iodine was reduced (purple solution to colorless). What was the H_2S concentration?

$S^{-2} + I_2 \rightarrow S_0 + 2 I^{-1}$ iodine is the oxidant. H_2S is the reducing agent.

(mL) (N) = milliequivalents
(10 mL) (0.001 N) = 0.01 milliequivalent of I_2
1 meq I_2 is equivalent to 0.5 meq of H_2S
0.01 meq of I_2 is equivalent to 0.005 meq of H_2S
1 meq H_2S is equivalent to 24.45 mL H_2S at 25°C and 760 mm Hg
0.005 meq $H_2S \approx 0.122$ mL H_2S

$$\frac{0.122 \text{ mL } H_2S}{12 \text{ L}} = \frac{0.010 \text{ mL}}{L}$$

$$\frac{10 \text{ mL}}{1000 \text{ L}} = 10 \text{ ppm } H_2S$$

H_2S is a potent chemical asphyxiant that, after CO gas, is the second leading cause of acute chemical asphyxiation fatalities in the world.

186. A direct-reading air-sampling instrument was calibrated with the following results:

Calibration (True) Concentration (ppm_v)	Instrument Reading (ppm_v)
10	7
30	63
70	91
200	320
500	570

What is the group correlation between the dependent and independent variables?

This can be computed by most standard hand-held scientific calculators or quite easily using computer spreadsheet analysis tools. The university where I am a part-time instructor provides a convenient software package with this capability. *MicroSoft Excel*® does as well.

Answer: Very high, at $r = 0.986$. Although group correlation is excellent, some individual readings are unsatisfactory. This instrument should be repaired or returned to the manufacturer for servicing and calibration.

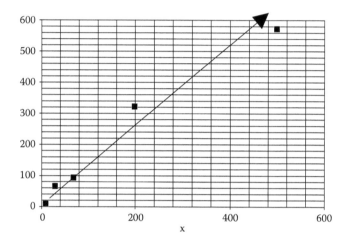

187. 1.3 pints of cyclohexane are uniformly evaporated from a large tray over a steam bath in an exhaust hood every 7 minutes. The exhaust hood face dimensions are 2.5 feet × 4.5 feet. The average hood face velocity is 150 feet per minute. The density of liquid cyclohexane is 0.78 g/mL. What is the cyclohexane vapor concentration in the exhaust air of the hood?

$$1.3 \text{ pints}/7 \text{ minutes} = 0.186 \text{ pint of cyclohexane evaporated per minute}$$

$$Q = A \times V = (2.5' \times 4.5') \times 150 \text{ fpm} = 1688 \text{ cfm}$$

$$\text{molecular weight } C_6H_{12} = 84 \text{ grams/gram-mole}$$

$$\frac{0.186 \text{ pint}}{\text{minute}} \times \frac{473 \text{ mL}}{\text{pint}} \times \frac{0.78 \text{ g}}{\text{mL}} \times \frac{1000 \text{ mg}}{\text{gram}} = \frac{68.623 \text{ mg}}{\text{minute}}$$

$$\frac{1688 \text{ ft}^3/\text{minute}}{35.3 \text{ ft}^3/\text{m}^3} = \frac{47.8 \text{ m}^3}{\text{minute}}$$

$$\frac{68,623 \text{ mg/minute}}{47.8 \text{ m}^3/\text{minute}} = \frac{1435 \text{ mg}}{\text{m}^3}$$

$$\text{ppm} = \frac{(\text{mg/m}^3) \times 24.45}{\text{molecular weight}} = \frac{(1435 \text{ mg/m}^3) \times 24.45}{84}$$

$$= 418 \text{ ppm cyclohexane}$$

Answer: 418 ppm$_v$ of cyclohexane vapor are in the exhaust air. This might exceed local air pollution regulations. A significant portion of these exhaust gases are water vapor because of the steam bath. Repeat the stack testing and calculate the emissions on a dry air basis.

188. Calculate air volume in Problem 182 if the discharge air temperature from the oven is 300°F.

   Use the ratio of absolute temperatures, that is, since the air expands with heat (but the mass remains the same), multiply the ratio of temperatures by a factor that is greater than 1.

$$Q = 5650 \text{ cfm}_{\text{stp}} \times \frac{460°F + 300°F}{460°F + 70°F} = 8102 \text{ cfm}$$

Answer: 8102 ft^3 of air at 300°F are discharged from this oven per minute. This is peculiar because the boiling point of water at sea level is 212°F. Is the 300°F from waste energy from the oven? Is there an additional source of needless sensible heat?

189. Which use of organic vapor monitors best characterizes a production worker's 8-hour TWAEs to toluene and methyl isobutyl ketone (MIBK) vapors?
   a. Use one OVM for the entire exposure period.
   b. Use four OVMs sequentially for 2 hours each.
   c. Use two OVMs sequentially for 3–4 hours each.
   d. Use one OVM full shift on Tuesday. Repeat air sampling on Thursday.
   e. Use two OVMs at the start of work. Remove one at lunch time and the other at the end of work. Repeat this process on another day of exposure.
   f. Sample randomly with one OVM for any 3-hour exposure period in the week. Repeat this on another day.

   Answer: e. This approach is powerful because the difference between results can help one determine which, if any, half of work shift contributes the highest exposure.

190. An electrician who infrequently works in a TDI polyurethane foam production area develops asthma and can no longer work there without experiencing dyspnea, chest tightness, and an anxious feeling. Seventeen air samples obtained in the area over the past 3 years were between 0.001 and 0.014 ppm$_v$ as 8-hour TWAEs. The TLV for TDI is 0.005 ppm$_v$ with a ceiling concentration of only 0.02 ppm$_v$. What appears most plausible?
   a. Since most of his exposures were well below the TLV, his illness must be due to nonoccupational factors (perhaps hobbies, avocations, home environment).

    b. He did not regularly work in the area, so it is unlikely he is sensitized to TDI.

    c. The TDI sample test results cannot be used as proof or disproof of an existing illness. It appears he might be sensitized to TDI vapors. He must no longer work in this area. To do so might risk his health and, indeed, his life.

    d. He is sensitized to the pyrolysis products released from hot insulation coating from smoldering electrical wires.

    e. There is insufficient information to reach any tentative conclusions. It is not known if the air samples were 8-hour averages or short-term samples. Obtain more data.

    f. There is no work relatedness of his condition to his exposures because no epidemiological studies have shown electricians are at an increased risk of isocyanate sensitization.

Answer: c. Study the *Preface* to the ACGIH *Threshold Limit Values®* book. Note these were area samples—not personal breathing zone samples. Area-sampling results are typically misleading and are lower than breathing zone samples. Apply the Precautionary Principle. Eliminate TDI emissions.

191. How many Pitot traverse readings should be taken when one measures velocity pressures in 6" to 40" round stacks or ducts?

    a. One carefully placed center line reading × 0.8

    b. From 5–30 as long as 50% of measurements are within 10% of each other provided that the gas or air is less than 100°F

    c. Six

    d. It depends upon the water vapor content, density, temperature, and dust load of the air or gas being tested

    e. Two 10-point traverses with 20 sampling points and both at 90° to each other

    f. Two five-point traverses with 10 sampling points and both at 90° to each other provided that the gas or air is less than 100°F

Answer: e. Refer to most recent edition of ACGIH's *Industrial Ventilation*.

192. A degreasing tank loses TCE vapor to the general plant atmosphere because this tank does not have a mechanical local exhaust ventilation system. Based on the solvent purchase records, 1.1 gallons evaporate from this tank and the draining parts every 8 h. Determine the emission rate (mg/min) to help design local exhaust ventilation for this 4 feet × 8 feet tank. The tank is covered to reduce evaporative losses except for 40 hours during the week when the process is operating. The density of TCE = 1.46 g/mL.

$$\frac{1.1\ \text{gallon}}{8\ \text{hours}} \times \frac{3785\ \text{mL}}{\text{gallon}} \times \frac{1.46\ \text{g}}{\text{mL}} \times \frac{\text{hour}}{60\ \text{minutes}} \times \frac{1000\ \text{mg}}{\text{g}} = \frac{12,664\ \text{mg}}{\text{minute}}$$

Answer: The average TCE evaporation loss is 12,664 milligrams per minute. Refer to OSHA's ventilation requirements for open surface tanks [29 *CFR* 1910.94 (d)].

193. A tracer gas (SF_6) was used to measure air infiltration rate into an empty building. Initial gas concentration was 0.01% (100 ppm$_v$). After 1 hour, it was 0.0012% (12 ppm$_v$). Dimensions of the building are 20' × 40' × 80'. What was the infiltration rate?

$$C = C_{oe}^{-(kt)/V}$$

where

C = tracer gas concentration after elapsed time, t
C_o = initial tracer gas concentration
e = 2.718
k = outside air infiltration rate
t = time
V = building volume
t = 1 hour = 60 minutes

$$V = 20' \times 40' \times 60' = 64{,}000 \, ft^3$$

$$k = \frac{-\ln[C/C_o]}{t/V} = \frac{-\ln[0.0012/0.01]}{60 \, \text{minutes}/64{,}000 \, ft^3} = \frac{-\ln 0.12}{0.0009375}$$

$$= \frac{-(-2.1203)}{0.0009375} = \frac{2263 \, ft^3}{\text{minute}}$$

Answer: The average air infiltration rate was 2262 ft³/minute.

194. Which tracer gases are used in building air exchange measurements?
    a. Sulfur dioxide, hydrogen, helium, nitrogen, methane
    b. Helium, phosgene, argon, hydrogen selenide, nitrogen
    c. Argon, nitrogen, helium, carbon dioxide, hydrogen
    d. Methane, SF_6, ethane, nitrogen dioxide, helium
    e. Sulfur hexafluoride, CO_2, helium, phosphine, arsine
    f. Nitrogen, CO_2, SF_6, helium, argon

Answer: d. Helium and SF_6 gases are used most frequently.

195. An Orsat analysis of flue gas was 12.9% CO_2 by volume in a combustion process using 40% excess air. Residual oxygen was 4.9% by volume in flue gas. What is the theoretical maximum concentration of CO_2 gas by volume in the flue gas under these conditions?

$$\text{Theoretical CO}_2 = \frac{\% \text{ CO}_2 \text{ in flue gas sample}}{1 - [\% \text{ O}_2 \text{ in flue gas sample}/21\%]}$$

$$= \frac{12.9\% \text{ CO}_2}{1 - [4.9\% \text{ O}_2/21\% \text{ O}_2]} = \frac{12.9\% \text{ CO}_2}{1 - 0.233} = 16.8\% \text{ CO}_2$$

Answer: The theoretical maximum concentration of CO_2 gas by volume was 16.8%. The combustion efficiency, therefore, was (12.9%/16.8%) × 100 = 76.8%.

196. Isokinetic sampling for dust in a cement plant stack will be done with an air pump operating near 1 cfm. The velocity of the dust entering the nozzle must equal the velocity of the dust passing the nozzle to ensure representative dust sampling. The velocity of the stack gas is uniform and laminar at 37.65 feet/second at dust sampling point. What collecting nozzle diameter is required?

$$\text{Sampling rate} = 60 \times V \times \frac{\pi \times d^2}{4 \times 144 \text{ in}^2/\text{ft}^2},$$

where
V = stack or duct velocity, feet per second
d = internal diameter of sampling probe, inches

Select a nominal meter stack sampling rate of 1 ft³/minute.
1 cfm = 0.327 × (37.65 feet/second) × d^2

$$d^2 = \frac{(1 \text{ ft}^3/\text{minute})}{(0.327)(37.65 \text{ feet/second})} = 0.0812 \text{ in}^2$$

$$d = \sqrt{0.0812 \text{ in}^2} = 0.285 \text{ inches}$$

The closest standard nozzle size is 0.25 inch internal diameter. Therefore, calculation of meter rate for the smaller nozzle is necessary to give the same probe-to-stack velocity (and collection of a representative dust sample):

$$Q = V \times A$$

$$(37.65 \text{ ft/s}) \times (60 \text{ seconds/minute}) = 2259 \text{ fpm}$$

Answer: Air-sampling pump meter rate = 2259 fpm × 0.00545 × (0.25 inch)² = 0.769 ft³/minute.

197. An Orsat stack gas analysis from burning natural gas is 10.4% CO_2, 2.9% O_2, and 86.7% N_2 by volume. The gas mixture is 89% methane, 5% nitrogen, and 6% ethane by volume. What is the maximum theoretical percent CO_2 and excess air? Use 90% as the approximation of ratio of dry products of combustion (ft³/ft³ of gas burned) to the air required for stoichiometric combustion (ft³/ft³ of gas burned).

$$\% \, CO_2 = \frac{10.4\% \, CO_2}{1 - [2.9\% \, O_2 / 21\% \, O_2]} = 12.06\% \, CO_2$$

$$\text{Excess air} = \frac{12.06\% \, CO_2 - 10.4\% \, CO_2}{10.4\% \, CO_2} \times 0.90 = 14.37\% \, \text{excess air}$$

Answers: 12.1% CO_2 and 14.4% excess air (oxygen). This calculation assists combustion engineers to achieve optimum oxidation of fuel ($ saved) and the production of minimum amounts of CO gas and other emissions in the flue and stack exhaust gases.

198. A pipe fitter working on a blast furnace inhaled 15,000 ppm$_v$ CO for 1 minute (his first inhalation caused immediate collapse, and he continued to breathe the CO gas, although unconscious, for another 50 seconds until he was rescued by a coworker). The background ambient CO concentration throughout the steel mill was 3 ppm$_v$. What was his 8-hour TWAE assuming that he had no other exposures to CO that day? Where do you think his SCBA was?

1 minute × 15,000 ppm$_v$ CO	=	15,000 ppm$_v$-minutes
479 minutes × 3 ppm$_v$ CO	=	+ 1437 ppm$_v$-minutes
		16,437 ppm$_v$-minutes

$$16,437 \, \text{ppm}_v\text{-minutes}/480 \, \text{minutes} = 34 \, \text{ppm}_v$$

Answer: 34 ppm$_v$ TWAE to CO, assuming he survived this massive, but brief, exposure. If he had not been rescued by his vigilant coworker, should we have told his widow that his 8-hour TWAE complied with OSHA's 35 ppm$_v$ PEL for CO?

199. Two adjacent gas lines rupture in a chemical plant exposing workers to an acrid aerosol. Five were hospitalized with severe pulmonary edema. A laboratory reconstruction of the incident revealed a maximum concentration of 6.3 milligrams of NH_4Cl/m^3 (STEL = 20 mg/m^3), 0.3 ppm$_v$ Cl_2 (STEL = 1 ppm$_v$), and 3.6 ppm$_v$ NH_3 (STEL = 35 ppm$_v$). Workers were not exposed for more than 10 minutes. What appears most plausible?
    a. Since all exposures were well below PELs, these workers appear to have some other effects, perhaps psychosomatic "illness" or mass hysteria
    b. There were mistakes in the laboratory simulation tests
    c. These air contaminants cannot cause pulmonary edema
    d. The workers instead most probably have chemical pneumonitis
    e. The ammonia gas reacted with the chlorine gas to form chloramines—gases that are potentially far more injurious to the lungs than chlorine or ammonia. The industrial hygienist failed to consider and detect the *de novo* chloramines

f. These workers must have had preexisting COPD (emphysema or bronchitis).

Answer: e.

200. The analytical detection limit for a hydrocarbon vapor with a molecular weight of 82 is 5 micrograms. The PEL is 150 ppm$_v$. What fraction of PEL could be reported with a 100 mL air sample?

$$\text{ppm} = \frac{(\text{mcg/L}) \times 24.45}{\text{molecular weight}} = \frac{(5 \text{ mcg/0.1 L}) \times 24.45}{82} = 14.9 \text{ ppm}$$

Answer: 14.9 ppm$_v$, or 10% of the PEL. Such a very small air sample might be obtained to measure peak exposures. 100 mL could be drawn through a small charcoal tube using a detector tube pump in 2 or 3 minutes. This method of air sampling could be used to augment full work shift air sampling using pumps on the worker's belts and breathing zone collection tubes.

201. The partial pressure of water vapor in an air sample at 22°C is 12.8 mm Hg. The saturation vapor pressure of water at this temperature is 19.8 mm Hg. What is the relative humidity of this air at 22°C and when heated to 28°C? The water vapor saturation pressure at 28°C is 28.3 mm Hg.

$$(12.8 \text{ mm Hg/19.8 mm Hg}) \times 100 = 64.6\% \text{ RH at } 22°C$$

$$(12.8 \text{ mm Hg/28.3 mm Hg}) \times 100 = 45.2\% \text{ RH at } 28°C$$

Answers: 64.6% and 45.2% relative humidities, respectively, at 22°C and 28°C. If this air sample is cooled to 15°C, the air becomes saturated and the relative humidity is 100% because the saturation vapor pressure of the water is also 12.8 mm Hg.

202. A 5.3 cubic foot cylinder containing 2.7% arsine gas in nitrogen gas under high pressure bursts inside a 10' × 18' × 38' room with no ventilation. What is the gas concentration of AsH_3 in ppm$_v$ after mixing? The molecular weight of AsH_3 is = 78 grams/gram-mole.

$$5.3 \text{ ft}^3 \times 0.027 = 0.143 \text{ ft}^3 \text{ of arsine gas}$$

$$10' \times 18' \times 38' = 6840 \text{ ft}^3$$

$$(0.143 \text{ ft}^3/6840 \text{ ft}^3) \times 10^6 = 20.91 \text{ ppm}_v$$

Answer: PEL = 0.05 ppm$_v$. This is an IDLH atmosphere. Molecular weight is irrelevant.

203. A typical hydrocarbon emission rate for a stationary source coal combustion unit exceeding 10^8 BTU/hour capacity is 0.2 lb per ton of coal burned. What is the approximate annual hydrocarbon emission from a power plant burning 275 tons of coal per day in a steam generator with a 450×10^6 BTU/hour capacity?

$$\frac{275 \text{ tons}}{\text{day}} \times \frac{365.25 \text{ days}}{\text{year}} \times \frac{0.2 \text{ lb HC}}{\text{ton}} = \frac{20,089 \text{ lb HC}}{\text{year}}$$

Answer: Over 10 tons of hydrocarbon vapors and gases are emitted per year.

204. The annual hydrocarbon emission for fluid catalytic units in petroleum refineries is typically 220 pounds for every 1000 barrels of fresh feed. What are annual HC emissions for a unit with no CO boiler operating 250 days per year and consuming 12,000 barrels of fresh feed per day?

$$\frac{12,000 \text{ barrels}}{\text{day}} \times \frac{220 \text{ pounds}}{1000 \text{ bbl}} \times \frac{250 \text{ days}}{\text{year}} = \frac{660,000 \text{ lb HC}}{\text{year}}$$

Answer: 660,000 pounds (330 tons) of hydrocarbon vapor emissions per year.

205. A worker leans into a large gasoline storage tank that was recently filled to the 90%$_v$ level. He collapses, but fortunately was seen by another worker and was revived by cardiopulmonary resuscitation (CPR). Analysis of the tank's head space shows 46.8 volume percent hydrocarbons. What was the most likely cause for his syncope and close brush with the "Grim Reaper?"

Answer: Since the tank's atmosphere contained 46.8% gasoline vapors, the balance was 53.2% air. 21% of this was oxygen (0.21 × 53.2% = 11.2% oxygen). The oxygen-deficient atmosphere plus the narcotic gasoline vapors most likely acted in a toxicologically additive way to cause his collapse. OSHA regards an atmosphere containing less than 19.5% O_2 at any expected ambient barometric pressure as oxygen deficient. Note that percent oxygen in an atmosphere does change with increases in elevation although oxygen's partial pressure decreases at higher altitudes.

206. A 2-inch wide by 4-foot-long slot hood exhausts air at a rate of 2000 feet per minute. How many pounds of air are exhausted per hour by this hood? Assume this hood is in an industrial plant in Tampa, Florida.

$$\frac{2'' \times 48''}{144 \text{ in}^2/\text{ft}^2} \times \frac{2000 \text{ ft}}{\text{minute}} \times \frac{60 \text{ minutes}}{\text{hour}} \times \frac{0.075 \text{ lb}}{\text{ft}^3} = \frac{6000 \text{ lb}}{\text{hour}}$$

Answer: 6000 pounds (3 tons) of air are exhausted every hour.

207. The driver of an LPG fork lift truck had the following CO exposures throughout a 10-hour work shift: 14 ppm$_v$ for 3-$\frac{3}{4}$ hours delivering parts to work stations, 16 ppm$_v$ for 2-$\frac{1}{2}$ hours returning empty pallets to the shipping and receiving dock, 2 ppm$_v$ for two 15-minute breaks, 135 ppm$_v$ for 1-$\frac{1}{2}$ hours inside railroad box cars, and 82 ppm$_v$ for 1-$\frac{3}{4}$ hours inside truck trailers. What was her TWAE to CO gas?

	ppm$_v$ CO	×	Time (h)	=	ppm$_v$-hours
Delivering parts	14	×	3.75	=	52.5
Returning pallets	16	×	2.5	=	40
Coffee breaks	2	×	0.5	=	1
Inside RR box cars	135	×	1.5	=	202.5
Inside truck trailers	82	×	1.75	=	143.5
			10.0		439.5

439.5 ppm$_v$-hours/10 hours = 43.95 ppm$_v$ TWAE.

Answer: Her TWAE to CO is 44 ppm$_v$ that exceeds the 35 ppm$_v$ TLV. Try to reduce her exposure time inside trucks and rail cars.

208. The Agency for Toxic Substances and Disease Registry regards 0.26 microgram inorganic mercury per cubic meter of air as a concentration below which adverse health effects should not occur with continuous exposure. At a detection level of 0.1 microgram, how long must one sample at 1 liter/minute through adsorbent tube to detect this concentration?

$$\frac{0.26 \text{ mcg}}{m^3} \times \frac{m^3}{1000 \text{ L}} \times \frac{1 \text{ L}}{\text{minute}} = \frac{0.00026 \text{ mcg}}{\text{minute}}$$

$$0.1 \text{ mcg} \times \frac{\text{minutes}}{0.00026 \text{ mcg}} = 384.6 \text{ minutes}$$

Answer: Sample for 385 minutes assuming a 100% collection and desorption efficiency.

209. Air was sampled at an average flow rate of 2.34 liters/minute through a tared PVC membrane filter for 3 hours and 27 minutes. The filter weight difference after air sampling was 2.32 milligrams. What was the airborne TSP (total suspended particulates) concentration?

$$207 \text{ minutes} \times 2.34 \text{ L/m} = 484.4 \text{ liters}$$

$$2320 \text{ mcg}/484.4 \text{ liters} = 4.79 \text{ mg/m}^3$$

Answer: 4.79 milligrams of TSP/m^3 of air.

210. A solvent blend is 20% ethyl acetate and 80% toluene (vol/vol). What is the vapor-phase concentration of each component?

	Ethyl Acetate	Toluene
Density	0.90 g/mL	0.87 g/mL
Vapor pressure	74 mm Hg	22 mm Hg
Molecular weight	88.1	92.1

Use 100 mL of the solvent blend as the basis for your calculations.

$$20 \, mL \times 0.90 \, g/mL = 18 \, g \text{ ethyl acetate}$$

$$80 \, mL \times 0.87 \, g/mL = 69.6 \, g \text{ toluene}$$

Partial pressure of ethyl acetate

$$= \frac{(18 \, g/88.1 \, g/mole) \times 74 \text{ mm Hg}}{(18 \, g/88.1 \, g/mole) + (69.6 \, g/92.1 \, g/mole)} = \frac{15.119}{0.96} = 15.75 \text{ mm Hg}$$

Partial pressure of toluene

$$= \frac{(69.6 \, g/92.1 \, g/mol) \times 22 \text{ mm Hg}}{(18 \, g/88.1 \, g/mol) + (69.6 \, g/92.1 \, g/mol)} = \frac{16.63}{0.96} = 17.32 \text{ mm Hg}$$

$$15.75 \text{ mm Hg} + 17.32 \text{ mm Hg} = 33.07 \text{ mm Hg total vapor pressure}$$

$$(15.75 \text{ mm Hg}/33.07 \text{ mm Hg}) \times 100 = 47.6\% \text{ ethyl acetate vapor}$$

$$(17.32 \text{ mm Hg}/33.07 \text{ mm Hg}) \times 100 = 52.4\% \text{ toluene vapor}$$

Calculations were made using Raoult's law. Note how the minor (20%), but more volatile, component in liquid phase increases and enriches in percent composition in vapor phase (47.6%). Refer to Problem 18 for discussion of application and deviations from Raoult's law.

211. A process evaporates 1.3 pints of benzene into the air of an empty room every 8 hours. The room is 10' × 16' × 35'. The density of benzene is 0.88 g/mL. The molecular weight of benzene is 78.1 grams/gram-mole. What volume of dilution air is required to maintain benzene vapor concentration below the NIOSH-recommended TWAE limit of 0.1 ppm$_v$?

Answers: Benzene is too hazardous and toxic to rely on ventilation by dilution air as a control method. Reduction in evaporation rate and the use of mechanical local exhaust ventilation are recommended to protect health of exposed workers. Substitute a less hazardous solvent, for example, water-based, or toluene with <1 ppm$_v$ benzene.

212. A gasoline-powered electrical generator engine produces 47 cfm of total exhaust gases that contain 1.2% (vol/vol) of carbon monoxide in a 12' (h) × 30' (w) × 40' (l) garage. Garage ventilation is 0.3 cfm/ft² of floor area. Assuming a negligible concentration of CO in the make-up air, what is the concentration of CO gas in the garage atmosphere after 30 minutes of engine operation?

$$\text{ppm}_v\, CO = \frac{G \times 10^6\ (1 - e^{-[Q/V]t})}{Q}$$

where
  G = rate of CO generation
  V = volume of garage
  Q = ventilation rate
  t = time
  e = 2.7813

$$47\,\text{cfm} \times 0.012 = 0.56\,\text{ft}^3 \text{ of CO generated/minute } (G)$$

$$0.3\,\text{cfm/ft}^2 \times 30' \times 40' = 360\,\text{cfm (ventilation, } Q)$$

$$V = 12' \times 30' \times 40' = 14{,}000\,\text{ft}^3$$

$$\text{ppm}_v\, CO = \frac{0.56 \times 10^6 \left[1 - e^{-[(360 \times 30/14{,}400)]}\right]}{360\ \text{cfm}} = 824\ \text{ppm}_v\, CO$$

Answer: 824 ppm$_v$ CO gas after 30 minutes. Evacuate the building, turn the generator off, ventilate with fresh air, do not permit occupancy until air samples for CO are the same as ambient air, and educate and train generator operator. Blood of exposed persons should be tested for carboxyhemoglobin (COHgB).

213. In the preceding example, if the generator is turned off after an industrial hygienist takes a CO detector tube air sample, how long will it take for the CO concentration to reduce to 20 ppm$_v$? Assume steady, uniform ventilation of the garage.

$$\text{Dilution ventilation time} = \frac{V}{Q} \times \left[\ln \frac{C_2}{C_1}\right] = \frac{14{,}400\,\text{ft}^3}{(360\,\text{ft}^3/\text{minute})} \times \left[\ln \frac{824\,\text{ppm}}{20\,\text{ppm}}\right]$$

$$= 40 \times \ln 41.2 = 40 \times 3.72 = 148.8\,\text{minutes}$$

Answer: 149 minutes ≅ 2.5 hours.

214. A chemical plant operator is exposed to 87 ppm$_v$ MEK and 23 ppm$_v$ toluene. Both are narcotic solvents with additive toxic effects. Their respective OSHA PELs are 200 and 100 ppm$_v$. What is worker's additive exposure to these solvent vapors?

$$87\,\text{ppm} + 23\,\text{ppm} = 110\,\text{ppm}$$

$$\frac{87\,\text{ppm}}{200\,\text{ppm}} + \frac{23\,\text{ppm}}{100\,\text{ppm}} = 0.44 + 0.23 = 0.67$$

Answers: 110 ppm$_v$ total solvent vapors. Exposure is below PEL for the mixture, however, it exceeds the action level (67% of PEL for the mixture of solvent vapors; the action level is 50%). Industrial hygiene controls are warranted. ACGIH TLV for toluene is 20% (20 ppm$_v$) of the OSHA PEL based on current toxicological evidence. One might be in regulatory compliance but still fail to protect workers' health and safety.

215. Many workers in a Wisconsin office building complain of eye irritation, sore throat, headache, and sinus "problems" every winter. Several air samples reveal vapor concentrations of mineral spirits from an adjacent printing operation ranging from 0.3 to 1.5 ppm$_v$. (The geometric mean is 0.5 ppm$_v$.) There are no other remarkable industrial hygiene issues. What would you do?
   a. Take more air samples.
   b. Disregard their symptoms because the vapor concentrations average less than 1% of the PEL.
   c. Measure the relative humidity and consider the additive effects of dry air plus solvent vapors on health and comfort. Suggest they be seen by a physician. Consult with the physician.
   d. Look for other air contaminants because mineral spirits' vapors cannot be the root cause of their mild complaints.
   e. Increase the air velocity flow through the work areas.
   f. Consider rotating some workers to other areas.
   g. Test the mineral spirits for benzene contamination.
   h. Their complaints are not statistically significant until 37% of the work force reports similar problems. It was only 28% here from day to day.

Answer: c.

216. A reactive gas is present in a plant atmosphere. This gas dissociates into a less hazardous gas at a predictable rate that is independent of the concentration that is present in the air. The gas is also removed by plant exhaust ventilation system. A direct-reading gas measurement instrument reveals 25 ppm$_v$. Thirty minutes later, the gas concentration is 10 ppm$_v$. What is the combined decay constant for this gas with this ventilation and dissociation decay? Assume no further release of the gas when measurements are made.

$$\ln\left[\frac{\text{final concentration}}{\text{initial concentration}}\right] = -T_{1/2} \times \text{time}$$

$$\ln\left[\frac{10\,\text{ppm}}{25\,\text{ppm}}\right] = -T_{1/2} \times 30\,\text{minutes}$$

$$\ln 0.4 = -T_{1/2} \times 30\,\text{minutes}$$

$$\frac{-0.9163}{30\,\text{minutes}} = -T_{1/2}$$

$$-T_{1/2} = -0.0305/\text{minute}$$

Answer: Combined decay constant is 0.0305 per minute (i.e., 3.05% decay every minute).

217. In the previous problem, the physical half-life of this gas is 40 minutes. Calculate the combined effective half-life (T_{eff}) and dilution ventilation half-life (T_v) of the gas.

$$\ln(1.2) = \frac{-0.0305}{\text{minute}} \times T_{eff}$$

$$T_{eff} = \frac{-0.693}{-0.0305/\text{minute}} = 22.7\,\text{minutes}$$

$$T_{eff} = \frac{T_p \times T_v}{T_p + T_v}, \quad \text{where } T_p = \text{physical half-life}$$

$$22.7\,\text{minutes} = \frac{40\,\text{minutes} \times T_v}{40\,\text{minutes} + T_v}$$

$$908 + 22.7 + T_v = 40 \times T_v$$

$$908/40 = (22.7 \times T_v/40) = T_v$$

$$22.7 + 0.568 \times T_v = T_v$$

$$22.7 = T_v\,(1 - 0.568)$$

$$T_v = 52.5\,\text{minutes}$$

Answers: Combined effective half-life is 22.7 min. Ventilation half-life is 52.5 min.

218. An industrial hygiene chemist wants to prepare a 25-liter gas-sampling bag with 35 ppm$_v$ CO to calibrate his instrument. What volume of 1760 ppm$_v$ CO gas must he dilute with CO-free air to obtain a 35 ppm$_v$ concentration?

$$C_1 \times V_1 = C_2 \times V_2$$

$$V_1 = \frac{C_2 \times V_2}{C_1} = \frac{35 \text{ ppm} \times 25 \text{ L}}{1760 \text{ ppm}} = 0.497 \text{ L}$$

Answer: Dilute 497 mL of 1760 ppm$_v$ CO to 25 liters with CO-free air.

219. A room with dimensions of 12' × 20' × 26' contains an ammonia compressor that leaks at a rate of 0.3 cfm. If 260 cfm is the volumetric rate of dilution air supplied to the room, by how much must air supply be increased to ensure that the ammonia concentration does not exceed 10 ppm$_v$.

$$12' \times 20' \times 26' = 6240 \text{ ft}^3$$

$$Q_o = 260 \text{ cfm}$$

$$C = 10 \text{ ppm}$$

$$C_o = \frac{0.3 \text{ ft}^3}{6240 \text{ ft}^3} \times 10^6 = 48 \text{ ppm NH}_3$$

$$Q = 260 \text{ cfm} \times \frac{48 \text{ ppm}}{10 \text{ ppm}} = 1248 \text{ cfm}$$

Answer: Increase the dilution air by 988 to 1248 cfm. Instead, quit sandbagging. Stop and repair the leak!

220. According to the olfactory perceptions by room occupants, the odor intensity in a work area has "doubled." Assuming the source of the odor has a constant generation rate, by how much was dilution ventilation apparently decreased? The dilution ventilation for the initial odor concentration was 16,400 cfm.
    The equation, with I = the odor intensity, is

$$\frac{I_1}{I_2} = \frac{\log \text{cfm}_2}{\log \text{cfm}_1}$$

This equation is similar to the Weber–Fechner law of physiological reactions in general, which is a physiological sensation is proportional to the logarithm of the stimulus.

$$\frac{1}{2} = \frac{\log \text{cfm}_2}{\log 16,400 \text{ cfm}}$$

$$0.5 = \frac{\log \mathrm{cfm}_2}{4.215}$$

$$\log \mathrm{cfm}_2 = 2.108$$

$$\mathrm{cfm}_2 = 128$$

Answer: Dilution air was reduced approximately from 16,400 to 128 cfm. Or, stated another way, relatively tremendous volumes of clean air are required to dilute an odor to 50% of its perceived olfactory intensity. See Problem 221.

221. In the preceding problem where the sensory odor intensity doubled with reduction of the dilution air from 16,400 to 128 cfm, by how much did the actual odorant concentration apparently increase?

$$16,400\,\mathrm{cfm}/128\,\mathrm{cfm} = 128 \text{ times}$$

Answer: As the odor concentration increases to the 2nd power, apparent intensity doubles. Odor intensity varies with logarithm of concentration. Another psychophysical bioreaction covering the senses (olfaction, hearing, vision, taste, touch, etc.) is Stevens' law:

$$I = k\,C^{\alpha}$$

where I = perceived intensity of the sensation, or odor; k = a constant; and C = the physical intensity of the stimulus, the odorant. As a rule, sensation varies as a power function of the stimulus. In olfaction, α is less than 1. Stevens' law has generally replaced the Weber–Fechner logarithmic law (refer to Problem 220). Since α is invariably less than 1, a change in concentration results in a smaller change in odor perception. Typically, α ranges from 0.2 to 0.7. With the exponent being 0.7, the odorant concentration must be reduced by 10-fold to reduce the odor perception by a factor of 5. When α equals 0.2, there must be a 3000-fold reduction in the concentration to achieve a 5-fold reduction in odor perception.

222. An instrument calibrated at normal temperature and pressure (NTP = 25°C and 760 mm Hg) indicated 57 ppm at 16°C and 630 mm Hg pressure. What is the corrected concentration and the correction factor to be applied to all readings?

$$\mathrm{ppm_v}\text{ at NTP} = \mathrm{ppm_v}\text{ meter reading} \times \frac{P}{760\text{ mm Hg}} \times \frac{298\text{ K}}{T}$$

$$= 57\text{ ppm} \times \frac{630\text{ mm Hg}}{760\text{ mm Hg}} \times \frac{298\text{ K}}{289\text{ K}} = 48.7\text{ ppm}$$

Answer: The correction factor = (49 ppm_v/57 ppm_v) is 0.86.

223. Pot room air containing caustic soda dust (NaOH) was bubbled through a midget impinger at 0.92 L/m for 17.5 minutes. There were 13.2 mL of dilute H_2SO_4 acid solution in the impinger. Each milliliter of the acid will neutralize 0.43 micrograms of NaOH as indicated by a color change from purple to green. Assuming 100% collection efficiency by the impinger, what was the average airborne NaOH dust concentration when the indicator color (methyl purple) changed?

$$0.92 \, \text{L/m} \times 17.5 \, \text{minutes} = 16.1 \, \text{liters of air were sampled}$$

$$13.2 \, \text{mL} \times 0.43 \, \text{mcg/mL} = 5.68 \, \text{mcg NaOH}$$

$$5.68 \, \text{mcg NaOH}/16.1 \, \text{liters} = 0.35 \, \text{mcg/L}$$

Answer: 0.35 mg NaOH dust per cubic meter of pot room air.

224. Sulfur hexafluoride, SF_6, was used to measure reentry of the exhaust air from a hood into a plant. The hood had an exhaust ventilation rate of 18,500 cfm. SF_6 gas was released at a rate of 0.05 cfm into the hood's exhaust system. The plant had a general replacement air flow rate of 77,600 cfm. What was the percent reentry if the SF_6 steady-state concentration in the plant was 0.004 ppm_v?

$$\text{Exhaust concentration} = \frac{0.05 \, \text{cfm}}{18,500 \, \text{cfm}} \times 10^6 = 2.7 \, \text{ppm}$$

$$\text{Ventilation dilution factor} = \frac{77,600 \, \text{cfm}}{18,500 \, \text{cfm}} = 4.19$$

Concentration of SF_6 measured inside the plant was 0.004 ppm_v. Contaminant dilution, therefore, is 2.7 ppm_v/0.004 ppm_v = 675 times.
Reentry of contaminants from the hood into the plant, given as a fraction of the released SF_6 = 4.19/675 = 0.0062.

Answer: 0.0062 × the released amount, or 0.62% reentry.

225. Contrast ratio of surface areas of 1-micron diameter particles to particles with a diameter of 3 microns.

$$\text{Surface area of a sphere} = \pi \times (\text{diameter})^2$$

$$\text{for a 1-micron Ø particle: } \pi \times (1 \, \mu)^2 = 3.1416 \, \text{square microns}$$

$$\text{for a 3-micron Ø particle: } \pi \times (3 \, \mu)^2 = 28.274 \, \text{square microns}$$

28.274 microns^2/3.1416 microns^2 = nine (the square of the diameter, e.g., for a 5-micron diameter particle, the surface area would be 25 times greater).

Answer: Nine times greater. As the particle diameter increases 3-fold, the surface area increases by 6-fold. If the diameter increases, for example, 6-fold, the surface area increases (6)^2-fold, or 36 times.

226. A virulent strain of infectious virus has a molecular weight of 55×10^6. This virus, resistant to desiccation, is in an aerosol containing 12 picograms of the virus per cubic meter of air. How many virus particles are inhaled with each liter of aerosol?

$$\frac{12 \text{ picogram}}{m^3} = \frac{0.000012 \text{ microgram}}{1000 \text{ L}} = \frac{0.000000012 \text{ microgram of virus}}{L}$$

$$\frac{1.2 \times 10^{-14} \text{ g}}{L} \times \frac{\text{mole}}{55 \times 10^6 \text{ gram}} \times \frac{6.023 \times 10^{23} \text{ molecules}}{\text{mole}}$$
$$= \frac{1.3 \times 10^2 \text{ molecues}}{L}$$

Answer: 130 infectious virus particles per liter of inhaled air.

227. A worker sprayed banana plants in Ecuador with a pesticide. A breathing zone air sample using a tared PVC membrane filter was used to collect spray aerosol mist. Assume 35% of the collected particulate was not pesticide and 7% of the pesticide evaporated from the filter during air sampling. If the air-sampling rate was 2.1 L/minute for 7 hours and 41 minutes, what was sprayer's pesticide exposure if initial filter weight was 45.46 mg, and the final filter weight was 49.13 mg?

7 hours and 41 minutes = 420 minutes + 41 minutes = 461 minutes

(2.1 L/m) × 461 minutes = 968.1 liters

49.13 mg – 45.46 mg = 3.67 mg total aerosol particulate on this PVC filter

3.67 mg × 0.65 = 2.39 mg pesticide/m^3

(2.39 mg pesticide/m^3) × 1.07 for the evaporation factor = 2.56 mg pesticide/m^3 of breathing zone air.

228. During a laboratory instrument calibration for airborne oxidant (I_2 vapor), 45.67 mg of iodine crystals were placed in a 0.17 ft^3/minute NTP air stream. I_2 crystals were removed after 176 minutes and now weighed 43.79 mg.

What was the average concentration of sublimed iodine vapor in air stream in ppm$_v$? Molecular weight of $I_2 = 253.8$ grams gram-mole^{-1}

$$\frac{0.17\,\text{ft}^3}{\text{minute}} \times \frac{28.3\,\text{L}}{\text{ft}^3} \times 176\,\text{minutes} = 846.7\,\text{L}$$

$$45.67\,\text{mg} - 43.79\,\text{mg} = 1880\,\text{mcg}$$

$$\frac{1880\,\text{mcg}}{846.7\,\text{L}} = \frac{2.22\,\text{mcg}}{\text{L}}$$

$$\text{ppm}_v = \frac{(2.22\,\text{mcg/L}) \times 24.45 \text{ L/gram-mole at NTP}}{253.8 \text{ gram/gram-mole}} = 0.213865$$

Answer: 0.21 ppm$_v$ sublimed I$_2$ vapor in the effluent air stream. As you collect instrument readings you will notice that iodine vapor concentrations will initially increase and then stabilize. Use the stabilized value as your calibration point. Then proceed to change the iodine vapor concentrations by increasing and decreasing the air flow. Four more data points should suffice to establish your instrument calibration curve by plotting instrument reading versus ppm$_v$ I$_2$ vapor. The instrument calibration apparatus must be slightly warmer than ambient temperature to ensure that iodine vapor does not condense on surfaces before the instrument insertion probe point.

229. A hospital requires humidification for occupant comfort. Engineering calculations are based on an outdoor temperature of 41°F and a relative humidity of 20%. An indoor temperature at a relative humidity of 50% is selected as a design objective. Additional increases in humidity and dry bulb temperature will be gained from the occupants, other water vapor sources, and process lighting and heating, and so on. How much H$_2$O vapor must be evaporated into the influent outside air to achieve this humidity? Water vapor saturation is 6.77 g/m³ at 41°F and 17.26 g/m³ at 68°F. This hospital is in New York City.

Density of water vapor in 41°F air $= 0.20 \times 6.77\,\text{mg/m}^3 = 1.354\,\text{g/m}^3$

Density of water vapor in 68°F air $= 0.50 \times 17.26\,\text{g/m}^3 = 8.63\,\text{g/m}^3$

$$41°F = 5°C$$

$$68°F = 20°C$$

One m³ of air at 41°F expands to: $\dfrac{273\,\text{K} + 20}{273\,\text{K} + 5} = \dfrac{293\,\text{K}}{278\,\text{K}} = 1.054$ m³ at 68°F.

$$\text{Water vapor in } 1.054\,\text{m}^3 \text{ at } 68°\text{F} = 1.054\,\text{m}^3 \times \frac{8.63\,\text{g}}{\text{m}^3} = 9.10\,\text{grams}$$

The amount of water vapor to be added to each cubic meter of air at $41°\text{F} = 9.10\,\text{g} - 1.35\,\text{g} = 7.75\,\text{grams of } H_2O$.

Answer: 7.75 grams of water vapor must be added to every cubic meter of $41°\text{F}$ outdoor air at 20% relative humidity. Protracted relative humidity beyond 60–70% will support mold growth on organic substrates. Humidity measurements should be made regularly to ensure occupants' comfort.

230. A pressure vessel contained CO compressed to 17 atmospheres at $43°\text{F}$. The tank ruptures at what was believed to be an internal CO temperature of $300°\text{F}$. People died from blast effects and chemical asphyxiation. Incident reconstruction attempts to determine the probable internal pressure of CO when the tank blew apart.
    Since the tank volume is constant, $V_i = V_f$, then:

$$\frac{P_i V_i}{T_i} = \frac{P_f V_f}{T_f} \text{ becomes } \frac{P_i}{T_i} = \frac{P_f}{T_f} = \frac{17\,\text{atmospheres}}{(43 + 460)°\text{R}} = \frac{P_f}{(300 + 460)°\text{R}}$$

$$P_f = 25.69\,\text{atmospheres}$$

Answer: Tank rupture occurred near an internal pressure of 26 atmospheres. Correction for the nonideal gas behavior at this high pressure might be required.

231. A mixture of gases at $25°\text{C}$ contains nitrogen at 160 mm Hg, methane at 212 mm Hg, hydrogen at 110 mm Hg, ethane at 210 mm Hg, and propane at 195 mm Hg. What is the total pressure of the gas mixture and the percent noncombustible gas in the mixture?

$$(160 + 212 + 110 + 210 = 195)\,\text{mm Hg} = 887\,\text{mm Hg}$$

Nitrogen is the only noncombustible, nonflammable component:

$$\frac{160\,\text{mm Hg}}{887\,\text{mm Hg}} \times 100 = 18.04\%\,N_2$$

Answer: 18% nitrogen. 82% flammable gases. 887 mm Hg total pressure.

232. A pressure vessel contains 3700 kilograms of ethane gas at 1 atmosphere and $10°\text{C}$. How much ethane could it contain at $30°\text{C}$ and 14 atmospheres?

$$\text{The tank volume is constant, therefore: } \frac{P_i}{T_i} \propto \frac{P_f}{T_f}$$

$$\frac{14 \text{ atm.}/(273 \text{ K} + 30°\text{C}) \text{ K}}{1 \text{ atm.}/(273 \text{ K} + 10°\text{C}) \text{ K}} = \frac{0.046205}{0.003534} = 13.074$$

$$13.074 \times 3700 \text{ kg} = 48,374 \text{ kg } C_2H_6$$

233. A boll weevil returns home from work in the fields with 600 nanograms of pesticide with a molecular weight of 378 on each foot. He dies from acute cholinesterase inhibition poisoning while being clutched by his wife. Their subterranean nest—that has no ventilation—is 11 cubic centimeters. What is the concentration of the pesticide vapor in the nest air after evaporating from his feet? If the STEL for adult weevils is 1 ppm$_v$ (15 minutes), is his wife in danger? She is not wearing a protective apron, impervious booties, or a whole-body respirator (insects "breathe" through their integument).

$$1 \text{ nanogram} = 10^{-9} \text{ gram} = 10^{-3} \text{ micrograms}$$

$$(600 \text{ nanograms/foot}) \times 6 \text{ feet} = 3600 \text{ nanograms}$$

Answer: 3.6 micrograms of pesticide evaporates from his feet.

$$11 \text{ cm}^3 = 11 \text{ mL} \times \frac{L}{1000 \text{ mL}} = 0.011 \text{ L}$$

$$\frac{3.6 \text{ mcg}}{0.011 \text{ L}} = \frac{327.3 \text{ mcg}}{L}$$

$$\text{ppm} = \frac{(327.3 \text{ mcg/L}) \times 24.45}{378} = 21.2 \text{ ppm}$$

Since wife weevil has continuous exposure, telephone 911 immediately to request EMS team. Ask them to bring atropine and 2-PAM for antidote injection. Instruct her to leave the nest (or remove her by stretcher). Keep patient warm and at rest until help arrives. CPR (with an insect whole body "intubator") might be needed. Give nothing by mouth if she is unconscious. Check their kids who are in school for signs and symptoms of OP pesticide poisoning. Bar them and all others from entering the nest.

234. OSHA's action level for asbestos fibers is 0.1 fiber per cc as an 8-hour TWAE. Assuming a worker inhales 10 cubic meters of air containing 0.1 f/cc during his work shift, how many asbestos fibers will he inhale?

$$10 \text{ m}^3 \times \frac{1000 \text{ L}}{\text{m}^3} \times \frac{1000 \text{ cm}^3}{L} \times \frac{0.1 \text{ fiber}}{\text{cm}^3} = 10^6 \text{ fibers}$$

Answer: At least 1,000,000 asbestos fibers are inhaled during his 8-hour work shift. This only includes those fibers with a length-to-width aspect ratio of 3-to-1 and with a length longer than 5 microns. If the fibers have the same aspect ratio but are shorter than 5 microns, the total number of inhaled asbestos fibers would be substantially higher, say, 10^3–10^6.

235. Thirty percent of the workers in a metal machining plant simultaneously develop an acute illness and are too ill to work. Their symptoms include fever, myalgia, headache, dyspnea, and extreme fatigue. Inhalation of aerosols from the metal working fluid is believed to be the source of an apparent epidemic bacterial or viral infection. Analysis of the coolant reveals the presence of 10^8 aerobic bacteria per mL and 130,000 *Legionella pneumophillae* per mL. Workers were exposed to a geometric mean concentration of 5.8 mg total airborne mist and aerosol/m³. If the coolant metal working fluid was 95% water and had a density of 1 g/mL, how many *L. pneumophillae* bacteria were inhaled in each liter of air at the geometric mean concentration?

Consider the number of bacteria in 1 liter of air.

$$5.8\,\text{mg/m}^3 = 5.8\,\text{mcg/L}$$

The volume of coolant mist generated and required to be filtered by the sampling device, V, to collect 5.8 mcg of particulates (including all bacteria) is

$$V \times (1.00 - 0.95) = 5.8\,\text{mcg}$$

$$V = \frac{5.8\,\text{mcg}}{0.05} = 116\,\text{mcg of total mist (includes the aqueous portion)}$$

$$116\,\text{mcg} \times \frac{\text{g}}{10^6\,\text{mcg}} \times \frac{1.0\,\text{mL}}{1.0\,\text{g}} = 0.000116\,\text{mL}$$

$$0.000116\,\text{mL} \times \frac{130,000\,L.\ pneumophillae}{\text{mL}} = 15.08\,L.\ pneumophillae$$

Answer: 15 *L. pneumophillae* bacteria are inhaled in each liter of air. It has been estimated that as little as one *L. pneumophillae* bacterium in 50 liters of inhaled air can be an infectious dose in a susceptible human host.

236. Derive an equation to calculate the half-life for an air contaminant in a room with a volume (V), uniform ventilation (Q), original concentration (C_o), and concentration at half-life time (C) after generation stops ($G = 0$). t = the time required to reach the half-life concentration.

The volume of the space and the ventilation rate of the space determine the half-life.

$$C = C_o \times \left(e^{-\frac{Qt}{V}} \right)$$

at one half-life: $C = 0.5 C_o$

$$\frac{0.5\, C_o}{C_o} = e^{-\frac{Qt}{V}}$$

$$t = -\left[\frac{V}{Q} \right] \ln 0.5 = 0.693 \left[\frac{V}{Q} \right]$$

Answers: Air contaminant half-life $= 0.693 \left[\dfrac{V}{Q} \right]$.

237. Evaluate a plant that is 20' (h) × 45' (w) × 60' (l) that has a general ventilation rate of 1.3 cfm per square foot of floor area. Ventilation is evenly
distributed. If the CO gas concentration in this plant initially is 200 ppm$_v$,
what is the concentration after three half-lives? What is half-life of CO gas
for this plant under these conditions?

$$\frac{1.3\, \text{ft}^3/\text{minute}}{\text{ft}^2} \times (45' \times 60') = \frac{3510\, \text{ft}^3}{\text{minute}}$$

Plant volume $= 20' \times 45' \times 60' = 54{,}000\, \text{ft}^3$

$$\text{Half-life} = 0.693 \left[\frac{V}{Q} \right] = 0.693 \left[\frac{54{,}000\, \text{ft}^3}{3510\, \text{cfm}} \right] = 10.66\, \text{minutes}$$

For three half-lives $= (10.66\, \text{minutes/half-life}) \times 3\, \text{half-lives} = 32\, \text{minutes}$

$$200\, \text{ppm}_v \times (0.5)^3 = 25\, \text{ppm}_v$$

Answer: The concentration of CO gas remaining in the plant atmosphere
after three half-lives (32 minutes) of ventilation is 25 ppm$_v$, the current TLV,
assuming that CO release and generation stops when the ventilation begins.
The half-life for this situation is 10.7 minutes.

238. A dry cleaner operator was exposed to 11 ppm$_v$ "perc" vapor for 3.5 hours
at a spotting bench, 15 ppm$_v$ for 1.5 hours while sorting and loading clothes,
49 ppm$_v$ for 2.5 hours while removing and hanging clothes, and 2 ppm$_v$ for
2.5 hours while she took orders and returning cleaned clothing. What was
her TWAE to perchloroethylene vapor?

Job	C	×	T	=	CT
Spotting	11 ppm$_v$	×	3.5 h	=	38.5 ppm$_v$-h
Sorting and loading	15 ppm$_v$	×	1.5 h	=	22.5 ppm$_v$-h
Removing and hanging	49 ppm$_v$	×	2.5 h	=	122.5 ppm$_v$-h
Taking orders	2 ppm$_v$	×	2.5 h	=	5.0 ppm$_v$-h
			10.0 h	=	188.5 ppm$_v$-h

TWAE = 188.5 ppm$_v$-hours/10 hours = 18.85 ppm$_v$ "perc" using Haber's law.

Answer: Her 10-hour TWAE to perchloroethylene vapor was 19 ppm$_v$. 8-hour = 23.6 ppm$_v$.

239. A silver membrane air filter taken from a topside coke oven worker was extracted with benzene. This tared filter weighed 78.56 mg before sampling and 82.97 mg after air sampling. The same filter weighed 80.76 mg after extracting with warm benzene. If air was sampled at 2.07 L/min for 463 minutes, what was the worker's exposure to total airborne particulates and to the benzene-soluble fraction of the airborne coke oven emissions?

$$(2.07 \, L/min) \times 463 \, minutes = 958.4 \, liters$$

$$82.97 \, mg - 78.56 \, mg = 4.41 \, mg$$

$$\frac{4410 \, mcg \, total \, particulates}{958.4 \, L} = \frac{4.60 \, mg \, TSP}{m^3}$$

$$82.97 \, mg - 80.76 \, mg = 2.21 \, mg \, COE/m^3$$

Answer: 4.60 mg TSP/m³ and 2.31 mg coke oven emissions/m³ (50% of total much of which is benzene insoluble coal dust and soot).

240. A balloon man—an evil person—sold 12 balloons filled with 20% ammonia and 80% helium instead of pure helium to kids at a birthday party. He instructed them to prick every balloon at the same time for a birthday surprise. Each balloon was filled with 6 liters of the mixed gas. He promptly left the party being held in an 8' × 16' × 18' room with no ventilation. What was the burst concentration of ammonia gas when the now-gleeful children pricked their balloons?

$$12 \, balloons \times \frac{6 \, liters}{balloon} \times 0.2 = 14.4 \, liters \, of \, ammonia$$

$$8' \times 16' \times 18' \times \frac{28.3 \, L}{ft^3} = 65,249 \, liters$$

$$\frac{14.4\,L}{65,249\,L} \times 10^6 = 220.6\,ppm\,NH_3$$

Answer: The average concentration in the room was about 221 ppm$_v$; however, the burst concentration near each balloon was less than 200,000 ppm$_v$—the ammonia concentration in each balloon. This would cause coughing and tearing but not so much incapacitation that one could not call an EMS unit at 911 and the police to report and apprehend this vicious balloon man—an evil clown.

241. Which OSHA-approved gases and vapors are used to odorize LPG?
    a. Ethyl mercaptan, thiophane, amyl mercaptan
    b. Thiophene, butyl mercaptan, methyl mercaptan
    c. Ethane thiol, isoamyl mercaptan, mercaptobenzothiazole
    d. SO$_2$, butyl mercaptan, acetylene
    e. Propane thiol, thiophene, thiophane, hydrogen sulfide

Answer: a.

242. Ethyl mercaptan is typically used as the odorant for natural gas and liquefied fuel gases at a concentration of about 8 pounds per million cubic feet of gas. What is concentration of C$_2$H$_5$-SH in the mixed gas in ppm$_v$? Density of ethyl mercaptan (ethanethiol) is 0.839 g/mL. Molecular weight of ethyl mercaptan is 62.13 grams per gram-mole.

$$\frac{8\,lb}{10^6\,ft^3} \times \frac{454\,g}{lb} \times \frac{1000\,mg}{g} \times \frac{35.3\,ft^3}{m^3} = \frac{128\,mg}{m^3}$$

$$\frac{(128\,mg/m^3) \times 24.45}{62.13} = 50\,ppm\,C_2H_5SH$$

Answer: 50 ppm$_v$ ethyl mercaptan. This greatly exceeds odor threshold of approximately 0.5 ppb$_v$ (0.0005 ppm$_v$) most people can detect. Source concentration must be much greater than odor thresholds because substantial dilution occurs between the leak and the person detecting the gas. An important consideration, of course, is the notion that "If I'm smelling this odorant at, say 1 ppb$_v$, and the leak is 50 or more feet away, could concentration of LPG exceed its LEL somewhere between the source and me?" Any ignition sources between the source and me?

243. What is the saturation concentration of "Heptachlor" in air at NTP? The vapor pressure of this pesticide is 0.0003 mm Hg, and its molecular weight is 373.4. Does the saturation concentration exceed the OSHA PEL of 0.5 mg/m^3 (skin)?

$$\frac{0.0003\,\text{mm Hg}}{760\,\text{mm Hg}} \times 10^6 = 0.395\,\text{ppm}$$

$$\frac{\text{mg}}{\text{m}^3} = \frac{\text{ppm} \times \text{molecular weight}}{24.45} = \frac{0.395 \times 373.4}{24.45} = \frac{6.03\,\text{mg}}{\text{m}^3}$$

Answer: Six mg/m^3, or a vapor saturation concentration of "Heptachlor" about 12 times the OSHA PEL. However, in the field, the typical exposures to this pesticide are to mist—not solely to vapor—and evaluations must include appraisal of mist, vapor, and skin exposures among other issues of industrial hygiene.

244. An explosion occurred in a tank where LPG was used as fuel for a heater to dry the tank's interior. In forensic accident investigation, the LPG cylinder was found outside of the tank, and propane gas was fed into the heater in the tank by a hose. The heater tipped over, the ignition flame was extinguished, and leaking propane gas accumulated inside the tank. From the tare weight and the water capacity stamped on the LPG container, reweighing indicated 2.8 gallons of LPG had vaporized inside the eight-foot diameter, 24' long cylindrical tank. The source of ignition appeared to be an ineffective exhaust fan located near the far end of the tank. The density of LPG is 0.51 g/mL. What was the maximum concentration of propane gas in the tank?

$$\text{Volume of cylinder} = \frac{(1400\,\text{mg/m}^3 - \text{hour})}{8\,\text{hours}} = \frac{175\,\text{mg}}{\text{m}^3},$$

clearly a need for a full-face airline respirator during spraying.

$$2.8\,\text{gallons} \times \frac{3.785.3\,\text{mL}}{\text{gallon}} \times \frac{0.51\,\text{g}}{\text{mL}} = 5405.5\,\text{grams}$$

$$\frac{5405.5\,\text{g}}{34,164\,\text{L}} = \frac{0.1582\,\text{g}}{\text{L}} = \frac{158.2\,\text{mg}}{\text{L}} = \frac{158,200\,\text{mg}}{\text{m}^3}$$

Molecular weight of $C_3H_8 = 44$ grams/gram-mole

$$\text{ppm} = \frac{(158,200\,\text{mg/m}^3) \times 24.45}{44} = 87,909$$

Answer: 87,909 propane gas in the tank (= 8.79%). LEL = 2.2% = 22,000 ppm$_v$. UEL = 96,000 ppm. 8.79% approaches the "richer" end of the LEL–UEL range and, therefore—while certainly explosive—the magnitude of any explosion of gas or vapor increases as the concentration approaches the stoichiometric mid-range of explosibility.

245. The industrial hygienist often must explain how small a ppm is, how tiny are these dust particles, "What's a microgram?," how short a 5 micron fiber is, and "What does 0.002 mg Be/m^3 mean?," and so on. One good approach that can be used is to relate the familiar to the unfamiliar. As an example, to share the notion of the OSHA PEL for lead of 50 micrograms per cubic meter, one could relate the weight of the artificial sweeteners (e.g., Equal®) found in the little paper packages we see along with packets of sugar in restaurants. This could be used to demonstrate the dust concentration inside a building the size of, say, a football field 300 feet × 160 feet × 14 feet high. The weight of artificial sweeteners in those little packets is 1 gram. Assuming that the sweetener is an ultrafine lead powder, what would be the airborne lead dust concentration if it was suspended in the air of this gridiron-sized building?

$$\frac{1\,\text{gram} \times (10^6\,\text{micrograms/gram})}{300' \times 160' \times 14'/(35.315\,\text{ft}^3/\text{m}^3)} = \frac{53\,\text{mcg Pb}}{\text{m}^3}.$$

Answer: 53 micrograms of lead per cubic meter, or slightly over the OSHA PEL of 50 and 23 mcg/m^3 above the action level for work place airborne lead. Such a finely divided dust at this concentration would be essentially invisible.

246. A 70-kilogram man inadvertently swallowed a 1-gram chunk of sodium cyanide falling from dusty overhead beam onto a sandwich he ate for lunch. Immediately, the soluble cyanide salt converted into HCN gas when contacting his gastric HCl. This caused him to belch 400 mL of gas containing 2800 ppm HCN gas. How much cyanide was left in his body after belching? Assume excess of hydrochloric acid in his stomach and that the belched gas was totally expelled from his body.

$$NaCN + HCl \rightarrow HCN \uparrow + NaCl$$

molecular weights = 49 grams/gram-mole for NaCN and 27 for HCN grams/gram-mole.

One gram NaCN/(49 grams/mole) = 0.02041 mole NaCN; therefore, 0.02041 mole of HCN was produced.

$$\frac{\text{mg}}{\text{m}^3} = \frac{\text{ppm} \times \text{molecular weight}}{24.45} = \frac{2800\,\text{ppm} \times 27}{24.45}$$

$$= \frac{3092\,\text{mg}}{\text{m}^3} = \frac{3092\,\text{mcg}}{\text{L}}$$

$$\frac{3092\,\text{mcg}}{\text{L}} \times 0.4\,\text{L} = 1237\,\text{mcg HCN}$$

$$\%\,\text{CN}^- \text{ in HCN} = \frac{26}{27} \times 100 = 96.3\%$$

$$\%\,\text{CN}^- \text{ in NaCN} = \frac{26}{49} \times 100 = 53.1\%$$

$$1237\,\text{mcg HCN} \times 0.963 = 1191\,\text{mcg CN}^- = 1.19\,\text{mg CN}^-$$

$$1000\,\text{mg NaCN} \times 0.531 = 531\,\text{mg CN}^-$$

$$531\,\text{mg CN}^- - 1.19\,\text{mg CN}^- = 529.8\,\text{mg CN}^-$$

529.8 milligrams of cyanide remain as HCN in his body. This equals 529.8 mg/ 70 kg = 7.6 mg CN^-/kg, or (7.6 mg/kg) × (1/0.531) = 14.3 mg NaCN/kg. A LDL_0 of 2.857 NaCN/kg has been reported for man. HURRY! HURRY!— quickly administer antidotes amyl nitrite and sodium nitrite! Keep the person warm. Give CPR, oxygen, and so on. Call a doctor, EMS, local Poison Control Center. Do not become a victim yourself by inhaling exhaled HCN of victim during mouth-to-mouth CPR.

247. An industrial hygienist evaluated a maintenance worker's exposures to methyl chloroform vapors during the removal of sludge from a solvent degreasing tank. A 45 minute TWAE was 350 ppm$_v$ (1900 mg/m^3, the 8-hour OSHA PEL). A few days later, the same worker was overcome from methyl chloroform at a spray degreasing operation and admitted to a hospital. Reconstruction of the exposure ironically showed that he was also exposed to 350 ppm for 45 minutes. Charcoal tubes and air-sample pumps were used for both studies. What could account for the man's illness from one exposure and not the other apparently identical exposure?

    The first exposure was solely to methyl chloroform vapors, while the second most likely included the inhalation of mist particles as well as vapors. There is no good air-sampling technique that can distinguish between both MC vapors and MC mist particles in an aerosol mixture. Inhalation of respirable MC particles is expected to be far more irritating and injurious to lungs than to the vapor alone. Compare, for example, eye irritation expected from vapor exposure to that resulting from direct contact of liquid solvent on the cornea. The glycolipid-covered membranes in the alveoli are more susceptible to injury by mist particles than from vapor molecules. In the evaluation of exposures to mixed aerosols, the industrial hygienist must be vigilant for variations in biological responses from what appear to be the "same" exposure concentrations. Observation of a mist cloud in workers' breathing zones is a clue that exposures are to a two-phase aerosol.

248. A calibrated rotameter indicated an air flow rate of 2.37 L/minute at 7:07 am. This rotameter—used in a sampling system for airborne dust—indicated 2.18 L/m at 4:02 pm. How much air was sampled?

$$\frac{2.37\,L/min + 2.18\,L/min}{2} = 2.28\ L/min\ average\ air\ flow\ rate$$

Elapsed time = 15:62 − 7:07 = 8 hours and 55 minutes = 535 minutes

$$(2.28\,L/m) \times 535\,minutes = 1219.8\,L$$

Answer: Approximately 1220 liters of air were sampled (1.22 m³).

249. The PEL for a dust is 0.3 mg/m³ as TSP. The detection limit is 5 micrograms. At an air-sampling rate of 1.06 L/m, how long must one sample take to detect 10% of the PEL?

$$minimum\ sampling\ time\ (in\ minutes) = \frac{analytical\ sensitivity\ (in\ mcg)}{0.1 \times PEL\ (mcg/L) \times L/min}$$

$$= \frac{5\ micrograms\ TSP}{0.1 \times 0.3\ mcg/L \times 1.06\ L/min}$$

$$= 157.2\ minutes$$

Answer: Sample at least 2 h and 38 minutes.

250. A crane used in construction of a skyscraper raised a compressed gas bottle containing 200 ft³ of 100% acetylene to welders on the 41st floor. The bottle burst in free space at the 20th floor. If the gas concentration was 100,000 ppm 4 ft from the cylinder microseconds after bursting, what concentration of gas could be expected 8, 12, and 16 feet away from bottle (i.e., spheres with 16, 24, and 32 feet diameters)? Assume no wind or thermal air currents immediately diluted the gas. Assume uniform energy dispersion as the compressed cylinder of acetylene ruptured.

The inverse square law would appear to apply here; that is, the concentration will decrease as the inverse of the square of the distance from the source. This is an expanding sphere of acetylene gas decreasing in concentration and energy as the inverse of the square of the distance from the epicenter.

In the gas cylinder: 100% acetylene (i.e., 1,000,000 ppm$_v$)

At 4 feet: 100,000 ppm$_v$

At 8 feet: $(1/2^2) \times 100,000\,ppm_v = 25,000\,ppm_v$ (i.e., 2 × 4 feet)

At 12 feet: $(1/32) \times 100,000\,ppm_v = 11,111\,ppm_v$ (3 × 4 feet)

At 16 feet: $(1/4^2) \times 100,000\,ppm_v = 6250\,ppm_v$ (4 × 4 feet)

Answers: About 25,000 ppm$_v$, 11,000 ppm$_v$, and 6000 ppm$_v$ of C_2H_2 at 8', 12', and 16', respectively.

251. Consider a 1 cm^3 solid crushed into 1 micrometer cubes. How many particles are created? How does the total surface area of the small particles compare to surface area of the original cube? What are the implications for alveolar surface activity?

$$\text{Original: 6 facets} \times (1 \times 1 \text{ cm})/\text{facet} = 6 \text{ cm}^2$$

$$1 \text{ cm} = 10^4 \text{ microns/side}$$

$$\text{After crushing to tiny particles: } 10^4 \times 10^4 \times 10^4 = 10^{12} \text{ particles}$$

10^{12} particles $\times$ 6 microns^2/particle $= 6 \times 10^{12}$ microns$^2 = 6$ m^2 (and biological activity $\uparrow$

Answers: One large particle becomes 1,000,000,000,000 tiny particles, and the surface area increases from 6 to 60,000 cm^2. Ouch!

252. NIOSH reports that the inhalation of a single tubercle bacillus can cause an active tuberculosis lesion (*Occupational Respiratory Diseases*, 1986). This bacterium weighs as little as 10^{-13} gram. A careless laboratory worker was exposed to an aerosol of tubercle bacilli estimated to be about 20,000 microbes per cubic meter. She inhaled at least 30 liters of bio-contaminated air before leaving the facility. How many bacilli could she have inhaled? What was the total inhaled mass of bacilli?

$$20,000 \text{ microbes/m}^3 = 20,000/1000 \text{ L} = 20 \text{ microbes/liter of air}$$

$$\frac{20 \text{ microbes}}{L} \times 30 \text{ L of inhaled air} = 600 \text{ microbes dose}$$

$$600 \text{ microbes} \times 10^{-13} \text{ gram/microbe} = 6 \times 10^{-11} \text{ gram/microbe}$$
$$= 60 \text{ picograms.}$$

Viruses, by comparison, are about 1 attogram $= 10^{-18}$ gram $= 10^{-13}$ femtogram.

Answer: 600 tubercle bacilli microbes. 60 picograms of tubercle bacilli.

253. Jerry returns home from the dry cleaners with 26 pounds of clothing and hangs it in a 3' × 4' × 8' closet with about 20% contents. One milliliter of perchloroethylene remains in the clothing after he removes the plastic garment bags and hangs the clothing. He shuts the door. The closet has no ventilation. Tom, the family cat, is working inside the closet making mouse

traps. What is Tom's exposure when all of the perchloroethylene evaporates? The molecular weight and "perc's" density are 165.8 grams/gram-mole and 1.62 g/mL, respectively.

$$3' \times 4' \times 8' = 96\,\text{ft}^3$$

$$96\,\text{ft}^3 - (0.20 \times 96\,\text{ft}^3) = 76.8\,\text{ft}^3$$

$$\frac{1.62\,\text{g "perc"}}{76.8\,\text{ft}^3} = \frac{1{,}620{,}000\,\text{mcg}}{76.8\,\text{ft}^3} \times \frac{\text{ft}^3}{28.32\,\text{L}} = \frac{744.8\,\text{mcg}}{\text{L}}$$

$$\frac{(744.8\,\text{mcg/L}) \times 24.5\,\text{L/g-mole}}{165.8\,\text{g/g-mole}} = 109.8\,\text{ppm "perc" vapor}$$

Answer: 110 ppm$_v$ perchloroethylene vapor. Consultation with a seasoned veterinary industrial hygienist appears warranted.

254. Three grams of liquid mercury were spilled on a carpet. Vacuuming removed the mercury but mercury vapors continued to be emitted from the sweeper's bag. If the evaporation rate during vacuum sweeper operation was 300 micrograms of Hg per minute, and the exhaust rate was 50 cfm, what was the concentration of mercury vapor in the air exhausted from the sweeper?

$$\frac{300\,\text{mcg}}{\text{minute}} \times \frac{\text{milligrams}}{1000\,\text{micrograms}} = \frac{0.3\,\text{mg Hg}}{\text{minute}}$$

$$\frac{50\,\text{ft}^3}{\text{minute}} \times \frac{\text{m}^3}{35.315\,\text{ft}^3} = \frac{1.416\,\text{m}^3}{\text{minute}}$$

$$\frac{0.3\,\text{mg Hg/minute}}{1.416\,\text{M}^3\text{/minute}} = 0.212\,\text{mg Hg/M}^3$$

Answer: 0.212 mg of mercury vapor per cubic meter of air. Would the carpet fleas in the vacuum sweeper bag develop mercurialism? Become as "mad and buggy as a hatter?"

255. A paint sprayer using an air line respirator painted the inside of a 14' diameter by 30' long chemical storage tank with an epoxy resin paint. The paint contained 28% methyl ethyl ketone (by weight) as the solvent. The tank had no ventilation. If 1 gallon of the coating covers 325 ft², what MEK vapor concentration could be expected in the tank after the painter leaves? The

resin coating weighs 12.3 pounds per gallon. MEK's molecular weight is 72.1 grams/gram-mole.

$$\text{Lateral tank surface area} = 2 \times \pi \times r \times h = (2)\,\pi\,(7')\,(30') = 1320\,\text{ft}^2$$

$$\text{End surface areas} = 2 \times \pi \times r^2 = (2)\,\pi\,(7\,\text{ft})^2 = 308\,\text{ft}^2$$

$$\text{Total tank surface area} = 1320 + 308 = 1628\,\text{ft}^2$$

$$1628\,\text{ft}^2 \times \frac{\text{one gallon}}{325\,\text{ft}^2} = 5.01\,\text{gallons}$$

This was verified by finding an empty 5-gallon pail of the resin coating outside of the tank after the painter finished.

$$(12.3\,\text{lb/gallon}) \times 5.01\ \text{gallons} \times 0.28 = 17.25\,\text{lb MEK}$$

$$\text{Tank volume} = \pi \times r^2 \times h = \pi(7')^2\,(30') = 4618\,\text{ft}^3$$

$$4618\,\text{ft}^3 \times 28.32\,\text{L/ft}^3 = 130{,}782\,\text{L} = 130.78\,\text{m}^3$$

$$\frac{(7{,}832{,}000\,\text{mg}/130.78\,\text{m}^3) \times 24.45}{72.1} = 20{,}308\ \text{ppm MEK vapor}$$

Answer: Oh, boy! 20,308 ppm$_v$ methyl ethyl ketone vapor (2.03%) is average concentration in the tank! The LEL is 1.7%. The UEL is 11.4%. Ignition sources? Explosion proof lighting in the dark tank? Blow air into the tank to dilute the vapors to <PEL. Discharge explosive vapors to a safe area. Prohibit entry. Check vapors with a calibrated CGI. Consider diluting MEK vapors with nitrogen first to get well below the LEL, then follow with air dilution to < PEL.

256. The air flow through a low flow air-sampling pump and small charcoal tube was calibrated with a 100 mL soap film bubble tube. The calibration conditions were at NTP. What was the average flow rate in L/min if two bubble traverses required 73.7 and 74.1 seconds?

$$\frac{74.1\,\text{seconds} + 73.7\,\text{seconds}}{2} = \frac{73.9\,\text{seconds}}{100\,\text{mL}}$$

$$\frac{60\,\text{seconds}/100\,\text{mL}}{\times\,\text{seconds}/100\,\text{mL}} = \frac{0.1\,\text{L/minute}}{\times\,\text{L/minute}} = \frac{60\,\text{seconds}}{73.9\,\text{seconds}}$$

$$= 0.812 \times \frac{0.1\,\text{L}}{\text{minute}} = \frac{0.0812\,\text{L}}{\text{minute}}$$

Answer: 0.081 liter of air sampled per minute.

257. An impinger containing 8.9 mL of collection solution was analyzed and found to contain 0.19 mcg TDI/mL. The sample of air had been collected at 1.04 L/min for 16.5 minutes. If the collection efficiency of the impinger was 85%, what was the concentration of TDI in the breathing zone of the polyurethane foam maker? The molecular weight of TDI is 174.2 grams/gram-mole.

$$(1.04\,\text{L/min}) \times 16.5\,\text{minutes} = 17.16\,\text{liters}$$

$$8.9\,\text{mL} \times \frac{0.19\,\text{mcg}}{\text{mL}} \times \frac{1}{0.85} = 1.99\,\text{mcg TDI}$$

$$\frac{(1.99\,\text{mcg}/17.16\,\text{L}) \times 24.45}{174.2} = 0.016\,\text{ppm TDI}$$

Answer: 0.016 ppm$_v$ of toluene 2,4- and 2,6-diisocyanate isomers.

258. Twelve and one-half pounds of methylene chloride quickly evaporate in a 50 foot diameter × 95 feet long storage tank. A maintenance worker inspecting the tank wearing a HEPA respirator remained in the tank for 2 hours, 40 minutes. What was his 8-h TWAE to vapor? The molecular weight of CH_2Cl_2 is 84.9 grams per gram-mole.

$$\text{Tank volume} = \pi \times (25')^2 \times 95' = 186{,}532.5\,\text{ft}^3$$

$$186{,}532.5\,\text{ft}^3 \times \frac{28.32\,\text{L}}{\text{ft}^3} \times \frac{\text{m}^3}{1000\,\text{L}} = 5282.6\,\text{m}^3$$

$$12.5\,\text{lb} \times \frac{453.59\,\text{g}}{\text{lb}} \times \frac{1000\,\text{mg}}{\text{g}} = 5{,}669{,}875\,\text{mg}$$

$$\frac{(5{,}669{,}875\,\text{mg}/5282.6\,\text{m}^3) \times 24.45\,\text{L/g-mole}}{84.9} = 309\,\text{ppm}$$

$$160\,\text{minutes} \times 309\,\text{ppm}_v = 49{,}440\,\text{ppm}_v\text{-minutes}$$

$$320\,\text{minutes} \times 0\,\text{ppm}_v = 0\,\text{ppm}_v\text{-minutes}$$

$$49{,}440\,\text{ppm}_v\text{-minutes}/480\,\text{minutes} = 103\,\text{ppm}_v$$

Answer: HEPA filter respirator is worthless for protection against solvent vapor. His TWAE to CH_2Cl_2 vapor was 103 ppm$_v$. ACGIH classifies CH_2Cl_2 as A2 carcinogen with an 8-hour TWAE TLV® of 50 ppm.

In the preceding problem, assume the worker was found dead in the tank—a victim of solvent vapor inhalation. What could explain his death? 309 ppm$_v$?

A 160-minute exposure to 309 ppm$_v$ CH$_2$Cl$_2$ would be an unlikely cause of death unless there was ventricular fibrillation by sensitization of the myocardium from endogenously released adrenaline. This solvent's 100% vapor density is much greater than air, and a pocket or stratified vapor layer in the air would have a much greater concentration. The average vapor or gas concentration is very misleading when investigating confined spaces where gas and vapor pockets with very high concentrations can occur. For example, if there were a spill, and the worker fell striking his head and became unconscious, his breathing zone is now in a pocket of highly enriched vapor. Death would ensue if he did not regain consciousness, and he was not rescued. Clearly, in this example, OSHA's confined space entry standard was not in effect. His foreseeable death was fully preventable.

259. A welder was exposed to 260 mcg Pb/m^3 for 75 minutes while burning bolts, 38 mcg Pb/m^3 for 3 hours and 19 minutes when welding painted steel, and 560 mcg Pb/m^3 for 27 minutes when using abrasive blaster to remove lead-containing paint from steel prior to welding. The balance of his 8-hour exposure was 1.5 mcg Pb/m^3. What was his 8-hour TWAE?

Job	C (mg Pb/m³)	×	T (minutes)	=	CT (mcg/m³-minutes)
Burning	260	×	75	=	19,500
Welding	38	×	199	=	7562
Blasting	560	×	27	=	15,120
Other	1.5	×	179	=	269
			480	=	42,451

$$\frac{42,451 \, \text{mcg Pb/m}^3 - \text{minutes}}{480 \, \text{minutes}} = \frac{88.4 \, \text{mcg Pb}}{\text{m}^3}$$

Answer: Eight-hour TWAE to lead dust and fume is 88.4 mcg Pb/m^3. This exceeds OSHA's PEL by about 77%. Implement the OSHA lead standard (29 *CFR* 1910.1025).

260. A 10-pound chunk of calcium carbide fell into an open 55-gallon drum of water in a room with inside dimensions of 10' × 12' × 12'. The room had no ventilation, but there was a small candle burning in a corner. There was quantitative conversion of calcium carbide into acetylene gas. Could there be an explosion? Disregard any solubility of the acetylene in water. Other than drum and candle, assume the room is empty. Respective molecular weights of acetylene and calcium carbide are 64.1 and 26.04 grams/gram-mole.

$$CaC_2 + 2H_2O \rightarrow Ca(OH)_2 + C_2H_2 \uparrow$$

10 lb × 453.59 g/lb = 4535.9 grams of calcium carbide

$$\frac{4535.9\,\text{grams\,CaC}_2}{64.1\,\text{grams/mole}} = 70.76\,\text{moles of CaC}_2$$

Therefore, 70.76 moles of acetylene gas were generated and released into the air.

$$70.76\,\text{moles C}_2\text{H}_2 \times 26.04\,\text{grams/mole} = 1842.67\,\text{grams C}_2\text{H}_2$$

$$10' \times 12' \times 12' = 1440\,\text{ft}^3 \times (28.32\,\text{L/ft}^3) = 40{,}781\,\text{L} = 40.78\,\text{m}^3$$

$$\frac{(1{,}842{,}670\,\text{mg/40.78 m}^3) \times 24.45\,\text{L/g-mol}}{26.04\,\text{g/g-mol}} = 42{,}427\,\text{ppm acetylene gas}$$

Answer: The average concentration of acetylene gas in the room is 4.24%. The LEL–UEL range for acetylene gas in air is one of the widest of all hydrocarbon gases: 2.5–80%. Although the concentration is toward the lower end of the explosive range for acetylene, this room will be blown to kingdom come. Leave pronto!

261. A railroad tank car containing 9 tons of liquid chlorine de-rails, tips, ruptures, and spills its contents into a drainage ditch that is parallel to the right-of-way. A very light, hot summer breeze evaporates some liquid Cl_2 into a gas cloud that is about 1.3 miles long, 0.67 mile wide, and 200 feet high. What is the average Cl_2 gas concentration in this cloud? Assume 20% of the liquid chlorine in the ditch evaporates before the remaining liquid chlorine is contained by the emergency response team.

$$1.3\,\text{miles} \times (5280\,\text{feet/mile}) = 6864\,\text{feet}$$

$$0.67\,\text{mile} \times (5280\,\text{feet/mile}) = 3538\,\text{feet}$$

$$200' \times 6864' \times 3538' = 4.857 \times 109\,\text{ft}^3 \times (28.32\,\text{L/ft}^3) = 1.376 \times 10^{11}\,\text{liters}$$
$$= 1.376 \times 10^8\,\text{m}^3$$

$$9\,\text{tons} \times \frac{2000\,\text{lb}}{\text{ton}} \times \frac{453.59\,\text{grams}}{\text{lb}} \times \frac{1000\,\text{mg}}{\text{gram}} \times 0.20 = 1.633 \times 10^9\,\text{mg Cl}_2$$

$$\frac{((1.633 \times 10^9\,\text{mg Cl}_2)/(1.376 \times 10^8\,\text{m}^3)) \times (24.45\,\text{L/gram-mole})}{70.9\,\text{grams/gram-mole}}$$
$$= 4.1\,\text{ppm}_\text{v}\,\text{Cl}_2$$

Answer: The mean concentration of chlorine gas in this cloud is 4.1 ppm$_\text{v}$. However, while this type of calculation is helpful for air pollution and community risk assessment purposes, a much greater concentration exists close to the spill that is subject to the prevailing meteorological conditions. This would help determine gas diffusion parameters in x, y, and z planes and expected concentrations at selected points away from the spill. A coefficient of prayer >1 should be factored into this equation if this gas release was upwind and near a residential community, school, nursing home, or hospital.

262. Air was to be sampled through a midget impinger containing 15 mL distilled water for a specified time at 1.09 L/min. The barometric pressure is 760 mm Hg. The ambient temperature is 14°F. The laboratory analytical detection limit is 320,000 particles/mL. How long should one sample take to determine 20% of a 30 mppcf airborne dust concentration?

Answer: Sample until the water freezes in the impinger (that will happen in a minute or so)! Add NaCl, MeOH, or IPA to lower the freezing point of air-sampling liquid. If this is done, the following calculations apply:

$$\frac{320,000\,particles}{mL} \times 15\,mL = 4,800,000\,particles$$

$$20\%\,of\,30\,mppcf = \frac{6 \times 10^6\,particles}{ft^3} \times \frac{ft^3}{28.32\,L} = \frac{211,684\,particles}{liter}$$

$$\frac{4,800,000\,particles}{211,864\,particles/liter} = 22.66\,liters$$

$$\frac{22.66\,liters}{1.09\,Lpm} = 20.7\,minutes$$

263. The formula $Q = 0.75\,V\,(10\,X^2 + A)$ is used for calculations involving which type of exhaust hood?
   a. Slot hood without a flange and an aspect ratio of 0.2 or less
   b. Slot hood with flange and aspect ratio of 0.2 or less
   c. Plain opening hood without flange with aspect ratio of 0.2 or greater or round without flange
   d. Plain opening hood with flange and aspect ratio of 0.2 or greater or round hood without flange
   e. Circular canopy hood for hot gases, smoke, fumes, steam
   f. Plain multiple slot opening hood with two or more slots and an aspect ratio of 0.2 or greater
   g. All sheet metal hoods, but not for flexible duct hoods

Answer: d.

264. What is the expected capture velocity 8 inches in front of a 2.5" × 3" flanged slot hood with a slot velocity of 2000 fpm?
   The basic equation for air flow for a flanged slot with an aspect ratio of 2.5"/36" = 0.07 is $Q = 2.6\,LVX$, where Q = air volume rate, L = slot length, V = capture velocity, and X = distance in front of the hood face.

$$Q = A \times V = \frac{2.5" \times 36"}{144\,\text{in}^2/\text{ft}^2} \times \frac{2000\,\text{feet}}{\text{minute}} = \frac{1250\,\text{ft}^3}{\text{minute}}$$

$$V = \frac{Q}{2.6 \times L \times X} = \frac{1250\,\text{cfm}}{2.6 \times 3' \times 0.67'} = \frac{239\,\text{feet}}{\text{minute}}$$

265. Five grams of mothball crystals (*p*-dichlorobenzene) remain in an empty dresser drawer after the sweaters had been removed. The drawer is $19" \times 11" \times 28"$. The molecular weight of DCB is 147 grams/gram-mole. DCB vapor pressure is 0.4 mm Hg at 75°F. What is the vapor concentration in the closed drawer expressed in mg/m³? How many milligrams of DCB vapor are in air of the drawer? The OSHA PEL for DCB is 75 ppm$_v$. Assuming toxicokinetics of DCB in humans are similar to those in a moth, is the moth excessively exposed?

$$\frac{0.4\,\text{mm Hg}}{760\,\text{mm Hg}} \times 10^6 = 526\,\text{ppm of DCB vapor at saturation}$$

$$\frac{\text{mg}}{\text{m}^3} = \frac{\text{ppm} \times \text{molecular weight}}{24.45\,\text{L/g-mole}} = \frac{526\,\text{ppm} \times 147}{24.45} = \frac{3162\,\text{mg DCB}}{\text{m}^3}$$

$$\frac{19" \times 11" \times 28"}{(1728\,\text{in}^3/\text{ft}^3)} \times \frac{28.32\,\text{L}}{\text{ft}^3} \times \frac{\text{m}^3}{1000\,\text{L}} = 0.0959\,\text{M}^3$$

$$\frac{3162\,\text{mg}}{\text{m}^3} \times 0.0959\,\text{m}^3 = 303\,\text{milligrams of DCB}$$

Answers: 3162 mg DCB/m³. 303 mg DCB vapor. Yes.

266. The weighted average velocity pressure in a duct exhausting standard air is 0.84" wg. What is the duct's air velocity?

$$V = 4005 \times \sqrt{VP} = 4005 \times \sqrt{0.84} = 3671\,\text{fpm}$$

Answer: 3671 feet per minute.

267. An exhaust duct static pressure is 1.34" wg. The total system pressure is −0.63" wg. What is the standard air velocity pressure? What is the duct velocity?

Since this is an exhaust duct, the static pressure is negative, or −1.34" wg.

$$SP + VP = TP$$

$$VP = TP - SP$$

$$VP = -0.63" - (-1.34") = 0.71"$$

Note: VP is always positive.

$$V = 4005 \times \sqrt{VP} = 4005 \times \sqrt{0.71} = 3375 \text{ feet per minute}$$

Answer: Velocity pressure is 0.71" wg. Duct velocity is 3375 fpm.

268. A 14" internal diameter duct has an average duct velocity of 3160 fpm. What is the duct's volumetric flow rate?

$$Q = A \times V = \frac{\pi \times (7" \times 7")}{144 \text{ in}^2/\text{ft}^2} \times \frac{3160 \text{ ft}}{\text{minute}} = \frac{3378 \text{ ft}^3}{\text{minute}}$$

Answer: 3378 cubic feet of air are exhausted each minute.

269. Standard dry air at 70°F and 29.92" barometric pressure has a mass of
    a. 0.13 lb/ft³
    b. 0.013 lb/ft³
    c. 22.4 lb/1000 ft³
    d. 0.003 grains/ft³
    e. 0.130 lb/1000 ft³
    f. 0.075 lb/ft³

Answer: f.

270. The LEL for most combustible gases and vapors is normally constant up to about 250°F. What approximation factor is usually applied above these temperatures to account for the change in the LEL?
    a. 0.3
    b. 0.5
    c. 0.7
    d. 0.9
    e. 1.3
    f. 1.5

Answer: c.

271. An empty 47' × 166' × 20' building is supplied with outdoor air at 7300 cfm. How many air changes occur hourly? How many minutes are required per air change? How many cubic feet are supplied per square foot of floor area? The ceiling height is 20 feet. Forty-seven people will work in this single-story building. What is the outdoor air ventilation rate per person if 90% of the air is recirculated?

$$7300 \text{ cfm} \times 60 \text{ minutes/hour} = 438,000 \text{ cfh}$$

$$\frac{438{,}000\,\text{cfh}}{47' \times 166' \times 20'} = 2.8 \text{ air changes/hour}$$

$$\frac{60\,\text{minutes/hour}}{2.8\,\text{air changes/hour}} = 21.4 \text{ minutes/air change}$$

$$\frac{7300\,\text{cfm}}{47\,\text{ft} \times 166\,\text{ft}} = 0.94 \text{ cfm/ft}^2$$

$$\frac{0.1 \times 7300\,\text{cfm}}{47\,\text{occupants}} = \frac{15.5\,\text{ft}^3/\text{minute}}{\text{occupant}}$$

Answers: 2.8 air changes per hour. 21.4 minutes per air change. 0.94 cfm per ft² of floor area. 15.5 cubic feet of outside air per minute per worker/occupant.

272. Industrial hygienists rarely prescribe canopy hoods because of their often poor performance, and because these hoods are vulnerable to cross drafts. One engineering equation used to design such hoods, however, is
    a. $Q = 1.4\ PDV$
    b. $Q = (LWH/2) \times V$
    c. $Q = C_e \times SP_h \times V$
    d. $Q = V\,(5\,X^2 + A)$
    e. $Q = \pi\,CLW$

Answer: a. Where $P =$ the tank's perimeter, in feet. D is the vertical distance between the tank edge and the bottom edge of the canopy, in feet. V is the design capture velocity (fpm) and can range from as low as 50 fpm in undisturbed locations to 500 fpm or higher when moderate cross drafts and thermal convection air currents are disruptive to capture. Canopy hoods are generally contraindicated because workers leaning over the tank are in a vapor and/or a mist cloud originating from the source and ascending to the canopy hood. They can be excellent for humidity control of steamy sources.

273. If 1000 cfm are needed to capture air contaminants 6 in from a hood's face, what exhaust air volume is needed to capture contaminants 12 inches from hood's face?
    a. 500 cfm
    b. 1000 cfm
    c. 2000 cfm
    d. 4000 cfm
    e. 9000 cfm
    f. 16,000 cfm

Answer: d. Air volume varies as the square of the distance from the air contaminant source. This points out the importance of placing the exhaust hood face as close as possible to the source of air pollutants. Exhaust volume is conserved, and more effective capture is achieved. Tempered make-up air requirements are greatly reduced.

274. A side draft exhaust hood with face dimensions of 32" × 50" has an average face velocity of 110 fpm. What volume of air is exhausted by this hood? What will be the average face velocity if the hood opening is reduced to 26" × 40"?

$$\frac{32" \times 50"}{144\,in^2/ft^2} \times 110\,fpm = \frac{1222\,ft^3}{minute}$$

$$New\,area = \frac{26" \times 40"}{144\,in^2/ft^2} = 7.2\,ft^2$$

$$V = Q \times A = \frac{1222\,cfm}{7.2\,ft^2} = 170\,fpm$$

Answer: 1222 cfm. The hood capture velocity increased to 170 fpm by slightly reducing the opening and improving the process enclosure.

275. Static pressure at fan inlet of an exhaust system is −2.75" H_2O. Static pressure at the fan outlet is 0.25". The internal diameter of the exhaust duct at the fan inlet is an odd size: 23". The system exhausts 4000 cfm. What is the fan's static pressure?

$$23"\,diameter\,duct = (\pi\,r^2)/144\,in^2/ft^2 = 2.885\,ft^2$$

$$Fan\,inlet\,velocity, V = \frac{4000\,cfm}{2.885\,ft^2} = \frac{1836\,ft}{minute}$$

$$VP = \frac{V^2}{(4005)^2} = \frac{(1386)^2}{(4005)^2} = 0.12"$$

$$Fan\,SP = SP_{in} + SP_{out} - VP_{in} = 2.75" + 0.25" - 0.12" = 2.88"$$

Answer: 2.88 inches of water. This is used to select appropriate fan from commercial fan tables and catalogues.

276. An exhaust system operates at 19,400 cfm. A hood is added to the system that requires a total system capacity of 23,700 cfm. By how much should fan speed be increased to handle the extra exhaust volume? Let Q_1 be original exhaust volume, and Q_2 is new exhaust volume.

Since CFM varies directly with fan rpm: $\dfrac{Q_2}{Q_1} = \dfrac{\text{rpm}_2}{\text{rpm}_1} = \dfrac{23,700\,\text{cfm}}{19,400\,\text{cfm}} = 1.22.$

Answer: Increase fan speed by 22%, that is, multiply the fan rpm × 1.22. Caution: Ensure that the maximum fan rpm and safety-rated tip speed are not exceeded to prevent the shattering of fan blades producing shrapnel. This information is available from the fan manufacturer or distributor. Normally, such safety information is located on a metal tag attached to the motor or fan housing.

277. In the preceding problem, what is the required increase in fan horsepower to handle the increased air volume?

Fan horsepower varies as the cube of the fan speed and cube of fan volume. Therefore, the required ratio of increase in the fan speed (rpm) is

$$\left[\frac{23,700\,\text{cfm}}{19,400\,\text{cfm}}\right]^3 = 1.82$$

Answer: The required ratio of the increase in the fan horsepower is 1.82. This is an 82% increase in the energy requirements for only a 22% increase in fan volume and speed. Caution: be certain that the fan horsepower rating is ample. Check the motor plate for specifications and maximum operating capacity. Consult a professional electrician.

278. A ventilation system was designed to exhaust 17,500 cfm at a static pressure of 3.46" at the fan inlet. A static pressure reading at the fan inlet is 2.66" 2 years after the system was installed. What is the present exhaust volume? What's up?

$$Q_{\text{now}} = Q_{\text{then}} \times \sqrt{\frac{\text{SP}_{\text{now}}}{\text{SP}_{\text{now}}}} = 17,500\,\text{cfm} \times \sqrt{\frac{2.66"}{3.46"}} = 17,500\,\text{cfm} \times 0.877$$

$$= 15,348\,\text{cfm}$$

$$17,500\,\text{cfm} - 15,348\,\text{cfm} = 2152\,\text{cfm}$$

Answer: System volume decreased 2152 cfm, or about 12%. Diagnosis could include dirty filters and ducts, slipping fan belt, sticking dampers, corrosion and holes in system (air shunting), and plugged or dirty heat exchange coils. Note the exhaust volume varies as square of static pressure at the fan inlet, whereas exhaust volume varies directly as the fan revolutions and as the cube of the horsepower.

279. What is the discharge air volume from a 50-foot high plant with a 30 ft² outlet in the roof and a 35 ft² inlet at the ground level if the outdoor temperature is 20°F and the discharge air temperature is 75°F ($\Delta t = 55°F$)?

ASHRAE provides a convenient formula to estimate "chimney" or "stack effect" ventilation due to temperature differentials existing between the floor of a building and its roof outlet:

$$Q(\text{cfm}) = 9.4 \times A \times h \times \sqrt{t_{in} - t_{out}},$$

where

    A = total inlet or discharge area (ft²) at ground or ceiling (use the smaller of the two)

    h = distance between inlet and outlet, ft

    t_{in} = air inlet temperature, °F

    t_{out} = air outlet temperature, °F

$$Q = 9.4 \times 30\,\text{ft}^2 \times 50' \times \sqrt{75°F - 20°F} = 14{,}788\,\text{cfm}$$

Answer: About 14,800 cfm are naturally exhausted due to thermal gradient demonstrating large volumes of natural ventilation result from differences in temperature forces alone. For example, it has been estimated, in some steel mills, 600 tons of air is naturally exhausted for every ton of steel produced.

280. A 10-inch diameter circular air supply duct is blowing air at 4200 feet per minute (fpm) as measured at the discharge point. What general quantitative statement applies regarding air velocity ("throw") at some distance away from the duct outlet?

    Answer: The velocity 30 duct diameters away is about 10% of the discharge velocity. That is, in this case, 30 duct diameters = 30 × 10" = 300" = 25 feet. The air velocity at this distance reduces to nearly 420 fpm if there are no significant cross drafts.

281. A low circular canopy hood is positioned over a circular tank containing water at 180°F. Distance between the tank and hood is less than 3 feet. If the room air temperature is 70°F and tank diameter is 4 ft, what is the total air flow rate required for this hood to capture the water vapor?

$$Q = 4.7 \times (D)^{2.33} \times (T)^{0.42},$$

where

    Q = total hood air flow, cfm

    D = hood diameter, feet (add 1 ft to source diameter)

    T = difference between the temperature of the hot source of air contaminants and the ambient temperature, °F

$$Q = 4.7 \times (4'+1')^{2.23} \times (180°F - 70°F)^{0.42} = 4.7 \times (5)^{2.23} \times (110)^{0.42}$$

$$= 4.7 \times 36.2 \times 7.2 = 1225\,\text{cfm}$$

Answer: 1225 ft³ of air per minute. Fan selection must be based on thermally expanded air volume and the air temperature existing at the inlet of the fan.

282. What is the approximate capture velocity 12 inches away from an un-flanged one-foot diameter duct exhausting air at a duct velocity of 3600 fpm?

   Answer: A rule of thumb is the capture velocity one duct diameter away from an un-flanged exhaust opening is about 10% of the duct velocity or, in this case, 360 fpm.

283. To ensure uniform capture and air distribution that is applicable to most exhaust hoods, a maximum plenum velocity should be no more than _____% of the slot velocity.
   a.   10%
   b.   30%
   c.   50%
   d.   75%
   e.   110%
   f.   150%

   Answer: c. For example, the maximum plenum velocity for a 2000 fpm slot hood should not exceed 1000 fpm. This is a good choice for uniform air flow across the slot and a moderate pressure drop. If, during design, calculated plenum velocity is too high (i.e., >50% of the design slot velocity), the size of the plenum must be increased—that is, the depth must be increased to slow the air velocity down in the plenum.

284. A standard bench grinder hood has a coefficient of entry of 0.78. A piece of string wrapped around the take-off exhaust duct to measure the circumference was 16". The throat static pressure of the hood is –2.75". What volume of air is exhausted by this hood—that is, what is the volumetric flow rate in cfm, standard air?

$$Q(\text{cfm}) = 4005 \times A \times C_e \times \sqrt{\text{SP}},$$

where
   A = cross section area of duct at throat, ft²
   C_e = coefficient of hood entry
   SP = static pressure at hood throat, inches of water

$$\text{Duct circumference} = 2 \times \pi \times r = 16"$$

$$r = \frac{16"}{2 \times \pi} = 2.55 \text{ inches}$$

   Therefore, the internal duct diameter = 5" (a standard size—the true duct radius = 2.5 inches).

$$A = \frac{\pi \times (2.5")^2}{(144 \, in^2/ft^2)} = 0.136 \, ft^2$$

$$\% \, outdoor \, air = \frac{ppm_{return \, air} - ppm_{mixed \, air}}{ppm_{return \, air} - ppm_{outdoor \, air}} \times 100$$

$$= \frac{750 \, ppm - 650 \, ppm}{750 \, ppm - 400 \, ppm} \times 100 = \frac{100 \, ppm}{350 \, ppm} \times 100 = 28.6\%$$

285. In general, the addition of a flange to an exhaust system hood opening improves air contaminant capture by about _____ percent.
   a. 15%
   b. 25%
   c. 40%
   d. 60%
   e. 75%

   Answer: b. Alternatively, at equal capture velocities, the flange decreases the air flow rate by about 25%. Typically, the width of the flange should be equal or exceed 25% of the square root of the hood's face area. In other words, a hood is 5' × 6' = 30 ft². 0.25 [(30 ft²)$^{0.5}$] ≅ 1.4 feet. This applies to a free-standing exhaust hood. If this hood rests on the floor, the floor acts as a boundary layer. Accordingly, exhaust volume might be reduced, and the flange can be 25% smaller, or 1.4' × 0.75 = 1.05'. Ensure the flange edges are guarded with a soft material to prevent cuts and other injuries. Splitting a garden hose longitudinally and placing over sharp metal or plastic edges works. Alternatively, consider a flexible soft rubber flange.

286. A duct velocity of 3740 feet per minute is equivalent to what velocity pressure?
   The basic equation is

$$V = 4005 \times \sqrt{VP}.$$

   V = duct velocity of standard air (0% humidity, 70" F, and 29.92" Hg), fpm
   VP = velocity pressure, inches water gauge

   Rearranging:

$$VP = \left(\frac{V}{4005}\right)^2 = \left(\frac{3740 \, fpm}{4005 \, fpm}\right)^2 = 0.872 \, inch \, of \, water$$

   Answer: This velocity pressure equals 0.872" H_2O assuming standard air is transferred. Standard air is defined as air at 0% relative humidity, 29.92" Hg, and 70°F. We often refer to the "volume" of air that is being transferred.

What we really should state is the "volumetric flow rate," for example, cubic feet of air per minute, cfm.

287. A ventilation system supplies air through five branch ducts. The system delivers 15,000 cfm, and another duct will be added to increase the total supply volume to 17,000 cfm. The static pressure at the fan discharge is 0.59". The static pressure in the main duct where the branch is to be added is 0.38". What are the new static pressures at the fan discharge and at the fan inlet plenum?

Calculate new SP drop in the main duct from the fan to the new connection:

$$\sqrt{\frac{17,000\,\text{cfm}}{15,000\,\text{cfm}}} \times (0.59" - 0.38") = 1.065 \times 0.21 = 0.22\,"\text{H}_2\text{O}$$

New SP at fan discharge = 0.38" H_2O + 0.22" $\text{H}_2\text{O}"$ = 0.6" H_2O

New static pressure at fan inlet plenum = $\dfrac{17,000\,\text{cfm}}{15,000\,\text{cfm}} \times 0.30" = 0.34"\text{H}_2\text{O}$

Answers: 0.22" is the new static pressure drop at new duct. 0.50" is new static pressure at the fan discharge. 0.34" is the new static pressure at the fan inlet. The required increase in fan rpm and horsepower can now be obtained from fan performance catalogues.

288. A five horsepower fan motor exhausts 3250 cfm. Amperes measured on the circuit are 9.8. The fan motor's rated amps are 13.2 for five horsepower at 220 volts. What will be the new air volume if the fan speed is increased to fully load the motor at five horsepower?

$$\text{Actual horsepower} = \frac{\text{measured amperage}}{\text{rated amperage}} \times \text{rated horsepower}$$

$$= \frac{9.8\,\text{amps}}{13.2\,\text{amps}} \times 5\,\text{HP} = 3.7 \text{ horsepower}$$

According to the fan laws, horsepower varies as the cube of the air volume (cfm) or the fan speed (rpm):

$$\frac{\text{HP}_1}{\text{HP}_2} = \left[\frac{Q_1}{Q_2}\right]^3$$

$$Q_2 = Q_1 \times \left[\frac{\text{HP}_2}{\text{HP}_1}\right]^{0.333} = 3250\,\text{cfm} \times \sqrt[3]{\frac{5\,\text{HP}}{3.7\,\text{HP}}} = 3592\,\text{cfm}$$

Answer: The new volume at full five horsepower is 3592 cubic feet per minute.

289. What is the density of dry air at 120°F? The density of dry air at 70°F is 0.75 lb/ft³.

Air density varies with the absolute temperature. That is, the density decreases as the air temperature increases:

$$\frac{0.075\,lb}{ft^3} \times \frac{460° + 70°F}{460° + 120°F} = \frac{0.075\,lb}{ft^3} \times \frac{530°A}{580°A} = \frac{0.0686\,lb}{ft^3}$$

Answer: The density of dry air at 120°F is 0.0686 lb/ft³.

290. In Problem 289, what is density of the air at 2000 feet elevation? The standard barometric pressure at 2000 feet is 27.80" of Hg.

The density of air (or any gas or vapor) is reduced by elevation:

$$\frac{0.075\,lb}{ft^3} \times \frac{530°\,A}{580°\,A} \times \frac{27.80"\,Hg}{29.92"\,Hg} = \frac{0.0637\,lb}{ft^3}$$

Answer: Therefore, it can be seen that as air is heated and reduced in pressure, its density is reduced by both forces. That is, air expands into a larger volume, but it retains its original mass. Mass is conserved.

291. What happens to air density when an inlet duct is used? For example, calculate the density correction for an inlet duct with a—20" inlet suction at an atmospheric pressure of 407" H_2O. Apply this to air at 120°F and 2000 feet elevation.

$$\frac{407" - 20"}{407"} = 0.951$$

$$Density = \frac{0.075\,lb}{ft^3} \times \frac{530°A}{580°A} \times \frac{27.80"Hg}{29.92"Hg} \times \frac{0.951 \times 407"}{407"} = \frac{0.0605\,lb}{ft^3}$$

292. What happens to the density of dry air as water vapor is added to it?

Answer: Density of air is reduced as water vapor is added to it (as in wet scrubber systems) because water vapor molecules weigh less than air molecules. Air's "apparent" molecular weight is 29; molecular weight of water is 18). As an example, consider water-saturated air at 120°F, 2000 feet elevation, and at 20" water suction (see Problems 289, 290, and 291):

$$\frac{0.0605\,lb}{ft^3} \times \frac{27.80"Hg}{29.92"Hg} \times \frac{387"}{407"} = \frac{0.0535\,lb}{ft^3}$$

293. A fan must be selected for a 20" suction pressure at the fan inlet at an air density of 0.0605 lb/ft³. How would you do this?

Since the pressure ratings of fans are based upon a standard gas density of 0.075 lb/ft³, the selection pressure must be adjusted to a density of 0.075 lb/ft³.

$$\text{Fan selection static pressure} = 20'' \times \frac{0.075\,\text{lb/ft}^3}{0.0605\,\text{lb/ft}^3} = 24.8''$$

Answer: 24.8 in of water static pressure.

294. Minimum basic design dilution ventilation rates, respectively, for propane fuel and gasoline fuel lift trucks are
   a. 300 and 600 cfm
   b. 600 and 300 cfm
   c. 1000 and 2000 cfm
   d. 6000 and 10,000 cfm
   e. 5000 and 8000 cfm

   Answer: e. This assumes regular engine maintenance program, less than 1% and 2% CO in the exhaust gas, truck operating times less than 50%, good distribution of dilution ventilation air flow, and at least 150,000 ft³ plant volume per lift truck. See ACGIH's *Industrial Ventilation* for other conditions of maintenance, truck operation, and ventilation parameters.

295. A fan exhausts 1000 cfm of air at 600°F at a static pressure of 10" of H_2O. What is the required fan horsepower? What will be the savings in horsepower (i.e., electricity costs) if the air is cooled to 100°F?

   $$\text{Fan horsepower} = 0.000158 \times 1000\,\text{cfm} \times 10''\,H_2O = 1.58$$

   Use Charles' law to calculate the new air volume at 100°F:

   $$\frac{Q\,(\text{at }100°F)}{Q\,(\text{at }600°F)} = \frac{100°F + 460°}{600°F + 460°}$$

   $$Q\,(\text{at }100°F) = \frac{1000\,\text{ft}^3}{\text{minute}} \times \frac{560°A}{1060°A} = 528\,\text{cfm},$$

   where Q (at 100°F) denotes volume flow rate at 100°F, and (100°F + 460°) is absolute temperature. Assuming the increase in density at the lower temperature results in a negligible reduction in pressure loss, then at 100°F:

   $$\text{Fan horsepower} = 0.000158 \times 528\,\text{cfm} \times 10 = 0.83\,\text{HP}$$

   Answer: 0.83 horsepower that = (0.83/1.58) × 100 = 52.5% of the horsepower required at 600°F.

296. A paint spray booth is 7 feet high × 10' wide. Spraying is often done as far as 5 ft in front of the booth. A nearly draft-free area requires 100 feet per minute capture velocity at the point of spraying. Determine the required exhaust rate.

$$Q = V \left[\frac{10\,x^2 + 2A}{2} \right]$$

This is a modification of classic ventilation equation that applies to a free-standing, unobstructed hood. The above equation is applied to rectangular hoods bounded on one side by a plane surface. The hood is considered to be twice its actual size with an additional portion being mirror image of the actual hood and the bounding plane being the bisector.

$$Q = 100\,\text{fpm} \times \frac{[(10)(5')^2] + (2)(7')(10')}{2} = \frac{19,500\,\text{ft}^3}{\text{minute}}$$

$$\text{Hood face velocity} = \frac{Q}{A} = \frac{(19,500\,\text{ft}^3/\text{minute})}{7' \times 10'} = \frac{280\,\text{feet}}{\text{minute}}$$

Answer: Hood's exhaust rate is 19,500 cfm. Average hood face velocity is 280 fpm.

297. A 28-inch diameter fan operating at 1080 rpm supplies 4700 cfm at 4.75 inches of static pressure. What size fan of the same type and series would supply 10,900 cfm at the same static pressure?

$$\text{Fan diameter} = \left[\frac{10,900\,\text{cfm}}{4700\,\text{cfm}} \right]^{0.333} \times 28'' = 37\,\text{inches}$$

Answer: 37-inch diameter is an odd dimension. A 36", 38", or 40" diameter fan would most likely be available from most fan manufacturers and distributors.

298. A fan exhausts 600°F air from a drier at 12,000 cfm. Density of this air is 0.0375 lb/ft³ (because of its high water vapor content). The static pressure at the dryer's discharge is 4 inches of water. The fan speed is 630 rpm. The fan uses 13 horsepower. What is the required horsepower if 70°F air is exhausted?

$$\text{HP} = 13\,\text{HP} \times \frac{0.075\,\text{lb/ft}^3}{0.0375\,\text{lb/ft}^3} = 26\,\text{HP}$$

Answer: 26 horsepower are required. If only a 15-horsepower fan, for example, was in this system requiring 26 HP, it would be necessary to use damper when starting "cold" to prevent electrically overloading the fan motor. Always ensure the installation of a fan with sufficient horsepower to handle the system requirements. The cost of installing a slightly larger fan can often be justified at initial installation since future system modifications can be anticipated (e.g., adding another hood to the exhaust system or a collector with a higher static pressure drop).

299. Assume that there are 63,000,000 automobiles in the United States and that each uses an average of 3 gallons of gasoline every day. Further assume that there are no gasoline vapor recovery systems when the car's fuel tank is filled and that each gallon of gasoline pumped to the fuel tank displaces a gallon of saturated gasoline vapor. If the average molecular weight of gasoline is taken as approximately 72 grams gram-mole^{-1}, and average vapor pressure of gasoline is 130 mm Hg, how many tons of gasoline vapors are evaporated and enter the atmosphere by filling automobile fuel tanks every year in the United States? Assume that the national average barometric pressure for high-density population areas is 740 mm Hg, and that the average temperature for these areas is 50°F.

$$\frac{130 \text{ mm Hg}}{740 \text{ mm Hg}} \times 10^6 = 175,676 \text{ ppm gasoline vapor at saturation}$$

$$\frac{\text{mg}}{\text{m}^3} = \frac{\text{ppm} \times \text{molecular weight}}{23.85 \text{ L/g-mole at 50°F}} = \frac{175,676 \times 72}{23.85} = \frac{530,343 \text{ mg}}{\text{m}^3}$$
$$= \frac{530.34 \text{ g}}{\text{m}^3}$$

$$\frac{530.34 \text{ g}}{\text{m}^3} \times \frac{\text{lb}}{454 \text{ g}} \times \frac{\text{m}^3}{35.315 \text{ ft}^3} \times \frac{0.1337 \text{ ft}^3}{\text{gallon}} = \frac{0.0044225 \text{ lb}}{\text{gallon}}$$

$$\frac{0.0044225 \text{ lb}}{\text{gallon}} \times \frac{3 \text{ gallons/day}}{\text{automobile}} \times \frac{365.25 \text{ days}}{\text{year}} \times 63,000,000 \text{ automobiles}$$
$$= \frac{3.053 \times 10^8 \text{ lb}}{\text{year}} \times \frac{\text{ton}}{2000 \text{ lb}} = \frac{152,648 \text{ tons}}{\text{year}}$$

Answer: Approximately 150,000 tons of gasoline vapor per year. This crude estimate does not consider evaporative hydrocarbon losses from trucks, airplanes, locomotives, and so on. Nor are hydrocarbon emissions that result from careless overfilling of tanks, refining, filling storage tanks, trucks, and so on considered in this simplified accounting. Noel De Nevers in his *Air Pollution Control Engineering* (McGraw-Hill, 1995, p. 287) estimated 400,000 gallons "lost" to the United States (Earth's) atmosphere daily from gasoline vapor displacement emissions.

The average density of gasoline is 6.073 lb/U.S. gallon.

$$152{,}648 \text{ tons/year (author's calculation)} \times 2000 \text{ lbs/ton}$$
$$= 305{,}296{,}000 \text{ lbs/year.}$$

(305,296,000 lbs/year)/6.073 lbs/gallon = 5,027,036 gallons/year. My estimate is much higher than De Never's pointing out the difficulties in estimating these and similar emissions on an annual basis for any number of stated assumptions.

300. A bizarre analytical chemist found that the average concentration of butyric acid in an adult tennis shoe was 4.7 micrograms after wearing it for a basketball game. What is the concentration of butyric acid existing in an unventilated empty gymnasium locker after evaporating from a pair of stinky shoes? The locker is 12" × 12" × 66". The molecular weight of butyric acid is 88.11 grams/gram-mole.

$$\frac{4.7 \text{ mcg BA}}{\text{shoe}} \times \text{two shoes} = 9.4 \text{ micrograms of butyric acid vapor}$$

$$\frac{12" \times 12" \times 66"}{(1728 \text{ in}^3/\text{ft}^3)} \times \frac{28.32 \text{ L}}{\text{ft}^3} = \text{closed locker volume} = 155.8 \text{ L}$$

$$\text{ppm} = \frac{(\text{mcg/L}) \times (24.45 \text{ L/g-mole})}{88.1 \text{ g/g-mole}} = \frac{(9.4 \text{ mcg/155.8 L}) \times 24.45}{88.1}$$
$$= 0.0167 \text{ ppm BA}$$

Answer: 16.7 parts of butyric acid vapor per billion parts of stale locker air. Phew!! Odor threshold for butyric acid ($0.0006 \text{ ppm}_v = 0.6 \text{ ppb}_v$—a stench) is considerably less. Butyric acid and butyraldehyde are found in rancid butter and putrefying animal fatty acids. Not surprisingly, neither have a TLV or a OSHA PEL because their horrid odor limits exposure to less than reversible toxic effects. Tibetan monks have burned animal butter fat candles in caves for centuries with apparently no adverse health effects.

301. Silica sand and steel shot are used in an abrasive blasting cabinet to remove lead-containing paint, cadmium plating, plutonium–mercury–beryllium alloy, and osmium tetroxide from arsenic–nickel castings! The interior dimensions of this cabinet are 3.5' × 3.5' × 4.5'. The total area of openings into the cabinet is 1.7 ft². What is the required exhaust rate if no less than 500 fpm face capture velocity is used? What is the exhaust rate if no less than 20 air changes per minute are necessary in the cabinet?

$$Q = A \times V = 1.7\,\text{ft}^2 \times 500\,\text{fpm} = 850\,\text{cfm}$$

Q required for 20 air changes/minute = 20 × booth volume = 20 × (3.5' × 3.5' × 4.5') = 103 cfm.

Recalculating face capture velocity using the larger of the two exhaust volumes:

$$V = Q/A = 1103\,\text{cfm}/1.7\,\text{ft}^2 = 649\,\text{fpm}$$

Answer: 850 and 1103 cfm. Use the larger of these. Give very careful attention to the engineering control of dust emissions in the exhaust air. Perhaps use of ultra-high efficiency bag filters connected in series (and in parallel for system maintenance) should be considered. Maybe the baghouse filter housings should be in separate exhausted enclosures each with their independent dust collection and absolute containment systems. Maintenance of these systems is a consummate challenge to ensure superior industrial hygiene control work and engineering practices.

302. Two duct branches in an exhaust ventilation system have greatly different static pressures at their union. Balance could be achieved by increasing the flow in the branch with the lower loss. The air volume for the smaller branch is 580 cfm. The static pressure calculated for branch is –1.75". The static pressure for combined branches at the junction is –3.5". What is corrected exhaust air volume for this branch?

Since pressure losses increase as the square of the volume, increase the air flow through the branch with the lower resistance:

$$Q_{\text{new}} = Q_{\text{calc}} \times \sqrt{\frac{SP_{\text{junction}}}{SP_{\text{branch}}}} = 580\,\text{cfm} \times \sqrt{\frac{-3.5"}{-1.75"}}$$
$$= 580\,\text{cfm} \times \sqrt{2} = \frac{820\,\text{ft}^3}{\text{minute}}$$

Answer: The new ventilation volume rate in the smaller branch is 820 cfm.

303. A sulfur-bearing fuel oil is burned in a combustion process using 20% excess air. Analysis of the oil is (% by weight): 88.3% carbon, 9.5% hydrogen, 1.6% sulfur, 0.10% ash, and 0.05% water. Assuming 4% conversion of sulfur dioxide into sulfur trioxide, what is the required amount of combustion air and the total concentration of sulfur oxides in the stack flue gases? What is the ash concentration in the flue gases at 12% CO_2 assuming complete combustion? This fuel oil has a theoretical dry air combustion requirement of 176.3 standard cubic feet (scf) per pound. The combustion requirement is 177.6 scf at 40% relative humidity and 60°F per pound of this fuel oil.

Use 1 pound of this fuel as the basis for your calculations.
Theoretical combustion air requirement:

$$C + O_2 \rightarrow CO_2 \uparrow \qquad (0.883\,lb)\,(32/12) = 2.35\,lb \text{ of } O_2$$
$$H_2 + 1/2\,O_2 \rightarrow H_2O \uparrow \qquad (0.095\,lb)\,(16/2) = 0.076\,lb\,O_2$$
$$S + O_2 \rightarrow SO_2 \uparrow \qquad (0.016)\,(32/32) = 0.016\,lb\,O_2$$

$2.35\,lb\,O_2 + 0.76\,lb\,O_2 + 0.016\,lb\,O_2 = 3.13\,lb\,O_2$ per pound of fuel

$$\text{for } CO_2: \quad \frac{2.35\,lb}{3.13\,lb} \times \frac{176.3\,ft^3}{lb\,oil} = 132.4\,scf \text{ of air}$$

$$\text{for } H_2: \quad \frac{0.76\,lb}{3.13\,lb} \times \frac{176.3\,ft^3}{lb\,oil} = 42.8\,scf \text{ of air}$$

$$\text{for } S: \quad \frac{0.016\,lb}{3.13\,lb} \times \frac{176.3\,ft^3}{lb\,oil} = 0.90\,scf \text{ of air}$$

Air requirements at 20% excess combustion air:

$$176.3\,scf \times 1.2 = 212\,scf \text{ dry air/lb of oil}$$

$$177.7\,scf \times 1.2 = 213\,scf \text{ moist air/lb of oil}$$

Combustion products:

$$(0.883\,lb\,CO_2)\,(44/12)\,(379\,scf/44\,lb/mole) = 27.9\,scf\,CO_2$$

$$(0.095\,lb\,H_2O)\,(18/2)\,(379\,scf/18\,lb/mole) = 18.0\,scf\,H_2O \text{ from combustion}$$

$$(0.0005\,lb\,H_2O)\,(379\,scf/18) = 0.011\,scf \text{ from water in fuel}$$

$$\text{Nitrogen: } (212\,scf)\,(0.79) = 167.5\,scf$$

Water in air at 40% relative humidity and 60°F:

$$(0.0072\,scf/scf \text{ air})\,(213\,scf) = 1.5\,scf\,H_2O$$

$$\text{Sulfur oxides as } SO_2: (0.016)\,(64/32)\,(379/64) = 0.19\,scf\,SO_2$$

$$\text{Oxygen: } (176.3\,scf)\,(1.20 - 1.00)\,(0.21) = 7.4\,scf$$

Total scf of combustion products: $27.9 + 18.0 + 0.011 + 167.5 + 1.5 + 0.19 + 7.4 = 222.3\,scf$

SO_2 concentration:

$$(0.016) \, (379/32) \, (1/222.3) \, (10^6) \, (0.96) = 818 \, ppm_v \, SO_2$$

SO_3 concentration:

$$(0.016) \, (379/32) \, (1/222.3) \, (10^6) \, (0.04) = 34 \, ppm_v \, SO_3$$

Flue ash concentration:

$$(0.001 \, lb) \, (7000 \, grains/lb) \, (1/222.3) = 0.0315 \, grain/scf$$

Answers: 213 scf of moist air per pound of fuel oil. 818 ppm_v SO_2, 34 ppm SO_3, and 0.0315 grain per standard cubic foot of air.

304. BZ air was sampled for total barley dust at 1.8 L/m for 5 hours, 40 minutes with a 37 mm MCE MF with respective pre-sampling and post-sampling weights of 33.19 and 38.94 mg. What was the grain silo filler's 8-hour TWAE exposure to respirable dust if 85 mass-percent was nonrespirable?

$$5 \, hours, 40 \, minutes = 340 \, minutes$$

$$1.8 \, L/m \times 340 \, minutes = 612 \, L = 0.612 \, m^3$$

$$38.94 \, mg - 33.19 \, mg = 5.75 \, mg$$

5.75 mg/0.612 m^3 = 9.4 mg/m^3 TWAE (assuming 340 minutes represented an 8-hour exposure)

$$340 \, minutes \times 9.4 \, mg/m^3 = 3196 \, mg/m^3\text{-minutes}$$

$$\underline{+ \, 140 \, minutes \times 0 \, mg/m^3 = \quad 0 \, mg/m^3\text{-minutes}}$$

$$480 \, minutes \qquad\qquad 3196 \, mg/m^3\text{-minutes}$$

$$3196 \, mg/m^3\text{-minutes}/480 \, minutes = 6.7 \, mg/m^3 \, 8\text{-hour TWAE}$$

15% mass respirable:

$$9.4 \, mg/m^3 \times 0.15 = 1.4 \, mg/m^3 \, respirable$$

$$6.7 \, mg/m^3 \times 0.15 = 1.0 \, mg/m^3 \, respirable$$

$$\text{Grain dust TLV (oats, wheat, barley)} = 4 \, mg/m^3$$

Answers: 9.4 mg/m^3 TWAE to total barley dust assuming the 340-minute dust exposure represented 8-hour exposure. 6.7 mg/m^3 assuming balance of loader's exposure was essentially dust-free. 1.4 mg/m^3 TWAE and 1.0 mg/m^3 TWAE as respirable dust for 340-minute and 8-hour exposures, respectively (total airborne BZ dust, i.e., grain, silica, silicates, pesticides, insect parts, microbes, endotoxins, spores, *ad nauseum*).

305. Air was sampled for HCl gas (MW = 36.45 grams/gram-mole) in 15 mL of impinger solution at 0.84 L/m for 17 minutes, 20 seconds. The HCl collection efficiency was 80%. A chemist analyzed 4.7 mcg Cl/mL in the sample and 0.3 mcg/mL in control blank impinger. What was the steel pickler's exposure in ppm_v?

$$17 \text{ minutes, } 20 \text{ seconds} = 17.33 \text{ minutes}$$

$$0.84 \text{ L/m} \times 17.33 \text{ minutes} = 14.56 \text{ L}$$

Correcting for collection efficiency: 4.7 mcg Cl/mL/0.8 = 5.88 mcg Cl/mL

Correcting for control blank: 5.88 mcg Cl/mL − 0.3 mcg Cl/mL
$$= 5.58 \text{ mcg Cl/mL}$$

$$5.58 \text{ mcg Cl/mL} \times 15 \text{ mL} = 83.7 \text{ mcg Cl}$$

$$83.7 \text{ mcg Cl} \times (36.45/35.45) = 86.06 \text{ mcg HCl}$$

$$\text{ppm} = \frac{(\text{mcg/L}) \times (24.45 \text{ L/g-mole})}{\text{molecular weight}} = \frac{(86.06 \text{ mcg/14.56 L}) \times 24.45}{36.45}$$
$$= 4.0 \text{ ppm}$$

Answer: 4 ppm_v HCl gas ACGIH TLV for HCl is 2 ppm_v (C). Control intervention is required.

306. Determine the 8-hour TWAE of a scrap metal processor to Pb dust and fume with exposures of 3 hours, 15 minutes to 17 mcg Pb/m³; 97 minutes to 565 mcg Pb/m³; and 2 hours, 10 minutes to 46 mcg Pb/m³. The worker wore an approved HEPA dust/fume/mist filter cartridge respirator for $5\text{-}\frac{1}{2}$ hours.

Concentration (C) × time (T) = dose (CT) (Haber's law)	
17 mcg Pb/m³	× 195 minutes = 3315 mcg/m³-minutes
565 mcg Pb/m³	× 97 minutes = 54,805 mcg/m³-minutes
46 mcg Pb/m³	× 130 minutes = 5980 mcg/m³-minutes
422	64,100 mcg/m³-minutes

$$\text{OSHA Pb PEL} = 50 \text{ mcg/m}^3 \times 480 \text{ minutes} = 24,000 \text{ mcg/m}^3\text{-minutes}$$

$$64,100 \text{ mcg/m}^3/\text{minutes}/24,000 \text{ mcg/m}^3\text{-minutes} = 2.67 \times \text{PEL}$$

$$64,100 \text{ mcg/m}^3/480 \text{ minutes} = 133.5 \text{ mcg Pb/m}^3 \text{ TWAE}$$

If we assume the balance of the 8-hour work shift was "zero" exposure:

(422 minutes/480 minutes) × 133.5 mcg/m³ = 117.4 mcg Pb/m³ TWAE

Answers: TWAE = 133.5 mcg Pb/m³, or 2.67 × the PEL. Assuming the balance of the work shift was "zero" exposure: TWAE = 117.4 mcg Pb/m³.

307. A 7000-gallon storage tank in Albuquerque where the barometric pressure is 640 mm Hg contains 2000 gallons of chlorobenzene. Chlorobenzene vapor pressure is 12 mm Hg at 70°F. The molecular weight of DCB is 112.6 grams/gram-mole. What is the vapor saturation concentration in ppm$_v$, %, and mg/m³? What does chlorobenzene smell like? What is its reported odor threshold?

$$\frac{12 \text{ mm Hg}}{640 \text{ mm Hg}} \times 10^6 = 18,750 \text{ ppm} = 1.875\%$$

$$\frac{\text{mg}}{\text{m}^3} = \frac{\text{ppm} \times \text{molecular weight}}{24.45} = \frac{18,750 \times 112.6}{24.45} = \frac{86,350 \text{ mg}}{\text{m}^3}$$

Answers: 18,750 ppm$_v$. 1.875%. 86,350 mg chlorobenzene vapor/m³. The odor threshold (detection, not necessarily recognition) for chlorobenzene vapor is reported as 1.3 ppm$_v$. Chlorobenzene is reported to smell like almonds. Almonds smell like what? Why, chlorobenzene, of course!

308. A chemical plant operator had the following 8-hour TWAEs on Monday: 32 ppm$_v$ toluene, 19 ppm$_v$ xylene, and 148 ppm$_v$ MEK. Their respective TLVs are 50, 100, and 20 ppm$_v$. By what percent is the additive exposure limit exceeded?

$$\frac{32 \text{ ppm}_v}{50 \text{ ppm}_v} + \frac{19 \text{ ppm}_v}{100 \text{ ppm}_v} + \frac{148 \text{ ppm}_v}{20 \text{ ppm}_v} = 8.23 \text{ (no units)}$$

Answer: 823% in excess of the TLV, or about 16.5 times the action level of 0.5.

309. Air in an empty room (20' × 38' × 12') contains 600 ppm$_v$ cyclohexene vapor. How long will it take to dilute this to 6 ppm$_v$ with a 1550 cfm vane-axial exhaust fan? K factor = 3.

$$t = ? \text{ to } C_0 \rightarrow C$$

600 ppm$_v$ diluted to 6 ppm$_v$

$$20' \times 38' \times 12' = 9120 \text{ ft}^3$$

2.3 room volume for 10% of C_0 (or 60 ppm$_v$)

another 2.3 room volume for 10% of previous dilution (diluted to 6 ppm$_v$), or

$$2.3 \text{ volume} + 2.3 \text{ volume} = 4.6 \text{ volume}$$

$$4.6 \text{ volume} \times 9120 \text{ ft}^3 = 41{,}952 \text{ ft}^3$$

$$41{,}952 \text{ ft}^3 \times \frac{\text{minutes}}{1550 \text{ ft}^3} = 27.1 \text{ minutes}$$

K estimate = 3

$$27.1 \text{ minutes} \times 3 = 81.3 \text{ minutes}$$

Solved another way:

$$t = \frac{-\ln[C/C_o]}{Q/V} = \frac{-\ln[6\,\text{ppm}/600\,\text{ppm}]}{1550\,\text{cfm}/9120\,\text{ft}^3} = \frac{-\ln 0.01}{0.17\,\text{minute}}$$

$$= \frac{-(-4.605)}{0.17\,\text{minute}} = 27.1 \text{ minutes}$$

Answer: With perfect mixing, the concentration of 600 ppm$_v$ would dilute to 6 ppm in 27.1 minutes. Applying the ventilation imperfect mixing factor of 3, the dilution time increases to over 81 minutes. Verify the cyclohexene vapor concentration is less than 10% of the TLV by air sampling before allowing reoccupancy. The TLV is 300 ppm$_v$.

310. 7.3 µL liquid styrene (MW = 104.2 grams/gram-mole, density = 0.91 g/mL) was evaporated in 21.6 L glass calibration bottle. What is styrene vapor concentration in ppm$_v$?

$$7.3 \,\mu L = 0.0073 \,\text{mL}$$

$$0.0073 \,\text{mL} \times 0.91 \,\text{g/mL} = 0.00664 \,\text{g} = 6.64 \,\text{mg}$$

$$21.6 \,\text{L} = 0.0216 \,\text{m}^3$$

$$\text{ppm} = \frac{(\text{mg/m}^3) \times (24.45 \,\text{L/g-mole})}{\text{molecular weight}} = \frac{(6.64 \,\text{mg}/0.0216 \,\text{m}^3) \times 24.45}{104.2}$$

$$= 72.1 \,\text{ppm}$$

311. A rotameter was calibrated at 25°C and 760 mm Hg (NTP). What is the corrected air flow rate when the rotameter indicates 2 L/min at 630 mm Hg and 33°C?

$$Q_{actual} = Q_{indicated} \times \sqrt{\frac{P_{cal}}{P_{field}} \times \frac{T_{field}}{T_{cal}}} = \frac{2 \,\text{liters}}{\text{minute}} \times \sqrt{\frac{760 \,\text{mm Hg}}{630 \,\text{mm Hg}} \times \frac{306 \,\text{K}}{298 \,\text{K}}}$$

$$= \frac{2.23 \,\text{L}}{\text{minute}}$$

Square root function must be used with orifice meters (e.g., rotameters and critical orifices). This derives from orifice theory and must be used when an air flow rate orifice functions at pressure and temperatures differing from calibration conditions. In such situations, Charles' and Boyle's laws must not be applied. Displacement meters such as wet or dry test meters do not require application of this correction equation.

312. What is the effective specific gravity of 13,000 ppm$_v$ of a gas in air when the gas has a specific gravity of 4.6? Will the mixture stratify with the denser gas at floor level?

$$13,000\,ppm_v = 1.3\% \text{ (i.e., 98.7\% air)}$$

$$0.987 \times 1.0 = 0.987$$

$$\frac{+\,0.013 \times 4.6 = +\,0.0598}{1.000 \quad\quad = \quad 1.0468}$$

Answer: 1.0468 is only 4.68% greater than the density of air. No, never, pointing out fallacy of placing exhaust air ducts at the floor level to capture vapors "heavier than air" or near the ceiling for gases that are "lighter than air." Of course, in those facilities, such as a paint mixing "kitchen," where large volumes of organic solvents might spill, exhaust air hoods near the floor are often advisable to capture solvent vapors as they evaporate and to promote good ventilation mixing near the areas where a combustible air and vapor mixture could be produced. Breathing zone ventilation must also be provided. Total reliance on the so-called "floor sweeps" is unacceptable.

313. Analysis of an 866 L MF air sample detected 2667 mcg Zn. How much zinc oxide (ZnO) fume does this represent in the welder's breathing zone? Molecular weights of Zn and O are 65 and 16, respectively.

$$ZnO \text{ molecular weight} = 65 + 16 = 81$$

$$81/65 = 1.246$$

$$2667\,mcg\,Zn \times 1.246 = 3323\,mcg\,ZnO$$

$$3323\,mcg\,ZnO/866\,L = 3.84\,mg\,ZnO/m^3$$

314. What is air flow rate through an 8-inch diameter duct with a transport velocity of 2900 fpm? What capture velocity is expected 8 inches in front of the duct inlet if there is a wide flange around the inlet? Without cross-drafts, what discharge velocity is expected 20 feet from the exhaust outlet? What is the expected reduction in capture velocity 8 inches in front of the exhaust inlet if this wide flange is removed?

$$\text{Area of circle} = \pi r^2 = \pi(4")^2 = 50.27\,in^2 = 0.349\,ft^2$$

$$Q = AV = (0.349\,\text{ft}^2)\,(2900\,\text{fpm}) = 1012\,\text{cfm}$$

$$\text{at } 8\text{": } 290\,\text{fpm (one duct diameter)}$$

$$\text{at } 20\text{': } 290\,\text{fpm (30 duct diameters)}$$

$$290\,\text{fpm} \times 0.75 = 218\,\text{fpm}$$

Answer: 1012 cfm. 290 fpm at one duct diameter on the exhaust side of the fan. 290 fpm at 30 duct diameters on exhaust air discharge side. Both assume no significant disruptive ventilation cross-drafts. Capture velocity on the suction side of the fan increases up to 25% with a wide flange. Therefore, with the absence of a wide flange, the capture velocity one duct diameter in front of the exhaust inlet would be about 75% of the nominal capture velocity with a flange, or 290 fpm − (290 fpm × 0.25) = 218 fpm.

315. Nine detector tube BZ air samples were obtained randomly during a work shift with results of 2, 10, 5, 6, 2, 4, 14, 3, and 6 ppm$_v$ SO_2. What are worker's arithmetic mean and median exposures? Assuming air samples are log-normally distributed, what are standard deviation and 95% confidence range of his or her exposures?

$$n = 9$$

SO$_2$, ppm$_v$

x	$\log x$	$(\log x)^2$
2	0.301	0.0906
2	0.301	0.0906
3	0.477	0.2275
4	0.602	0.3264
5	0.699	0.4886
6	0.778	0.6053
6	0.778	0.6053
10	1.000	1.0000
14	1.146	1.3133
$\Sigma = 52$	$\Sigma = 6.082$	$\Sigma = 4.7836$

Arithmetic mean (the average) = 52 ppm$_v$ SO_2/9 = 5.8 ppm$_v$ SO_2

Median = antilog 6.082/9 = antilog 0.6758 = 4.5

$$\text{standard deviation} = \text{antilog} \sqrt{\frac{4.7836 - (6.082^2/9)}{9 - 1}}$$

$$= \text{antilog} \sqrt{\frac{4.7836 - 4.1101}{8}}$$

$$= \text{antilog} \sqrt{0.0842} = \text{antilog } 0.2902 = 1.95$$

Values of the Student's t-distribution for 95% confidence range (bilateral test) are

Number of Measurements	Degrees of Freedom	t-Value
2	1	12.706
3	2	4.303
4	3	3.182
5	4	2.776
6	5	2.571
7	6	2.447
8	7	2.365
9	**8**	$\Rightarrow$**2.306**
10	9	2.262
11	10	2.228
21	20	2.086
31	30	2.042
51	50	2.009
101	100	1.984
501	500	1.965
1001	1000	1.962
∞	∞	1.960

$$95\% \text{ confidence range} = \text{antilog}\left[0.6758 \pm 2.306\sqrt{\frac{0.0842}{9}}\right]$$

$$= \text{antilog}\left[0.6758 \pm 2.306\,(0.09670)\right] = \text{antilog}\,(0.6758 \pm 0.2230)$$

$$\text{Upper limit} = \text{antilog}\,0.8988 = 7.9\,\text{ppm}_v\,SO_2$$

$$\text{Lower limit} = \text{antilog}\,0.4528 = 2.8\,\text{ppm}_v\,SO_2$$

Answers: Arithmetic mean = 5.8 ppm$_v$ SO_2. Median = 4.5 ppm$_v$ SO_2. Standard deviation = 1.95. The 95% confidence interval range of the exposures = 2.8 to 7.9 ppm$_v$ SO_2.

316. A sealed 55-gallon drum containing 2 gallons of n-butylamine was in an empty room (20' wide × 40' long × 10' high) for 3 weeks. A process operator wearing a full-face airline respirator and protective clothing finishes filling the drum with n-butylamine. The air-tight room has no operating ventilation during the drum filling. This plant is located in Montana at an elevation where the barometric pressure is 680 mm Hg. Vapor pressure and molecular weight of n-butylamine are 82 mm Hg and 73.2 grams/gram-mole, respectively. There is a single 1550 cfm exhaust fan in 20' wall with negative pressure-activated make-up air louver located in opposite wall. The ventilation mixing factor, K, is estimated at 3 (unitless).

- What is the saturation concentration of *n*-butylamine vapor in the drum in ppm_v and mg/m^3 before it is filled?
- What is the average *n*-butylamine vapor concentration (in ppm_v) in the room after the drum has been filled and the bung has been tightened?
- After the full drum is sealed, how long will it take to dilute *n*-butylamine vapor in the room to less than $1\,ppm_v$ by operating the exhaust fan?

$$\frac{82\,mm\,Hg}{680\,mm\,Hg} \times 10^6 = 120,588\,ppm \cong 12\%$$

$LEL_v = 1.7\%$
$UEL_v = 9.8\%$
$FP = 10°F$
$IDLH = 2000\,ppm_v$

ACGIH TLV, OSHA PEL, and NIOSH REL
$= C\ 5\,ppm_v$ (C is $15\,mg/m^3$.) SKIN

$$\frac{mg}{m^3} = \frac{ppm \times molecular\ weight}{24.45} = \frac{120,588 \times 73.2}{24.45} = \frac{361,024\,mg}{m^3}$$

55 gallons − 2 gallons = 53 gallons = 200.6 liters (the volume of saturated air that is displaced into the room)

$$200.6\,L \times 361\,mg/L = 72,417\,mg\ \textit{n}\text{-butylamine vapor}$$

$$20' \times 40' \times 10' = 8000\,ft^3 = 226.6\,m^3$$

$$72,417\,mg/226.3\,m^3 = 320.0\,mg/m^3$$

$$ppm = \frac{(mcg/L) \times 24.45}{molecular\ weight} = \frac{(320\,mcg/L) \times 24.45}{73.2} = 106.9\,ppm$$

$8000\,ft^3 \times 2.3$ room volumes $= 18,400\,ft^3$ (yields 10% of $106.9\,ppm_v$
$= 10.7\,ppm_v$)

$8000\,ft^3 \times 4.6$ room volumes $= 36,800\,ft^3$ (yields 1% of $106.9\,ppm$
$= 1.07\,ppm_v$)

$$36,800\,ft^3/1550\,cfm = 24\,minutes\ (approximately)$$

24 minutes × K (= 3) = 72 minutes. Therefore, operate the exhaust fan for at least 72 minutes. Verifying by another equation:

$$t = \frac{-\ln[C/C_o]}{Q/V} = \frac{-\ln[1\,ppm/107\,ppm]}{1550\,cfm/8000\,ft^3} = \frac{-\ln 0.00935}{0.194\,minute}$$

$$= \frac{4.673}{0.194\,minute} = 24\,minutes$$

$$24\,minutes \times K(=3) = 72\,minutes$$

Answer: 120,588 ppm$_v$. 361,024 mg/m^3 107 ppm$_v$. > 72 min. Vapor concentration is actually lower than this because the high vapor pressure of this amine produces back pressure on the liquid somewhat hindering evaporation. See Part I for further explanation of calculating the saturated vapor concentration of very volatile materials—in general, those with a vapor pressure exceeding 20 mm Hg at 20°C.

317. An industrial hygienist determines a worker's peak exposure to isopropylamine by drawing the BZ vapor through a small charcoal tube using a 100 mL detector tube pump. Since pump's orifice samples at a critical rate of 33 mL/minute, this method ensures maximum sampling rate for this size charcoal tube is not exceeded. TLV for isopropylamine is 5 ppm$_v$. If the true peak BZ concentration is 0.5 ppm$_v$, how much vapor will industrial hygienist collect? Molecular weight of isopropylamine is 59.08 grams/gram-mole.

$$\frac{mg}{m^3} = \frac{mcg}{L} = \frac{ppm \times molecular\,weight}{24.45} = \frac{0.5 \times 59.08}{24.45} = \frac{1.208\,mcg}{L}$$

$$\frac{1.208\,mcg}{L} \times 0.1\,L = 0.1208\,mcg$$

Answer: 0.12 microgram of isopropylamine. Check with the industrial hygiene chemist to ensure that enough vapor has been collected to satisfy the minimum analytically detectable concentration. Otherwise, a larger air sample is required.

318. Calculate the flash point of an aqueous solution containing 75% methyl alcohol by weight. The flash point of 100% methanol is 54°F. Methanol's vapor pressure at this temperature is 62 mm Hg.

   Calculate based on 100 pounds of solution. The mole fractions for each solution component are needed to apply Raoult's law. Remember that the number of moles = mass/molecular weight.

	Pounds	Molecular Weight	Moles	Mole Fraction
Methanol	75	32	2.34	0.63
Water	25	18	1.39	0.37
			3.73	1.00

Raoult's law is used to calculate the vapor pressure (P_{sat}) of pure methanol based on the partial pressure required to flash, where x is the mole fraction and p is the vapor pressure of the 100% flammable component at its flash point.

$$p = x\, P_{sat}$$

$$P_{sat} = \frac{p}{x} = \frac{62\,\text{mm Hg}}{0.63} = 98.4\,\text{mm Hg}$$

Answer: Using graph of the vapor pressure of methanol versus temperature, the flash point of aqueous solution is approximately 67.5°F. Addition of water, if miscible with the flammable solvent, not surprisingly, raises the flash point. See Problem 18 for the discussion of applications and the deviations from Raoult's law.

319. The LEL and UEL for a flammable gas are 2.2% and 7.8% (volume/volume), respectively. At the mid-point between the LEL and the UEL ($\cong 5\%$), the explosion pressure is about _____ times greater than an explosion occurring just above the LEL or just below the UEL, respectively.
    a. 2
    b. 4
    c. 10
    d. 50
    e. 100

Answer: c. This obviously varies from explosive gas to explosive gas, vapor to explosive vapor, and dust to explosive dust, but at the stoichiometric mid-point, the explosion pressure generated is typically an order of magnitude greater than that at the LEL or about an order magnitude greater than blast pressure at the UEL.

320. Generally, with excellent mixing of clean dilution ventilation with contaminated air, _____ complete air changes are necessary to ensure confined space atmosphere equals the ambient atmosphere concentration.
    a. 2
    b. 5
    c. 10
    d. 20
    e. 53

Answer: d. Assumption made of the confined space atmosphere is 100% contaminant (i.e., 1,000,000 ppm$_v$). After 2.3 air changes, the concentration is reduced to 100,000 ppm$_v$. After another 2.3 air changes (4.6 total), the level is reduced to 10,000 ppm$_v$. After a total of 6.9 air changes, the concentration is now 1000 ppm$_v$. After another 2.3 air changes (9.2 total), level is 100 ppm$_v$. After another 2.3 changes (11.5 total), the initial 100% concentration has been reduced to 10 ppm$_v$. After 18.4 total air changes, the concentration is reduced to 0.01 ppm$_v$, and after 20.7 air changes, the concentration is finally at 0.001 ppm$_v$ (1 ppb$_v$)—level well below most TLV®s and PELs (*bis*-chloromethyl ether and osmium tetroxide are current notable exceptions for gases and vapors). The above of course, requires that there is excellent mixing of the ambient air with the contaminated air, and there is recirculation of exhaust air into the make-up air inlet.

321. In testing atmosphere of a confined space for air contaminants, what is the proper sequence (first, second, third)? Consider flammables and combustibles the same.
   a. Toxics, oxygen, flammables
   b. Oxygen, flammables, toxics
   c. Combustibles, toxics, oxygen
   d. Oxygen, toxics, flammables
   e. Flammables, oxygen, toxics
   f. Toxics, flammables, oxygen
   g. None of the above because the sequence is not important

   Answer: b. Remember the mnemonic "**OFT**" = **O**xygen, **F**lammables, **T**oxics—and done very **OFT**en.

322. VICI Metronics, Inc. (Santa Clara, California), a high-quality manufacturer of gas and vapor diffusion vials, gives the following example to produce 10 ppm$_v$ toluene vapor in a 1000 cubic centimeters/minute air stream at 30°C:
   a. Calculate the required vapor generation rate:

   $$r = \frac{F\,C}{K}, \quad \text{where K} = \frac{24.47\,\text{L/g-mole}}{92.13 = \text{molecular weight}} = 0.266$$

   $$r = \frac{(1000\,\text{cc/minute})(10\,\text{ppm}_v)}{0.266} = \frac{37,594\,\text{nanograms}}{\text{minute}}$$

   b. Calculate the vapor diffusion rate:

   known: molecular weight of toluene = 92.13 (g/mole)

   Diffusion vial length = 7.62 cm = L

   T = 30° + 273 K = 303 K

   p = 36.7 mm Hg (toluene vapor pressure at 30°C)

   D_o = 0.0849 cm²/s (diffusion coefficient at 25°C)

   P = 750 mm Hg (atmospheric pressure)

$$A = 0.1963 \, \text{cm}^2 \, (\text{5-mm diameter diffusion vial cross-section})$$

$$r = (1.9 \times 10^4)(303\,\text{K})(0.0849)(92.13)\left[\frac{0.1963}{7.62}\right]x$$

$$\log\left[\frac{750}{750 - 36.7}\right] = \frac{25,276 \, \text{nanograms}}{\text{minute}}$$

c. Length of capillary tube can be shortened to give a higher diffusion rate, for example:

$$L_2 = L_1 \times \frac{r_1}{r_2} = 7.62\,\text{cm} \times \frac{25,276 \, \text{ng/min}}{37,594 \, \text{ng/min}} = 5.1\,\text{cm}$$

d. To estimate diffusion rate of a given volatile compound, use the equation:

$$r = 1.90 \times 10^4 \times T \times D_o \times M \times \left[\frac{A}{L}\right] \times \log\left[\frac{P}{P - \rho}\right]$$

Capillary diffusion tubes can provide a constant source for dynamic gas and vapor calibration systems generating ppb to high ppm concentrations. Generation rates are easily calibrated and verified by simple gravimetric procedures. The principle is based on fact gases and vapors will diffuse at a steady rate through a capillary tube held at constant temperature and pressure. The vapor pressure will remain constant and serves as the constant driving force for diffusion through a capillary tube. The bore diameter and the diffusion path length then determine the rate for a specific volatile material. Variations in atmospheric pressure and temperatures and the carrier gas composition affect the diffusion rate. The actual rate is verified by simply pre- and post-weighing the diffusion tube during the period of use.

323. One gallon of gasoline is accidentally spilt into a 10′ × 10′ × 10′ press pit below a large metal stamping machine. The pit does not have a fixed forced mechanical ventilation system. Hours go by. What is the average concentration of gasoline vapor? What are the hazards? The density of gasoline is 0.75 g/mL. Average molecular weight of gasoline is approximately 73 grams/gram-mole.

$$\frac{10' \times 10' \times 10'}{35.3 \, \text{ft}^3/\text{m}^3} = 28.32 \, \text{m}^3$$

$$1\,\text{gallon} \times \frac{3785\,\text{mL}}{\text{gallon}} \times \frac{0.75\,\text{g}}{\text{mL}} \times \frac{1000\,\text{mg}}{\text{g}} = 2,838,750\,\text{mg}$$

$$\text{ppm} = \frac{(\text{mg/m}^3) \times (24.45 \, \text{L/g-mole})}{\text{molecular weight}} = \frac{(2{,}838{,}750 \, \text{mg/28.32 m}^3) \times 24.45}{73}$$

$$= 33{,}572 \, \text{ppm}$$

Note the LEL and UEL for gasoline are 1.4% and 7.6% (by volume), respectively. The concentration is in the highly dangerous mid-explosive range. TLV = 300 ppm$_v$ (0.03%) with a ceiling of 500 ppm$_v$ and an IDLH concentration of 5000 ppm$_v$ (0.5%, or $\cong$ 1/3 of the LEL). The benzene vapor concentration alone in this vapor mixture is near 300 or more ppm$_v$. What if a welder later performs repairs on the press above the pit without taking "hot work" precautions? Does the vapor mixture have a "built-in match": a volatile explosive liquid that, at saturation concentration, is between the LEL and its UEL?

324. A compressed gas cylinder contains hydrogen at 25°C and at a gauge pressure of 2200 psig. The cylinder volume is 45 liters. What is the mass of hydrogen in this cylinder?

$$P_i \, V_i = P_f \, V_f$$

$$V_f = \frac{P_i \, V_i}{P_f} = \frac{(2200 \, \text{psig} + 14.7 \, \text{psia}) \, (45 \, \text{L})}{14.7 \, \text{psia}} = 6780 \, \text{L}$$

$$6780 \, \text{L} \times \frac{\text{mole}}{24.45} \times \frac{2 \, \text{grams H}_2}{\text{gram-mole}} \times \frac{\text{kg}}{1000 \, \text{g}} = 0.555 \, \text{kg}$$

Answer: 0.555 kilogram of hydrogen = 1.1 pounds of H_2.

325. Determine the volume that 1.5 moles of diethylsulfide [$(C_2H_5)_2S$] would occupy at 275°C and 12.33 atmospheres. The critical pressure (P_c) is 39.08 atmospheres. The critical temperature (T_c) for (C_2H_5)$_2$S is 283.8°C.

At this temperature and pressure, this chemical, like many other gases, does not behave like an ideal gas. Corrections, therefore, are required in the calculations.

$$P_r = \frac{12.33 \, \text{atm}}{39.08 \, \text{atm}} = 0.316$$

$$T_r = \frac{275°\text{C} + 273 \, \text{K}}{283.8°\text{C} + 273 \, \text{K}} = 0.983$$

From tables in physical chemistry handbooks, we obtain Z from the above correction factors = 0.87. Z is the compressibility factor, an empirical correction factor for the nonideal behavior of real gases.

$$V = \frac{(0.87)(1.50\,\text{moles})\,(0.0821\,\text{L-atm/mole-K})\,(548\,\text{K})}{12.33\,\text{atm}} = 4.76\,\text{L}$$

326. The vapor pressure (P_v) can be measured by passing an inert gas over a sample of the material and analyzing the composition of the gaseous mixture. Calculate the vapor pressure of mercury at 23°C and 745 mm Hg if a 50.40 gram sample of nitrogen plus mercury vapor contains 0.702 milligram of mercury.

The basic equation is

$$\frac{P_v}{P_t} = \frac{n}{n + n_{\text{inert}}}.$$

The amount of each gas in the mixture is

$$n\,(\text{Hg}) = \frac{7.02 \times 10^{-4}\,\text{g}}{200.59\,\text{g mole}^{-1}} = 3.50 \times 10^{-6}\,\text{mole}$$

$$n\,(\text{N}_2) = \frac{[50.40 - (7.02 \times 10^{-4})]\,\text{g}}{28\,\text{g mole}^{-1}} = 1.8\,\text{mole}$$

$$P_v = (745\,\text{mm Hg}) \times \frac{3.5 \times 10^{-6}}{(3.5 \times 10^{-6}) + 1.8} = 1.45 \times 10^{-3}\,\text{mm Hg}$$

Answer: 0.00145 mm Hg.

327. A compressed gas cylinder contains 75 liters of CO at 215 psig and 25°C. If the room atmospheric pressure is 14.4 psi, what mass of CO is vented to the laboratory when the valve is opened?

$$\text{CO originally in cylinder} = n_1 = \frac{PV}{RT}$$

$$= \frac{(215 + 14.4)\,\text{psi}[6895\,\text{Pa}]/(1\,\text{psi})]\,(75\,\text{L})}{(8314\,\text{L Pa K}^{-1}\,\text{mole}^{-1})(298\,\text{K})}$$

$$= 48.0\,\text{moles}$$

$$\text{CO remaining at 0 psig} = n_2 = \frac{(0 + 14.4)\,\text{psi}(6895\,\text{Pa})(75\,\text{L})}{(8314)(298\,\text{K})} = 3.0\,\text{moles}$$

$$[(48.0 - 3.0)\,\text{moles}](28 \times 10^{-3}\,\text{kg/mole}) = 1.3\,\text{kg}$$

Answer: 1300 grams of CO gas are released into the laboratory air.

328. A liquid with a molecular weight of 86 evaporates into the air of a work-place from an 8 feet × 2 feet open surface tank. The vapor pressure of this liquid is 30 mm Hg. The air and liquid temperature are both 25°C. The room air passing over the liquid is 100 feet per minute. What is the vapor generation rate?

The vapor generation rate, G (in lb/hour), is estimated from the EPA equation:

$$G = \frac{13.3792\,MPA}{T} \times \left[\frac{D_{ab}V_z}{\Delta Z}\right]^{0.5},$$

where
  M = molecular weight (lb/lb-mole) = 86
  P = vapor pressure (inches of mercury) = 30 mm Hg = 1.18" Hg
  A = liquid surface area (ft²) = 8' × 2' = 16 ft²
  D_{ab} = diffusion coefficient (ft²/sec of a through b in air) = ? (solved below)
  V_z = air velocity (feet/minute) = 100 ft/minute
  T = temperature (K, kelvin) = 25°C = 298.15 K
  ΔZ = pool or tank length along flow direction (feet) = 8 feet

$$D_{ab} = \frac{4.09 \times 10^{-5}\,(T^{1.9})\left[(1/29) + (1/M)\right]^{0.5}\,(M^{-0.33})}{P_t},$$

where
  T = temperature (K, kelvin) = 25°C + 273.15 K = 298.15 K
  P_t = pressure (in atmospheres) = 30 mm Hg = 0.03947 atm

$$D_{ab} = \frac{4.09 \times 10^{-5}\,(298.15\,K)^{1.9}\,[(1/29) + (1/86)]^{0.5}\,(86^{-0.33})}{0.03947\,\text{atmospheres}}$$

$$= 2.57\,cm^2/second = 0.00277\,ft^2/second$$

$$G = \frac{(13.3792)\,(86\,\text{lb/lb-mole})\,(1.18''Hg)(16\,ft^2)}{298.15\,K}$$

$$\times \sqrt{\frac{(0.00277\,ft^2/second) \times (100\,ft/minute)}{8\,feet}} = 13.56\,\text{lb/hour}$$

Answer: Approximately 13–14 pounds of liquid evaporates per hour. From this estimate, one can design industrial hygiene controls (e.g., enclosures, less volatile and/or toxic solvents, local exhaust ventilation (general dilution ventilation), etc.).

329. Assume from the preceding problem that general ventilation (10,000 cfm) is used to dilute vapors, and the ventilation mixing factor is 3. What is the contaminant concentration in the workplace?

$$C = \frac{(1.7 \times 10^5)(K)(G)}{M \times Q \times k},$$

where
　　C = air contaminant concentration (ppm$_v$)
　　K = ambient air temperature (degrees kelvin)
　　G = vapor generation rate (gm/sec)
　　M = molecular weight (grams/gram-mole)
　　Q = ventilation rate (cfm)
　　k = ventilation mixing factor (dimensionless, based on subjective judgment)

$$\frac{13.56\,\text{lb}}{\text{hour}} \times \frac{453.6\,\text{grams}}{\text{lb}} \times \frac{\text{hour}}{60\,\text{minutes}} \times \frac{\text{minute}}{60\,\text{seconds}} = \frac{1.709\,\text{grams}}{\text{second}}$$

$$C = \frac{(1.7 \times 10^5)(298.15\,\text{K})(1.709\,\text{g/sec})}{(86\,\text{g/g-mole})(10,000\,\text{cfm})\,(3)} = 33.6\,\text{ppm}_v$$

330. The saturation pressure of water vapor in air at 22°C is 19.8 mm Hg. What is the mass concentration of water vapor in air at this temperature when the barometric pressure is 725 mm Hg, and the relative humidity is 50%?

$$0.50 \times \left[\frac{19.8\,\text{mm Hg}}{725\,\text{mm Hg}}\right] \times 10^6 = 13,655\,\text{ppm}_v = 1.3655\%\,\text{water vapor in air}$$

$$\frac{\text{mg}}{\text{m}^3} = \frac{\text{ppm}_v}{((22.4\,\text{L/g-mole})/\text{molecular weight})}$$
$$\times\,(\text{absolute temperature}/273.15\,\text{K}) \times (760\,\text{mm Hg}/725\,\text{mm Hg})$$
$$= \frac{13,655\,\text{ppm}_v}{(22.4/18) \times (295.15/273.15) \times (760/725)} = \frac{9687\,\text{mg H}_2\text{O}}{\text{m}^3}$$

Answer: 13,655 ppm$_v$ = 9687 mg H$_2$O vapor/m^3

331. An industrial hygienist using a direct-reading instrument is measuring the mercury vapor in the atmosphere of a chloralkali plant. When the meter indicates a level of 0.04 mg Hg/m^3, there is a release of chlorine gas that, measured with a detector tube, was about 0.7 ppm$_v$ Cl$_2$. At this time, the mercury vapor meter reading fell to 0.01 mg/m^3. What could explain this apparent reduction?

Answer: Mercury vapor detector instruments are only responsive to elemental mercury vapor and not to mercury salts, oxides, or organomercury compounds. Most likely, since chlorine is a strong oxidizer, there was a gas-phase reaction of the mercury vapor and chlorine gas to produce mercurous chloride and mercuric chloride salts both of which are not detected by UV mercury vapor meter. The astute industrial hygienist is constantly vigilant for such possibilities in anomalous results (high or low).

$$\frac{mg}{m^3} = \frac{ppm_v \times molecular\,weight}{24.45} = \frac{0.7\,ppm_v \times (2 \times 35.5)}{24.45} = \frac{2.03\,mg\,Cl_2}{m^3}$$

Therefore, the excess chlorine molecules help ensure stoichiometric reactions in sufficient time.

$$Hg_0 + Cl_2 \rightarrow HgCl_2$$

$$2\,Hg_0 + Cl_2 \rightarrow Hg_2Cl_2$$

332. A worker inhales $1000\,ppm_v$ ethyl alcohol vapor continuously throughout his 8-hour work shift. He alleges ethanol intoxication. Is this likely?

$$Molecular\,weight\,CH_3CH_2OH = 46.07\,grams\,gram\text{-}mole^{-1}$$

$$Density\,of\,EtOH = 0.789\,g/mL$$

$$\frac{mg}{m^3} = \frac{ppm_v \times molecular\,weight}{24.45\,L/g\text{-}mole} = \frac{1000\,ppm_v \times 46.07}{24.45} = \frac{1884\,mg}{m^3}$$

During an 8-hour work shift, an average worker inhales approximately 10 cubic meters of air, so

$$10\,m^3 \times \frac{1884\,mg}{m^3} = 18,840\,mg\,EtOH = 18.84\,g\,EtOH$$

Assuming this worker absorbed 100% of inhaled EtOH (a reasonable assumption since EtOH is readily soluble in mucous membranes and readily absorbed into the systemic circulation from the respiratory tract):

$$18.84\,g\,EtOH \times \frac{mL}{0.789\,g} = 23.9\,mL\,EtOH$$

This equates to 0.8 ounces of ethyl alcohol, or nearly 1.5 12-ounce bottles of 4.5% ethanol beer—distributed over 8 hours while, concurrently, detoxification is occurring following zero-order kinetics.

Answer: Although absorption of ethyl alcohol vapors through respiratory tract is complete, and the absorbed ethanol does not initially pass through hepatic portal circulation, the amount absorbed over 8 hours (while there is detoxification at a rate high enough to prevent bioaccumulation to toxic levels) will not yield clinical intoxication. One on Antabuse® therapy, however, is compromised. Exposure of alcoholics on Antabuse to EtOH vapor is contraindicated because of the possibility of a violent adverse reaction. Furthermore, 1000 ppm$_v$ of ethanol vapor are, in the author's experience, highly irritating to upper respiratory tract mucous membranes. Few workers would willingly tolerate this vapor exposure for more than a few minutes.

333. One-micron particles are the optimal size for penetration into, and retention in, the terminal airways, the alveoli. What is the settling velocity of 1-micron particles of silica (density of α-quartz, $SiO_2 = 2.65$ g/cm³) in still air? Will such particles fall from the air and settle to the floor and the ground?

Apply Stoke's law:

$$V_s = \frac{g\,d^2(\rho - \rho_a)}{18\,\eta},$$

where

v_s = the particle settling velocity (in centimeters per second)
g = gravitational attraction of particle (981 centimeter/second)
d = particle diameter (centimeter)
ρ = particle density (g/cm³)
ρ_a = air density at 25°C (= 0.0017 g/cm³)
η = coefficient of air viscosity (= 1.828×10^{-4} poise) at 25°C

$$n_s = \frac{[981\,\text{cm/second}](0.0001\,\text{cm})^2(2.65\,\text{g/cm}^3 - 0.00117\,\text{g/cm}^3)}{(18)(1.828 \times 10^{-4}\,\text{poises})}$$

$$= \frac{0.0079\,\text{cm}}{\text{second}}$$

$$\frac{0.0079\,\text{cm}}{\text{second}} \times \frac{60\,\text{sec}}{\text{min}} \times \frac{60\,\text{min}}{\text{hour}} = \frac{28.4\,\text{cm}}{\text{hour}}$$

Answer: 28.4 cm/hour = 0.93 feet/hour. Normal air currents tend to keep such very small particles permanently suspended in air until they serve as condensation nuclei for atmospheric moisture or until they flocculate with other dust particles. Practically speaking, lung-damaging dust will not normally settle out of work place air. These invisible particles remain airborne. *Note*: Particles below 1 μm in diameter require application of Cunningham's factor to calculate settling rates in still air. This factor, or coefficient, $= C = C'[1 + K[1/r]]$, where $C' =$ the calculation (as above) using Stoke's law, $K = 0.8$–0.86, and $r =$ particle radius in centimeters.

Particles with a radius less than 0.1 micron behave like gas molecules and "settle" according to Brownian motion and the equation: $A = \sqrt{(RT/N) \times (t/3\pi\eta r)}$, where $A =$ the distance of motion in time, t; $R =$ the universal gas constant (8.316×10^7); $T =$ absolute temperature; $N =$ number of molecules in 1 mole (6.023×10^{23}); $\gamma =$ the viscosity of air in poises (1.828×10^{-4} at 70°F); and $r =$ the particle radius in centimeters.

334. The smallest particles visible to the unaided eyes under the best of viewing conditions (lighting, contrast, color, steadiness, etc.) are about 50–100 microns in diameter. The period at the end of this sentence is about 60–65 microns. Such large particles ("rocks" to industrial hygienists and air pollution engineers) do not penetrate far into respiratory tract. A 100-micron particle, for example, will not reach the alveolar region of the lungs. Less than 1% of all 100 micron particles penetrate as far as the tracheobronchial region. Close to 50% of the 100 micron particles are deposited in the nasopharyngeal region (nostrils and upper throat). The remaining 50% are not deposited and tend to be exhaled. Compare the settling rate of a 100 micron particle in still air to that of a 1 micron particle as calculated in Problem 333.

$$n_s = \frac{[981\,\text{cm/second}]\,(0.001\,\text{cm})^2\,(2.65\,\text{g/cm}^3 - 0.00117\,\text{g/cm}^3)}{(18)\,(1.828 \times 10^{-4}\,\text{poises})}$$

$$= \frac{0.789\,\text{cm}}{\text{second}}$$

$$\frac{0.789\,\text{cm}}{\text{second}} \times \frac{60\,\text{sec}}{\text{min}} \times \frac{60\,\text{min}}{\text{hour}} = \frac{2840\,\text{cm}}{\text{hour}}$$

Answer: 2840 centimeters per hour (= 93 feet per hour = 1.55 feet per minute). Even such "large" particles (still with relatively small mass) tend to be buoyant in normal air currents (e.g., 50 fpm or less is often taken as "still" air). It is easy to see how these larger, nonrespirable particles settling at a rate of 1.55 fpm in quiescent air tend not to settle in an atmosphere having perceptible air motion. That is, imagine a 100 micron particle "settling" at 1.55 fpm being tossed around by a 50 fpm air current—not unlike a cork bobbing in a turbulent sea.

335. In the estimation of a worker's total daily work place exposure to toxic agents (i.e., dose), the industrial hygienist must account not only for inhalation exposures, but those occurring from the percutaneous, or dermal, route as well as from ingestion and intraocularly. In the author's experience, toxicant absorption routes beside inhalation are often not adequately addressed and tend to be discounted. This is no doubt due, more often than

not, to the difficulty in quantitatively assessing the contribution of these routes of absorption to the overall body burden of a toxicant. With skin-absorbed contaminants, one must not only consider the toxicological legacy of a material to pass through intact skin and its dermal absorption rate (in units, e.g., of $\mu g/cm^2$ of skin surface area/hour), but also the skin contact time, skin surface area contacted, and the type of skin (e.g., eyelids have the thinnest skin, and the soles of the feet and the palms have the thickest integument).

Consider a worker falling into a tank and becoming totally immersed in a liquid at room temperature (the aromatic amine, Methyl Ethyl Death®) that has a dermal absorption rate of $2.3\,\mu g/cm^2/hour$. The elapsed time from when he fell into the tank until he was extracted and totally decontaminated was less than 17 minutes. This worker, Joe B. Lunchbox, was 6'-2" tall and weighed 200 pounds. Estimate his total maximum dermal absorption (using worst-case assumptions).

The Du Bois formula is used to calculate total body surface area of adult humans:

$$\text{BSA in m}^2 = (\text{weight in kg})^{0.425} \times (\text{height in cm})^{0.725} \times 0.007184$$

$$= \left[200\,\text{lb} \times \frac{\text{kg}}{2.204\,\text{lb}} \right]^{0.425} \times \left[74\,\text{in} \times \frac{2.54\,\text{cm}}{\text{inch}} \right]^{0.725} \times 0.007184$$

$$= 2.173\,\text{m}^2$$

$$2.173\,\text{m}^2 \times (10{,}000\,\text{cm}^2/\text{m}^2) = 21{,}730\,\text{cm}^2$$

$$\frac{(2.3\,\mu g/cm^2)}{\text{hour}} \times 17\,\text{minutes} \times \frac{\text{hour}}{60\,\text{minutes}} \times 21{,}730\,\text{cm}^2 = 14{,}160\,\mu g$$

$$\cong 14.1\,\text{mg}$$

Answer: $\cong 14.1$ milligrams of aromatic amine was the maximum dermally absorbed dose to which should be added the estimated inhalation and ingestion exposure doses.

336. The velocity pressure in a duct is 0.48 inches of water. The barometric pressure is 640 mm Hg. The air temperature is 190°F. Assume dry air. What is the duct air velocity?

$$P_b = 640\,\text{mm Hg} = 25.197"\,\text{Hg}$$

$$460°A + 190°F = 650°A$$

$$\text{Air density}, D = 1.325 \times \frac{P_b}{T_{abs}} = 1.325 \times \frac{25.197}{650} = \frac{0.05136\,\text{lb}}{\text{ft}^3}$$

$$\text{Air velocity, fpm} = 1096.2 \times \sqrt{\frac{P_v}{D}} = 1096.2 \times \sqrt{\frac{0.48"\,H_2O}{0.05136}} = \frac{3351\,\text{ft}}{\text{min}}$$

Note the air density decreases, in this case, by about one-third as air temperature increases and barometric pressure decreases. This should be intuitive for those familiar with the gas laws.

337. "Standard dry air" (at 70°F and 29.9" Hg) has a mass density of:
    a. 0.055 lb/ft^3
    b. 0.065 lb/ft^3
    c. 0.075 lb/ft^3
    d. 0.085 lb/ft^3
    e. 0.095 lb/ft^3
    f. None of the above

    Answer: c.

338. An industrial process generates 17,300 mg of contaminant hourly into a workplace atmosphere. It is desired to limit the workers' exposures to this solvent vapor to no more than 10 mg/m^3 as an 8-hour TWAE by the use of general dilution ventilation. If ventilation imperfection mixing factor is 4, how much ventilation is needed assuming make-up air is contaminant-free?

$$17,300\,\text{mg/hour} = 288.3\,\text{mg/minute}$$

$$V_{req} = 4\left[\frac{288.3\,\text{mg/minute}}{10\,\text{mg/m}^3}\right] \times \frac{35.3\,\text{ft}^3}{\text{m}^3} = \frac{4071\,\text{ft}^3}{\text{minute}}$$

Attempt to reduce the ventilation requirements by reducing generation rate (always the first priority), improving the mixing of fresh air with the contaminated air, using a less volatile, lower toxicity solvent and a lower amount of solvent, and improving the industrial hygiene work practices. Dilution ventilation is unacceptable for high toxicity chemicals, including carcinogens (e.g., benzene, carbon disulfide, dioxane, nitrobenzene), teratogens. (Refer to the latest edition of *Industrial Ventilation—a Manual of Recommended Practices* (the American Conference of Governmental Industrial Hygienists) for general dilution ventilation-restricted chemicals.)

339. Direct-reading, "real time" random spot breathing zone air sampling of a worker for NO_2 gas during 8 hours gave the following results: 1.3, 0.2, 0.6, 8.1, 15.6, 1.9, 0.5, 0.1, and 27.3 ppm$_v$. What is the geometric mean of these test results? You may assume the air-sample test results are log-normally distributed.

Test Result (ppm$_v$)	Log of Test Result
1.3	0.1139
0.2	−0.6990
0.6	−0.2218
8.1	0.9085
15.6	1.1931
1.9	0.2788
0.5	−0.3010
0.1	−1.0000
27.3	1.4361
	$\Sigma = 1.7086$

Arithmetic average (mean) of Σ of logs = 1.7086 ppm$_v$ NO$_2$/9 = 0.1898 ppm$_v$ NO$_2$. This, however, is not the arithmetic mean of the test results.

antilog of 0.1898 = geometric mean = $10^{0.1898}$ = 1.55 ppm$_v$ NO$_2$

Answer: Geometric mean = 1.6 ppm$_v$ NO$_2$ gas. That is, 50% of the NO$_2$ gas concentrations are expected to be above 1.6 ppm$_v$ and 50% below. Note that the arithmetic mean of the nine results is 6.2 ppm$_v$. The arithmetic mean (average) is not normally a meaningful and predictive statistic for air pollutant concentrations. The geometric mean is one of the most valuable descriptive statistics.

340. A low-flow personal air-sampling pump was calibrated using a 100 mL burette and required 51.3 seconds for the soap film bubble to traverse 78 milliliters. What was the air-sampling rate in liters/minute?

$$\frac{51.3\,\text{seconds}}{60\,\text{seconds/minute}} = 0.855\,\text{minute}$$

$$\frac{78\,\text{milliliters}}{1000\,\text{mL/liter}} = 0.078\,\text{liter}$$

$$\text{Air flow rate} = \frac{0.078\,\text{liter}}{0.855\,\text{minute}} = \frac{0.091\,\text{L}}{\text{minute}} = \frac{91\,\text{mL}}{\text{minute}}$$

Answer: 0.091 liter of air per minute. This low air flow rate does not exceed the recommended maximum rate of 100 mL air/minute for small charcoal tubes.

341. 6.6 liters of chlorine dioxide gas are inadvertently released into an empty, sealed 12' × 30' × 10' laboratory. After complete mixing, what is ClO$_2$ gas concentration?

$$(12' \times 30' \times 10') \times \left[\frac{28.32\,\text{L}}{\text{ft}^3} \right] = 101{,}952\,\text{L}$$

$$\text{ppm}_v = \frac{6.6\,\text{liters}}{101{,}952\,\text{liters}} \times 10^6 = 64.7\,\text{ppm}_v\ \text{ClO}_2$$

Since this is a highly reactive oxidizing gas, the concentration will decay in time as this gas reacts with reducing agents and organic materials in the room and in air. Do not enter this room! It is too dangerous without SCBA or until exhaust and dilution ventilation and confirmatory air sampling demonstrate that the ClO_2 gas level is <0.05 TLV or, better, at nondetectable levels.

342. The air velocity in an 8-inch diameter duct is 2500 fpm. What is the new duct velocity as this air passes into a constricted 6-inch diameter duct opening?

Area of circle $= \pi\, r^2$
6" Ø area $= 0.1963\,\text{ft}^2$
8" area $= 0.3491\,\text{ft}^2$
$V_1 \times A_1 = V_2 \times A_2$, where $V_1 =$ air velocity in duct with area A_1, and $V_2 =$ air velocity in duct with area A_2

$$V_2 = \frac{V_1 \times A_1}{A_2} = \frac{2500\,\text{fpm} \times 0.3491\,\text{ft}^2}{0.1963\,\text{ft}^2} = \frac{4446\,\text{feet}}{\text{minute}}$$

343. An open-top beaker containing 100 mL of liquid chlorine rests on the edge of a laboratory bench. The Cl_2 gas concentration 9 inches below the edge of the bench is 10,000 ppm$_v$ (1%), and the air and gas mixture at this point has a specific gravity of 1.015 (see Problem 8). What is the settling rate of this Cl_2 gas and air mixture?

$$V_s = \sqrt{\frac{2\,g\,(\text{SG} - 1)\,h}{\text{SG}}},$$

where
   $V_s =$ the settling rate of the gas or vapor mixture, ft/sec
   $g =$ gravity, 32.2 ft/sec^2
   $h =$ distance from source, ft
   SG $=$ specific gravity of the air and gas or vapor mixture relative to the specific gravity or density of air (unitless)

$$V_s = \sqrt{\frac{2 \times (32.2\,\text{ft}/\text{sec}^2) \times (1.015 - 1) \times 0.75\,\text{ft}}{1.015}} = \frac{0.845\,\text{feet}}{\text{second}}$$

Answer: 0.845 feet per second = 50.7 feet per minute, essentially identical to air velocity that is characteristic of "still" air (<50 fpm). *Note*: this equation cannot be used for gases or vapors farther than 1 ft from their emission source or when air currents disrupt the vapor or gas cascade.

344. An amateur photographer carelessly poured 1 liter of 3% sulfuric acid solution (vol/vol) into an open developing tray in a home basement darkroom. She leaves not recognizing her mistake. The tray contained 500 mL of 60 mg sodium sulfide/mL solution. Her unventilated darkroom is 5' × 10' × 8'. Which toxic gas evolved, and in what quantity, if 30% of the gas dissolves in the solution? What is the average concentration of the gas in the darkroom's air after mixing?

$$Na_2S + xs \ H_2SO_4 + H_2O \rightarrow Na_2SO_4 + H_2S\uparrow + H_2O$$

Observation of relative amounts of reactants (an excess of H_2SO_4) indicates there will be a stoichiometric reaction: that is, quantitative conversion of sodium sulfide into a strong acid solution to hydrogen sulfide gas.

$$5' \times 10' \times 8' = 400 \, ft^3 = 11.33 \, m^3$$

$$500 \, mL \times (60 \, mg \ Na_2S/mL) = 30,000 \, mg \ Na_2S$$

$$Molecular \ weight \ Na_2S = 78.04 \, grams/mole$$

$$Molecular \ weight \ H_2S = 34.08 \, grams/mole$$

$$Moles = \frac{grams}{molecular \ weight} = \frac{30 \, g \ Na_2S}{78.04} = 0.384 \, mole$$

Therefore, $0.7 \times 0.384 = 0.269$ mole of H_2S was released (i.e., 30% is dissolved).

grams H_2S released = molecular weight of H_2S × moles released

$$= \frac{34.08 \, grams}{mole} \times 0.269 \, mole = 9.17 \, g \ H_2S$$

$$ppm = \frac{(mg/m^3) \times 24.45}{molecular \ weight} = \frac{(9170 \, mg \ H_2S/11.33 \, m^3) \times 24.45}{34.08}$$
$$= 581 \, ppm \ H_2S$$

H_2S is a potent chemical asphyxiant that produces respiratory paralysis. This gas concentration greatly exceeds the 15 ppm$_v$ ACGIH STEL TLV® for H_2S. The IDLH for H_2S is 100 ppm$_v$; a few inhalations of 581 ppm$_v$ H_2S could be fatal. Large open pans of dangerous liquid chemicals in unventilated darkrooms are an invitation for a toxic gas release incident.

This example demonstrates the release of a *de novo* toxicant, that is, one that is generated and released by inadvertent mixture of two or more other chemicals. The generation is usually unforeseen by those not educated and trained in this hazard recognition and risk management techniques and practices. Careless mixing of household chemicals has resulted in inhalation fatalities, severe burn injuries, and life-long respiratory disability.

Examples of *de novo* toxicants are

Chemical Reaction	*De Novo* Gas and/or Mist Toxicant
Water-soluble sulfide salt + acid	H_2S (hydrogen sulfide)
Water-soluble cyanide salt + acid	HCN (hydrogen cyanide)
Water-soluble hypochlorite + acid	Cl_2 (chlorine), ClO_2 (chlorine dioxide)
Water-soluble hypochlorite + ammonia	1–3 chloramines (e.g., NH_2Cl)
Chlorinated solvent + high heat	$COCl_2$ (phosgene), HCl, Cl_2, ClO_2
Organic material + low oxygen	CO (carbon monoxide)
Adding H_2O to concentrated H_2SO_4	Violent eruption of hot acid + contents

345. A flammable solvent with toxic properties has a molecular weight of 78. What is its 100% vapor density in relation to air?

$$\text{vapor density, a ratio} = \frac{\text{molecular weight of vapor or gas}}{\text{composite molecular weight of air}} = \frac{78}{29} = 2.69$$

Answer: This vapor is approximately 2.7 times denser than air. The "apparent molecular weight" of gases that comprise air is nearly 29. Vapor density ratios are reported at equilibrium temperature under atmospheric conditions. Unequal or changing conditions can appreciably change density of any vapor or gas and their mixtures. As the vapor mixes with air, the relative density of the mixture approaches that of air itself (1.00, unitless).

346. A standard plumber's torch contains 1 pound of propane. An explosion leveled a mobile home after installation of bathroom sink's plumbing. An empty torch was found in the rubble. Litigation ensued with one side claiming propane gas was the explosion fuel (resident), and other side (plumber) alleging vapors from one ounce of fingernail polish remover (acetone) was the fuel. What appears to be the most plausible cause of explosion? Assume the ignition source is unknown, explosion originated in a closed unventilated bathroom (6' × 8' × 8'), and no one was present at the time of the explosion.

Assume worst-case scenarios, starting with propane:

$$6' \times 6' \times 8' = 288\,\text{ft}^3 = 8.16\,\text{m}^3$$

Assume an empty room with cabinets, counter, sink, and toilet not yet installed.

$$1\,\text{lb propane} = 454.6\,\text{grams}$$

$$\text{ppm} = \frac{(\text{mg/m}^3) \times 24.45}{\text{molecular weight}} = \frac{(454{,}600 \text{ mg/8.16 m}^3) \times 24.45}{44}$$

$$= 30{,}957 \text{ ppm} \cong 3.1\% \text{ gas in air}$$

Since the LEL for propane in air is 2.4%, explosion of propane gas is plausible if almost an entire cylinder was discharged into the air of the unventilated bathroom.

Molecular weight and density of acetone, respectively = 58.1 and 0.79 g/mL

$$\text{ppm} = \frac{(\text{mg/m}^3) \times 24.45}{\text{molecular weight}}$$

$$= \frac{((790 \text{ mg/mL})/8.16 \text{ m}^3) \times (29.57 \text{ mL/ounce}) \times 24.45}{58.1}$$

$$= 1205 \text{ ppm} \cong 0.12\%$$

Since the concentration of acetone vapor was far below its LEL of 2.5%, the fingernail polish remover solvent cannot be the cause of this explosion. We can speculate on the source of ignition in the unoccupied residence. If we postulate the pilot light on a remote water heater or kitchen stove, we see the propane gas concentration is now below its LEL. Ignition sources are many, and, since there is no dispute that there was an explosion, the ignition source is moot.

Answer: In absence of other information, the most likely cause of explosion was accumulation of propane gas above its LEL and below its UEL in the presence of an ignition source. Moreover, since Bob, the plumber, said he could not recall if he closed the torch valve on the fresh cylinder, and Sally, the mobile home owner, stated in her deposition that she never used more than $\frac{1}{4}$ ounce of polish remover at a time, explosion fuel, to a reasonable degree of scientific certainty, was propane gas.

347. An isokinetic stack sample of 98.4 cubic feet of air was collected for sulfuric acid mist in 100 mL of slightly acidic (nonvolatile acid), unbuffered, distilled H_2O in a Greenburg–Smith impinger. If the pH of the solution decreased from an initial pH of 4.3 to pH of 2.1 after sampling, how much sulfuric acid mist was collected? Assume 100% ionization of sulfuric acid (*see notes on next page).

$$98.4 \text{ ft}^3 \times \frac{28.32 \text{ L}}{\text{ft}^3} = 2786.7 \text{ L} = 2.787 \text{ m}^3$$

$$H_2SO_4 \leftrightarrow 2H^+ + SO_4^=$$

A solution of pH 2.1 contains $10^{-2.1}$ moles of hydrogen ion per liter. Similarly, since the initial pH was 4.3, the collection solution hydrogen ion concentration was $10^{-4.3}$ moles H^+ per liter before air sampling began.

$$10^{-2.1} = 0.00794 \text{ mole H}^+/\text{L}$$

$$10^{-4.3} = 0.00005 \text{ mole H}^+/\text{L}$$

Increase in hydrogen ion = $(0.00794 - 0.00005)$ moles/L = 0.00789 mole H$^+$/L

$$(0.00789 \text{ mole H}^+/\text{L}) \times 0.1 \text{ L} = 0.000789 \text{ mole H}^+$$

Since 1 mole of H$_2$SO$_4$ produces 2 moles of H$^+$:

$$(0.000789 \text{ mole}/2) = 0.0003945 \text{ mole H}_2\text{SO}_4 \text{ was collected.}$$

Molecular weight H$_2$SO$_4$ = 98.07 grams/mole

$$(98.07 \text{ grams/mole}) \times 0.0003945 \text{ mole} = 0.03869 \text{ gram H}_2\text{SO}_4$$

$$\frac{38.69 \text{ mg H}_2\text{SO}_4}{2.787 \text{ m}^3} = \frac{13.88 \text{ mg H}_2\text{SO}_4}{\text{m}^3}$$

*Actually, the result is low since sulfuric acid, as a diprotic acid, ionizes in two stages. The first stage: H$_2$SO$_4$ $\leftrightarrow$ H$^+$ + HSO$_4^-$ is essentially 100% ionization, that of a very strong acid. The second stage: HSO$_4^-$ $\leftrightarrow$ H$^+$ + SO$_4^=$ is that of a weak acid where ionization constant is 1.3×10^{-2}. Using quadratic equation and exotic calculations, one can determine precisely what sulfuric acid mist concentration was. Practically, it would be far easier and maybe more accurate to use a specific ion electrode for sulfate ion and then calculate the amount of sulfuric acid by using the ratio of the molecular weights of sulfate ion to sulfuric acid. Total hydrogen ion concentration at equilibrium is the sum of the concentrations due to both ionization stages. For example, for a 0.01 molar solution of sulfuric acid, hydrogen ion is 0.0147 molar (0.01 from the first stage of ionization, and 0.0047 from the ionization of hydrogen sulfate ion). The pH of this solution would be—log [H+] = −log (0.01 + 0.0047) = −log 0.0147 = pH 1.83.

If this is unprotected steel or "white metal" stack and pollution control equipment, severe corrosion is imminent. Perhaps fiberglass-reinforced PVC stack leading to a high efficiency caustic scrubber should be quickly considered and installed.

348. A process releases sulfur dioxide gas at a steady rate into an occupied work area. What is the generation rate of the gas in cubic feet per hour if 88,000 cubic feet of air per minute is needed to dilute the SO$_2$ to 2 ppm$_v$?

$$\frac{88,000 \text{ ft}^3}{\text{minute}} = \left[\frac{\text{generation rate, cfm}}{2 \text{ ppm}} \right] \times 10^6, \quad \text{or}$$

$$\text{generation rate, cfm} = \frac{(88,000\,\text{cfm}) \times 2\,\text{ppm}}{10^6} \times \frac{60\,\text{minutes}}{\text{hour}}$$

$$= \frac{10.56\,\text{ft}^3\,SO_2}{\text{hour}}$$

Answer: 10.56 cubic feet of sulfur dioxide gas are released per hour. If better control of the leak could be achieved, substantial reduction in the dilution air can be considered. Control at the source is always desirable over "capturing the horse once she/he's out of the barn." Good enclosure and confinement certainly beats using a lasso. Since gaseous molecules move around, they are difficult to capture once free. Tempering 88,000 scfm is pricey in hot and cold climates.

349. A composite mixture of shredded plastic waste contains 9% chlorine primarily from polyvinyl chloride polymers and copolymers. How much hydrochloric acid gas is released from 100% quantitative, stoichiometric combustion of this PVC waste?

$$RH\text{-}Cl + O_2 \rightarrow RO_2 + H_2O + HCl$$

Correct for conversion of Cl to HCl:

$$\frac{\text{Molecular weight HCl}}{\text{Molecular weight Cl}} = \frac{36.5}{35.5} = 1.028$$

Base the calculations on 100 pounds of plastic waste: 100 pounds × 0.09 × 1.028 = 9.25 pounds of HCl.

Answer: 9.25 pounds of HCl gas are released from 100% combustion of every 100 pounds of composite plastic waste assuming stoichiometric pyrolysis.

350. The vapor pressure of the very highly volatile diethyl ether is 401 mm Hg at 18°C. What is its vapor pressure at 32°C?

The molar heat of vaporization (ΔH_{vap}) of a chemical is defined as the energy (in kilojoules, normally) required to vaporize 1 mole of liquid phase of the chemical. For diethyl ether ("ether"), $\Delta H_{vap} = 26.0\,\text{kJ/mole} = 26,000\,\text{Joules/mole}$. A solution requires rearrangement of the Clausius–Clapeyron equation:

$$\ln P = \frac{\Delta H_{vap}}{RT} + C, \quad \text{where } C \text{ is a constant}$$

$$\ln \frac{P_1}{P_2} = \frac{\Delta H_{vap}}{R} \times \frac{T_1 - T_2}{T_1 \times T_2}$$

$P_1 = 401\,\text{mm Hg}$
$P_2 = ?$
$T_1 = 18°C = 291\,\text{K}$
$T_2 = 32°C = 305\,\text{K}$

$$\ln\frac{401\,\text{mm Hg}}{P_2} = \frac{26{,}000\,\text{Joules/mole}}{8.314\,\text{Joules/K-mole}} \times \left[\frac{291\,\text{K} - 305\,\text{K}}{(291\,\text{K})\,(305\,\text{K})}\right]$$

Taking the antilog of both sides: $\dfrac{401\,\text{mm Hg}}{P_2} = 0.6106$

$$P_2 = 657\,\text{mm Hg}$$

Answer: The vapor pressure of diethyl ether at 32°C is 657 mm Hg, close to its boiling point of 94°F at 760 mm Hg (sea-level barometric pressure).

351. The half-life of a chemical in air due to atmospheric oxidation is 4.7 hours. If the initial air concentration of this chemical was $19.8\,\text{mg/m}^3$, how much is left in the air after 12 half-lives?

    We can generalize the fraction of air contaminant left after n half-lives as $[1/2]^n$.

$$\frac{19.8\,\text{mg}}{\text{m}^3} \times \left[\frac{1}{2}\right]^{12} = \frac{19.8\,\text{mg}}{\text{m}^3} \times 0.000244 = \frac{0.00483\,\text{mg}}{\text{m}^3}$$

Answer: 4.83 micrograms per cubic meter after ($4.7\,\text{hours} \times 12 =$) 56.4 hours.

352. What is the half-life for an unstable air contaminant that decays by following first-order kinetics if the initial gas concentration is $367\,\text{ppm}_v$ and the concentration after 39.3 hours is $1.6\,\text{ppm}_v$? Assume that there is no further gas generation when timing begins, and that the loss is due entirely to chemical change and not due to loss by dilution ventilation or other means.

    This problem is similar to the preceding Problem 351.

$$367\,\text{ppm}_v,\ C_o \rightarrow 1.6\,\text{ppm}_v,\ C,\ \text{after 39.3 hours}$$

$$(367\,\text{ppm}) \left[\frac{1}{2}\right]^n = 1.6\,\text{ppm}_v$$

$$(0.5)^n = \frac{1.6\,\text{ppm}_v}{367\,\text{ppm}_v} = 0.00436$$

$$n = \frac{\log 0.00436}{\log 0.5} = \frac{-2.3605}{-0.301} = 7.84 \text{ half-lives}$$

$$\frac{39.3 \text{ hours}}{7.84 \text{ half-lives}} = \frac{5.01 \text{ hours}}{\text{half-life}}$$

Answer: $T_{1/2} = 5.0$ hours or, after 1 hour, 184 ppm$_v$ contaminant remains.

353. The OSHA PEL for lead (on a mass-to-volume basis) is 50 micrograms of lead per cubic meter of air. What is the PEL for lead if it was expressed on a mass to mass basis?

     Use the mass of lead to the mass of air at NTP containing the lead aerosol.

$$\frac{50 \text{ mcg Pb}}{m^3} \times \frac{M^3}{35.3 \text{ ft}^3} \times \frac{\text{ft}^3}{0.075 \text{ lb}} \times \frac{\text{lb}}{454 \text{ g}} \times \frac{g}{10^6 \text{ mcg}}$$

$$= \frac{50 \text{ mcg Pb}}{1202 \times 10^6 \text{ mcg air}} = \frac{1 \text{ mcg Pb}}{2.4 \times 10^7 \text{ mcg air}} = \frac{1 \text{ Pb}}{2.4 \times 10^7 \text{ air}} = 4.2 \times 10^{-8}$$

Answer: 1 part of lead by weight in approximately 24,000,000 parts of air by weight. This equals 42 parts of lead per billion parts of air, both by weight.

354. Determine the internal pressure exerted by 498 grams of sulfur pentafluoride gas at 95°F in a 6.4 L steel pressure vessel. Molecular weight of S_2F_{10} is 254.1 grams per gram-mole. IDLH is 1 ppm$_v$.

$$\frac{498 \text{ grams}}{254.1 \text{ grams/mole}} = 1.96 \text{ moles}$$

$$95°F = 35°C$$

$$P = \frac{nRT}{V} = \frac{(1.96 \text{ moles})(0.0821 \text{ L-atm/K-mole})(35°C + 273 K)}{6.4 L}$$

$$= 7.74 \text{ atmospheres}$$

355. Air bags in automobiles are rapidly detonated during a crash from decomposition of the propellant explosive sodium azide, NaN_3, according to the equation:

$$NaN_3 \text{ (solid)} \xrightarrow{\textbf{BOOM!}} 2\,Na \text{ (solid aerosol)} \uparrow + 3\,N_2 \text{ (g)} \uparrow$$

The released nitrogen gas rapidly inflates the "air" bag. Determine the volume of N_2 gas generated at 30°C and 640 mm Hg upon the decomposition of 70 grams of NaN_3 assuming quantitative stoichiometry.

Two moles sodium azide $\Rightarrow$ three moles nitrogen

$$30°C + 273\,K = 303\,K$$

$$\text{Moles}\,N_2\,\text{gas} = 70\,g\,NaN_3 \times \frac{1\,\text{mole}\,NaN_3}{65.02\,g\,NaN_3} \times \frac{3\,\text{mole}\,N_2}{2\,\text{moles}\,NaN_3}$$

$$= 1.615\,\text{moles of}\,N_2$$

$$V = \frac{nRT}{P} = \frac{(1.615\,\text{moles})\,(0.0821\,\text{L-atm/K-mole})\,(303\,K)}{640\,\text{mm Hg}/760\,\text{mm Hg}} = 47.7\,L\,N_2$$

The literature reports 100 g of sodium azide produces 1.64 moles of nitrogen gas. Because 70 grams sodium azide theoretically generates 1.615 moles of nitrogen gas, 100 grams should release 2.31 moles of N_2. However, since 1.64 moles of nitrogen are generated, the reaction is about 71% quantitative. That is, it appears there is about 29% undetonated or partial, nongaseous explosion reaction products (most likely, NaN_3 aerosol + ambient water vapor $\Rightarrow Na \Rightarrow NaOH \Rightarrow Na_2CO_3$ and $NaHCO_3$).

356. In the previous problem, assume the released sodium quickly reacts with moisture in the air to form a sodium hydroxide aerosol. Assuming dispersion into two cubic meters of car's interior atmosphere, calculate the average concentration of NaOH in the aerosol phase. Any unreacted sodium aerosol in contact with moist mucous membranes will also produce caustic, irritating NaOH powder. In the risk–benefit analysis, balance inhalation and ocular exposure to this caustic aerosol with main safety benefits of automobile air bags.

1 mole of sodium azide produces 1 mole of sodium:

$$\frac{70\,g\,NaN_3}{65.02\,g\,NaN_3/\text{mole}} = 1.077\,\text{mole}\,NaN_3 \Rightarrow 1.077\,\text{mole}\,Na$$

$$2Na + 2H_2O \rightarrow 2NaOH + H_2$$

Therefore, a maximum of 1.077 moles of NaOH are produced.

$$\frac{40\,\text{grams}\,NaOH}{\text{mole}} \times 1.077\,\text{moles} = 43.08\,\text{grams of}\,NaOH$$

$$\frac{43.08\,g\,NaOH}{2\,m^3} \times \frac{1000\,mg}{\text{gram}} = \frac{21,540\,mg\,NaOH}{m^3}$$

Answer: 21,540 mg $NaOH/m^3$ is momentarily intensely irritating, but one's life might have been saved by deployment of the "air" bag, and the duration of exposure is very brief. Regardless, Na and NaOH particulate aerosol in the eyes, especially, and in contact with moist mucous respiratory tract membranes could cause significant injuries if not immediately flushed with water. Those wearing contact lenses are particularly susceptible to corrosive alkaline chemical burns of the conjunctivae, corneas, and eyelids.

357. Consider the hydrogen gas generated in the previous problem. Could this, under the worst of circumstances, result in an explosive atmosphere? Assume that the driver was smoking as the air bag detonated and that his or her cigarette was not extinguished (or that there were other sources of ignition). Further assume that the released hydrogen gas is mixed with ambient air into in a 500 L volume.

   Each mole of Na produced yields 0.5 mole H_2. Regardless, it is unlikely that sodium aerosol would react completely, in even the most humid atmosphere, to provide a stoichiometric, rapid conversion to NaOH and hydrogen gas. Assuming the direst:

$$1.077 \text{ mole sodium} \Rightarrow 0.539 \text{ mole hydrogen gas}$$

$$\frac{2 \text{ grams } H_2}{\text{gram-mole}} \times 0.539 \text{ mole} = 1.078 \text{ grams of hydrogen gas}$$

$$ppm_v = \frac{(mg/m^3) \times 24.45}{\text{molecular weight}} = \frac{(1078 \text{ mg}/0.500 \text{ m}^3) \times 24.45}{2}$$
$$= 26,357 \text{ ppm}_v = 2.64\% \text{ hydrogen}$$

   LEL for $H_2 = 4\%$. UEL for $H_2 = 75\%$. The hydrogen gas concentration, during one of the worst scenarios, is below the LEL and far below stoichiometric mid-point for producing a maximum explosion blast pressure. A hydrogen gas explosion does not appear likely under these conditions. Note that this calculation is based on a detonation at sea-level barometric pressure whereas the original problem (355) was based on an atmospheric pressure of 640 mm Hg. One can easily make any corrections for various altitudes.

358. A 10,000 gallon airtight carbon steel tank is tightly sealed when the atmospheric pressure is 760 mm Hg. The tank's interior is not protected by paint, an oil film, or any other coating. The tank contains only air and atmospheric moisture. After 10 weeks, the interior pressure of the tank is 630 mm Hg without imploding. Assume the temperature remains constant at 25°C. What is happening?

Atmospheric oxygen is reacting with iron in the steel to form iron oxides. As O_2 is consumed, its partial pressure is reduced, and the overall pressure in the tank is reduced. Rust forms on the steel surfaces.

$$3Fe \text{ (s)} + 2O_2 \text{ (g)} \rightarrow Fe_3O_4 \text{ (s)}$$

$$4Fe \text{ (s)} + 3O_2 \text{ (g)} \rightarrow 2Fe_2O_3(s)$$

There is a concurrent percentage increase in inert gases (nitrogen and argon) in the atmosphere as oxygen is consumed. The amount of rust formed would not appreciably change the gas volume of this tank and may be disregarded in the following calculations. We want to know what the new oxygen concentration is and if it presents an inhalation health hazard if unprotected workers enter the tank.

The number of moles of oxygen consumed is related to the drop in pressure, or:

Initial tank pressure = (760 mm Hg/760 mm Hg) = 1.000 atmosphere

Final tank pressure = (630 mm Hg/760 mm Hg) = 0.829 atmosphere

1.000 atmosphere − 0.829 atmosphere = 0.171 atmosphere that corresponds to the chemical consumption of O_2 (oxidation of iron).

$$25°C + 273\,K = 298\,K$$

$$10,000 \text{ gallons} = 37,854 \text{ liters}$$

$$n = \frac{PV}{RT} = \frac{(0.171\,atm)\,(37,854\,L)}{(0.0821\,\text{L-atm/K-mole})\,(298\,K)}$$

$$= 264.6 \text{ moles of } O_2 \text{ consumed}$$

$$264.4 \text{ moles of } O_2 \times \frac{32 \text{ grams } O_2}{\text{mole}} = 8461 \text{ grams of } O_2 \text{ were consumed.}$$

The tank originally contained 10,000 gallons of air at 21% oxygen by volume. The original molar concentration of oxygen in the tank was

$$n = \frac{PV}{RT} = \frac{(0.21\,\text{atmosphere})\,(37,854\,L)}{(0.0821\,\text{L-atm/K-mole})\,(298\,K)} = 324.9 \text{ moles of } O_2$$

The oxygen in the tank was reduced to $[1 − (264.4\,\text{moles}/324.9\,\text{moles})] \times 100 = 18.6\%$ of the initial oxygen concentration.

Answer: The tank atmosphere has become substantially oxygen deficient, that is, 21% $O_2 \times 0.186 = 3.9\%$ O_2 by volume. Inhalation of this atmosphere would cause immediate collapse and death within minutes if the entrant was not rescued and given immediate CPR and EMS care. Carefully follow OSHA confined space entry procedures (29 *CFR* 1910.146) and respiratory protection (29 *CFR* 1910.134).

359. A diver wearing a self-contained under water breathing apparatus (SCUBA) is 20 feet deep in sea water. If, without breathing, he quickly rose to the water surface, what would happen to the air in his lungs?

Sea water is denser than fresh water: 1.03 g/mL versus 1.00 g/mL. Therefore, the pressure exerted by a 33-feet sea water column is equivalent to 1 atmosphere pressure. Since pressure increases with increasing depth, at 66 feet the pressure of the water is equivalent to 2 atmospheres, 99 feet $\approx$ 3 atmospheres, and so on. As ascent started at 20 feet below the surface, the total decrease in the pressure for this depth is [20 feet/33 feet] $\times$ 1 atmosphere = 0.606 atmosphere. When the diver reaches the surface, the air volume trapped in his lungs would have increased by a factor of $(1 + 0.606)$atm/1 atm = 1.606 times. This rapid expansion can fatally rupture the lung's delicate membranes. Development of air embolism is another serious possibility where expanded air in lungs is squeezed into the pulmonary capillaries.

Answer: Air in diver's lungs would expand, without his breathing, 1.6 times with a possible air embolism, pulmonary rupture, coma, and death.

360. A diver will descend to a water depth where the total pressure is equivalent to 2 atmospheres. What should the oxygen content of his SCUBA air be?

When P_T is the total gas pressure, the oxygen partial pressure, P_{O_2}, is given by

$$P_{O_2} = X_{O_2} P_T = \frac{n_{O_2}}{n_{O_2} + n_{N_2}} \times P_T.$$

n_{N_2} is the partial pressure from inert gases (nitrogen, argon, CO_2). However, since the gas volume is directly proportional to the number of moles of gas present (at constant temperature and pressure): $P_{O_2} = V_{O_2}/(V_{O_2} + V_{N_2}) \times P_T$. Thus, composition of air is 21% oxygen gas by volume and 79% inert gases by volume. When a diver is submerged, composition of air must be changed. At a depth equivalent to 2.0 atmospheres, the oxygen content of the air should be reduced to 10.5% by volume to maintain the same partial pressure of 0.21 atmosphere:

$$P_{O_2} = 0.21\,\text{atmosphere} = \frac{V_{O_2}}{V_{O_2} + V_{N_2}} \times 2.0\,\text{atmospheres,} \quad \text{or}$$

$$\frac{V_{O_2}}{V_{O_2} + V_{N_2}} = \frac{0.21 \text{ atmosphere}}{2.0 \text{ atmosphere}} = 0.105, \quad \text{or } 10.5\% \text{ oxygen gas by volume.}$$

361. A stainless-steel tank leaks formaldehyde gas at the rate of 0.3 mL/hour when the internal gas pressure is 3 atmospheres. What would the leakage effusion rate be if the tank contained vinyl chloride gas?

$$\frac{\text{HCHO leakage rate}}{\text{VC leakage rate}} = \sqrt{\frac{\text{VC molecular weight}}{\text{HCHO molecular weight}}} = \sqrt{\frac{62.49 \text{ grams/mole}}{30.03 \text{ grams/mole}}}$$

$$= \sqrt{2.0809} = 1.443$$

$$\text{VC leakage rate} = \frac{\text{HCHO leakage rate}}{1.443} = \frac{0.30 \text{ mL/hour}}{1.443} = \frac{0.208 \text{ mL}}{\text{hour}}$$

Answer: 0.208 mL of vinyl chloride gas effuses per hour.

362. Dinitrogen pentoxide, N_2O_5, decomposes according to first-order reaction kinetics. At 50°C, the rate constant is approximately 0.00054/second. This linear reaction dissociation of 1 mole of nasty gas into 2 moles of another evil gas is

$$2N_2O_5 \rightarrow 4NO_2 + O_2$$

What is the N_2O_5 concentration after 17.3 minutes if the initial gas concentration was 3.9 ppm_v? How long does it require for the initial gas concentration to decay to 0.4 ppm_v? How long will it take to convert 50% of the initial gas concentration?

$$\ln \frac{\text{conc}_{\text{initial}}}{\text{conc}_{\text{final}}} = kt = \ln \frac{3.9 \text{ ppm}}{\text{conc}_{\text{final}}}$$

$$= (0.00054/\text{second})\left[17.3 \text{ minutes} \times \frac{60 \text{ seconds}}{\text{minute}} \right]$$

$$\ln \frac{3.9 \text{ ppm}}{\text{conc}_{\text{final}}} = 0.5605$$

$$\frac{3.9 \text{ ppm}}{\text{conc}_{\text{final}}} = e^{0.5605} = 1.75$$

$$\text{conc}_{\text{final}} = \frac{3.9 \text{ ppm}}{1.75} = 2.23 \text{ ppm } N_2O_5$$

$$\ln \frac{3.9 \text{ ppm}}{0.4 \text{ ppm}} = (0.00054/\text{second})t$$

$$\ln 9.75 = (0.00054/\text{second}) \, t$$

$$t = \frac{\ln 9.75}{0.00054/\text{second}} = \frac{2.277}{0.00054/\text{second}} = 4217 \text{ seconds} = 70.3 \text{ minutes}$$

$$t = \frac{1}{k} \ln \frac{\text{conc}_{\text{initial}}}{\text{conc}_{\text{final}}} = \frac{1}{0.00054/\text{second}} \times \ln \frac{1}{0.5} = 1852 \text{ seconds} \times \ln 2$$

$$= 1852 \times 0.693 = 1283 \text{ seconds} = 21.4 \text{ minutes}$$

Answers: 2.23 ppm. 70.3 minutes. 21.4 minutes.

363. An industrial hygienist and a safety engineer collaborate to determine the bursting temperature of a steel gas vessel. If this vessel contains a fixed mass of gas at 25°C at 3 atmospheres, and it can withstand a pressure of 20 atm, what is the maximum temperature to which this vessel can be increased before it bursts?

Apply Gay–Lussac's law: at constant volume, the pressure of a mass of gas varies directly with the absolute temperature:

$$\frac{P_i}{T_i} = \frac{P_f}{T_f}, \quad \text{or } T_f = \frac{P_f T_i}{P_i} = \frac{(20 \text{ atm})(273 \text{ K} + 25°\text{C})}{3 \text{ atm}}$$

$$= \frac{(20 \text{ atm})(298 \text{ K})}{3 \text{ atm}} = 1987 \text{ K}$$

Answer: 1987 kevin, well above the melting point of steel at approximately 1380°C (1653 kevin). However, a safety factor is needed: either reduce maximum temperature rating and/or increase bursting strength of the gas vessel. Use higher temperature alloy.

364. Some commercial drain cleaners contain aluminum powder and NaOH. The following reaction occurs when a dry mixture of these two powders is poured into the water of a greasy, clogged drain:

$$2\text{NaOH} + 2\text{Al} + 6\text{H}_2\text{O} \rightarrow 2\text{NaAl(OH)}_4 + 3\text{H}_2 \uparrow + \Delta$$

Generated heat aids in melting grease. NaOH saponifies grease and fats. As the hydrogen is evolved, gas bubbles stir up solids that plug the drain. If 4 g of aluminum are added to a drain containing excess NaOH, could sufficient hydrogen gas be evolved in a 1-cubic foot volume above the drain to cause explosion if there was an ignition source? Atomic weight of Al is 26.98 grams/mole. Assume that the temperature above the drain is 25°C. Disregard any solubility of hydrogen gas into the standing water in the clogged drain.

$$1 \text{ mole of aluminum} \Rightarrow 1.5 \text{ moles of hydrogen gas.}$$

$$\frac{4.0 \text{ grams Al}}{26.98 \text{ grams/mole}} = 0.148 \text{ mole of Al}$$

$$0.148 \text{ mole Al} \times 1.5 = 0.222 \text{ mole of hydrogen gas}$$

$$0.222 \text{ mole} \times \frac{2 \text{ grams hydrogen}}{\text{mole}} = 0.444 \text{ gram H}_2$$

$$1 \text{ ft}^3 = 28.32 \text{ L}$$

$$\text{ppm} = \frac{(\text{mg/m}^3) \times 24.45}{\text{molecular weight}} = \frac{(444 \text{ mg}/0.02832 \text{ m}^3) \times 24.45}{2}$$
$$= 191,663 \text{ ppm} \cong 19.2\%$$

Answer: Under this nearly worst-case scenario, an explosion would occur because LEL of hydrogen gas is 4%. The UEL is 75%. Suppliers of these drain cleaners recommend immediately placing a container over the drain after the chemicals are added to contain any eruption of corrosive, hot alkali into one's face and eyes. This is good advice; however, chemical splash goggles and rubber gloves and a face shield must be worn before chemicals are added. Since an explosion could occur, ignition sources must be prohibited, and ventilation is necessary to dilute any hydrogen gas to at least 20% below its LEL. Furthermore, one might not be able to place a container over the drain quickly enough since the chemicals could react instantaneously.

Of all chemicals used to clean drains, concentrated sulfuric acid, in the author's view, must never be used because of its "wild card" behavior and incredibly high heat of dilution. Mechanical aids such as a plunger ("plumber's helper"), a drain snake, boiling water, and compressed air should always be tried before using corrosive chemicals. Reliance upon plumbers might be more prudent than using concentrated H_2SO_4, HCl, NaOH, and KOH. The average homeowner does not have sufficient chemical sophistication to use such highly hazardous agents.

365. The ozone gas molecules present in the stratosphere absorb much of the harmful radiation emitted from the sun. The typical pressure and temperature of ozone in the stratosphere are 0.001 atm and 250 kevin, respectfully.

How many ozone molecules are in each liter of air in these upper atmospheric conditions?

$$PV = nRT$$

$$n = \frac{PV}{RT} = \frac{(0.001\text{atm})(1.0\text{L})}{(0.0821 \text{ L-atm/mole-K})(250\text{K})} = 0.0000487 \text{ mole O}_3$$

$$0.0000487 \text{ mole O}_3 \times \frac{6 \times 10^{23} \text{ molecules O}_3}{\text{mole}} = 2.9 \times 10^{19} \text{ molecules O}_3/\text{L}$$

Answer: 2.9×10^{19} molecules of ozone gas per liter—a bunch.

366. What is the partial pressure of dioxane (in atm) at $12.0\,\text{ppm}_v$ vapor in air by volume if total barometric pressure is 730 mm Hg and temperature is 16°C?

$$P_{\text{dioxane}} = X_{\text{dioxane}} \times P_{\text{total}} = \frac{12}{10^6} \times 730 \text{ mm Hg} \times \frac{1 \text{ atm}}{760 \text{ mm Hg}}$$

$$= 1.15 \times 10^{-5} \text{ atm}$$

Answer: 0.0000115 atm partial pressure from dioxane vapor. The air temperature does not enter into these calculations. Refer to Dalton's law of partial pressures.

367. If an average adult inhales 320 mL of dioxane vapor at concentration in previous problem with each breath, how many molecules of dioxane are inhaled?

$$PV = nRT$$

$$n = \frac{PV}{RT} = \frac{(1.15 \times 10^{-5} \text{ atm})(0.32\text{L})}{(0.0821 \text{L-atm/mole-K})(289 \text{ K})} = 1.55 \times 10^{-7} \text{ mole}$$

$$1.55 \times 10^{-7} \text{ mole} \times \frac{6 \times 10^{23} \text{ molecules}}{\text{mole}} = 9.3 \times 10^{16} \text{ molecules of}$$
dioxane vapor per inhalation

368. A 55-gallon drum is splash filled with a volatile solvent. For each gallon of solvent added to the drum, 1 gallon of air that is nearly saturated with solvent vapors will be displaced. To reduce solvent vapor emissions into the ambient air, a dip pipe extending to within 2 inches of the bottom of the drum is used for filling. What percent of vapor saturation can be expected in the displaced air?

a. 130–150%—Slower filling produces super-saturation, that is, vapor plus mist.
b. 100%—There is no difference in vapor emissions over splash filling.
c. 80%
d. 50%
e. 30%
f. 10%

Answer: d. Several empirical studies have demonstrated reductions of vapor emissions by approximately 50% whenever fill tubes are used in place of splash filling of volatile solvents.

369. A breathing zone air sample was obtained from welder exposed to fume of barium fluoride, BaF_2, a fluxing agent. Air was sampled at an average rate of 1.64 L/m for 473 minutes of worker's 8-hour work shift. Filter contained 593 micrograms of barium. If this worker had no other exposure to fluoride fume, what were his TWAEs to barium and fluoride that day? Molecular weights of BaF_2 and Ba = 175.33 and 137.33 grams gram-mole^{-1}, respectively.

   473 minutes $\cong$ 8 hours (98.5% of the full work shift). Reduce exposure calculation by 1.5% assuming the remainder of the work shift was barium and fluoride free air.

$$\frac{1.64 \text{ liters}}{\text{minute}} \times 473 \text{ minutes} = 775.7 \text{ liters of air were sampled}$$

$$\frac{593 \text{ mcg Ba}}{775.7 \text{ L}} = \frac{0.764 \text{ mg Ba}}{\text{m}^3}$$

$$0.764 \text{ mg Ba/m}^3 \times 0.985 = 0.753 \text{ mg Ba/m}^3$$

$$\frac{\text{mol.wt.BaF}_2}{\text{mol.wt.Ba}} = \frac{175.33}{137.33} = 1.277$$

$$593 \text{ mcg Ba} \times 1.277 = 757.3 \text{ mcg BaF}_2$$

$$757.3 \text{ mcg BaF}_2 - 593 \text{ mcg Ba} = 164.3 \text{ mcg F}$$

$$\frac{164.3 \text{ mcg F}}{775.7 \text{ L}} = \frac{0.212 \text{ mg F}}{\text{m}^3}$$

$$0.212 \text{ mg F/m}^3 \times 0.985 = 0.209 \text{ mg F/m}^3$$

Clearly, exposure exceeds the TLV of 0.5 mg/m^3 for barium (soluble compounds). The TLV of 2.5 mg/m^3 for soluble fluoride compounds is not exceeded. The toxic effects of both are most likely not additive. Barium fluoride is sparingly soluble in water (0.12 g/100 cc at 25°C), so whether there is truly an overexposure would be a tough judgment call. Prudence would dictate that, since little is known about the chronic toxicity of barium fluoride, better industrial hygiene is needed to reduce his exposure to well below 0.5 mg Ba/m³.

370. Use the thermal mass balance calculation method to estimate outdoor air quantity when the temperature of the HVAC system return air is 76°F, the temperature of the mixed air is 72°F, and the temperature of the outdoor air is 42°F.

$$\text{Outdoor air (percent)} = \frac{T_{\text{returnair}} - T_{\text{mixedair}}}{T_{\text{returnair}} - T_{\text{outdoorair}}} \times 100$$

$$= \frac{76°F - 72°F}{76°F - 42°F} \times 100 = 11.8\%$$

371. Use the CO_2 measurement calculation method to estimate the outdoor air quantity when concentrations of CO_2 are 440 ppm$_v$ in the mixed air, 490 ppm$_v$ in the return air, and 325 ppm$_v$ in the outdoor air.

$$\text{Outside air (percent)} = \frac{C_S - C_R}{C_O - C_R} \times 100 = \frac{440\,\text{ppm} - 490\,\text{ppm}}{325\,\text{ppm} - 490\,\text{ppm}} \times 100$$

$$= \frac{-50\,\text{ppm}}{-165\,\text{ppm}} \times 100 = 30.3$$

Answer: 30.3% outside air.

372. What is the minimum free area for introducing outside air into a combustion device per 2000 BTU input?
  a. One square inch
  b. Two square inches
  c. Five square inches
  d. 10 square inches
  e. 27 square inches

Answer: a.

373. Painters sometimes, perhaps foolishly, mix a liquid insecticide mixture into paint that they apply on walls, ceilings, and trim. Presumably, any insects that later contact the dry paint absorb the pesticide and subsequently die. Chlorpyrifos has been used for this purpose. The TLV for this organophosphate insecticide, that has a "Skin" notation by the ACGIH TLV Committee, is 0.2 mg/m³. The molecular weight of Chlorpyrifos is 350.57, and its vapor pressure is 1.87×10^{-5} mm Hg at 25°C. A painter added three liquid ounces of an 11% mixture (weight/volume) of Chlorpyrifos to 1 gallon of paint

and painted the four walls and the ceiling of an infant's bedroom. The wall dimensions of the room were 8' × 10'.

If we assume the bedroom had no ventilation after the paint dried, what could have been maximum saturation concentration of Chlorpyrifos vapor? Was sufficient Chlorpyrifos in the paint to saturate the air with vapor? If the painter sprayed this paint, instead of brushing and rolling, and the average total mist concentration in his breathing zone was 1 mL of paint/m³ for the 1 hour it required to paint the room, what was his 8-hour TWAE to Chlorpyrifos mist? Assume that the paint was completely used to paint this bedroom; that is, 1 gallon of paint typically covers about 400 square feet {in this case: [4 (8' × 10')] + (10' × 10') = 420 ft²}. Assume that during paint spraying 1 mL of paint existed as a mist aerosol in each cubic meter of his or her breathing zone air, and the density of the paint was 1.4 g/mL.

$$\frac{1.87 \times 10^{-5}\,mm\,Hg}{760\,mm\,Hg} \times 10^6 = 0.0246\,ppm\ \text{Chlorpyrifos vapor at saturation}$$

$$mg/m^3 = \frac{ppm \times molecular\ weight}{24.45\,L/g\text{-mole}} = \frac{0.0246\,ppm \times 350.57\,g/mole}{24.45\,L/g\text{-mole}}$$

$$= 0.35\,mg/M^3$$

This concentration of Chlorpyrifos vapor did not exceed the TLV when calculated as an 8-hour TWAE; however, since the painter was in the room for only 1 hour, and it is highly unlikely that the Chlorpyrifos would yield a vapor saturation concentration in this brief time, she/he most probably, was not excessively exposed to vapor (if she/he had no skin contact).

$$\frac{11\,g}{100\,mL} \times 3\,ounces \times \frac{29.57\,mL}{ounce}$$

$$= 9.76\ \text{grams of Chlorpyrifos per gallon of paint}$$

$$Bedroom\,volume = 10' \times 10' \times 8' \times \frac{m^3}{35.3\,ft^3} = 22.67\ m^3$$

$$22.67\,m^3 \times \frac{0.35\,mg}{m^3} = 7.93\,mg\ \text{Chlorpyrifos vapor in the}$$
$$\text{bedroom atmosphere (at vapor saturation)}$$

The 9.76 grams of Chlorpyrifos in the gallon of paint greatly exceeds the 0.00793 gram of total vapor in the bedroom's saturated atmosphere. Therefore, a vapor saturation concentration was theoretically achievable, although highly unlikely.

$$\frac{9.76\,g}{gallon} \times \frac{1000\,mg}{g} \times \frac{gallon}{3785\,mL} = \frac{2.58\,mg}{mL}$$

$$\frac{(2.58\,mg/m^3 - hours)}{8\,hours} = \frac{0.32\,mg}{m^3} = 8\,hour\,time\text{-}weighted\,average\,exposure\,to\,Chlorpyrifos\,mist$$

Since the paint density was 1.4 g/mL, the painter's 8-hour TWAE to total mist = 1.4 g/mL × 1.0 mL × 1000 mg/g = 1400 mg total paint mist (includes Chlorpyrifos):

$$\frac{(1400\,mg/m^3 - hour)}{8\,hours} = \frac{175\,mg}{m^3}, clearly\,a\,need\,for\,a\,full\text{-}face\,airline\,respirator\,during\,spraying.$$

Answer: 0.0246 ppm$_v$ Chlorpyrifos vapor at saturation conditions. There was a sufficient amount of Chlorpyrifos in the paint to achieve vapor saturation in this bedroom. 0.044 mg Chlorpyrifos vapor/m^3 was the painter's calculated 8-hour TWAE. His/her total mist exposure, as an 8-hour TWAE, was 175 mg/m^3. In author's opinion, insecticides, including organophosphates, should not be used in home interiors especially if infants and those with compromised health could be in these enclosed rooms for long periods (e.g., sleeping 8 or more hours, those with cholinesterase enzyme deficiencies, handicapped persons perpetually confined to bed, those with neurological impairments, pregnant women, etc.).

374. Which of the following situations is the most reasonable indication of a significant indoor air quality issue?
  a. Increase in relative humidity from 30% to 55%
  b. Decrease in relative humidity from 55% to 30%
  c. Hourly fluctuations in dry bulb temperatures from 70 to 55°F back to 70°F
  d. Decrease of air flow velocity at work stations from 200 to 100 fpm
  e. Absence of thermophilic actinomycetes in exhaust ducts and plenums
  f. Steady-state air concentration of radon
  g. Increases of CO_2 from 300 to 400 ppm$_v$ in the morning to >1000 ppm$_v$ by noon
  h. A stench of CO gas in the air supply to employee work stations >1000 ppmv CO_2

Answer: g. Greater than 1000 ppm$_v$ CO_2.

375. The vapor pressure of TDI is close to 0.01 mm of Hg at 77°F. The vapor pressure of methylene *bis*-phenyl diisocyanate (MDI) is about 0.001 mm of Hg at 104°F. Compare the relative volatility of these diisocyanates.

Answer: Since the vapor pressure of a chemical approximately doubles with each 10°C (18°F) increase in temperature, it would be reduced by a factor of about 2 with each 18°F decrease in temperature. Therefore, as the temperature of liquid MDI is reduced by 18°F from 104°F to 86°F, MDI's vapor pressure drops from 0.001 mm Hg to about 0.0005 mm of Hg. If MDI's temperature is further lowered by another 18–68°F, the vapor pressure of the liquid drops to nearly 0.00025 mm of Hg. So, by approximate interpolation, the vapor pressure of MDI at 77°F is nearly 0.00038 mm of Hg.

In other words, TDI is about 0.01 mmHg/0.00038 mm Hg = 26.3 times more volatile than MDI.

This is an oversimplification of relative volatility since other factors determine the evaporation rate of a chemical besides vapor pressure. Note that MDI is a solid below 99°F although it still exerts a vapor pressure in the solid phase. If the MDI covered a surface area approximately 26.3 times greater than the surface area covered by liquid TDI, one could say that the overall emission rate (flux) of the two diisocyanates would be nearly equal.

376. The American Industrial Hygiene Association Workplace Environmental Exposure Level (WEEL) for titanium tetrachloride is 500 mcg per cubic meter of air. $TiCl_4$ rapidly hydrolyzes in moist air forming titanium dioxide fume and hydrogen chloride gas. Assuming stoichiometric conversion of the WEEL concentration to acid gas with an excess of water vapor in the atmosphere, what is the hydrogen chloride concentration? The molecular weight of titanium tetrachloride is 189.73.

$$TiCl_4 + 2H_2O \rightarrow TiO_2 + 4HCl$$

$$\frac{500 \times 10^{-6}\,g}{189.73\,g/mole} = 2.64 \times 10^{-6}\,mole\ of\ TiCl_4$$

Therefore, 4 (2.64 × 10⁻⁶ mole) = 10.56 × 10⁻⁶ mole of HCl is produced.

$$10.56 \times 10^{-6}\,mole\ HCl \times \frac{36.5\,g}{mole} = 385.4 \times 10^{-6}\,g\ HCl = 385.4\,mcg\ HCl$$

$$ppm = \frac{(mg/m^3) \times 24.45}{molecular\ weight} = \frac{(0.3854\,mg/m^3) \times 24.45}{36.5} = 0.26\,ppm\ HCl$$

Answer: 0.26 ppm$_v$ HCl gas. This is below the ACGIH TLV (ceiling) of 2 ppm$_v$. HCl gas is highly soluble in water and, therefore, it tends to be more of an upper respiratory irritant than a deep lung toxicant. However, because the gas might adsorb on the tiny TiO_2 particles, there could be deep lung penetration because the fume is likely sub-micron in size.

377. Many finely divided atmospheric dusts (e.g., flour, sugar, metals, starch, plastics) that settle and deposit on internal industrial plant structures can be a significant explosion hazard if a sufficient amount of dust becomes airborne and an ignition source of ample temperature is present. The amount of dust is generally considered to be significant if the settled amount exceeds:
    a. Any grossly visible amount removed by a piece of transparent adhesive tape
    b. 1/32 inch
    c. 1/16 inch
    d. 3/32 inch
    e. 1/8 inch
    f. 1/4 inch
    g. None of the above

    Answer: e. The settled dust layer thickness must not exceed 1/8 inch. If the settled dust approaches this thickness, better process enclosure, exhaust ventilation, and plant housekeeping are required. "Cleaning" with compressed air wands must never be permitted because the discharged air can accumulate a static charge sufficiently high (in electron volts) to ignite a suspended dust cloud at its lower explosion limit.

378. Inert gases are often used to exclude oxygen from systems where combustible and flammable vapors could accumulate to explosive concentrations. Nitrogen, although not truly chemically inert, is the most common gas used for this purpose. For organic hydrocarbon vapors and gases (but not inorganic explosive gases and vapors such as hydrogen, carbon monoxide, and carbon disulfide), what is the minimum oxygen concentration for combustion?
    a. 21%
    b. 15%
    c. 10%
    d. 5%
    e. 3%
    f. None of the above

    Answer: c. 10%, or slightly less than 0.5 atm. In typical engineering practice, a safety factor is applied so that the actual oxygen concentration is 6% or less in carefully controlled atmospheres (such as drying ovens) and 2% or less in systems where poor mixing of diluent inert gas with combustible gas, for example, could occur. Besides nitrogen, other "inert" gases used for such purposes include argon, carbon dioxide, steam, and helium.

379. A test tube containing 3 mL of a culture of *L. pneumophillae* (720,000 organisms/mL) shatters in a centrifuge. It is estimated that no more than 0.5 mL of this solution was released as respirable aerosol into a 30' × 40' × 12' laboratory that has, for purposes of estimating the worst-case respirable dose, no ventilation. What is the average concentration of

L. *pneumophillae* in the air? Assume this laboratory is essentially empty for these calculation purposes.

$$\frac{720,000 \text{ organisms}}{mL} \times 0.5 \, mL = 360,000 \text{ organisms}$$

$$30' \times 40' \times 12' \times = 14,400 \, ft^3 = 407,763 \, liters$$

$$\frac{360,000 \text{ organisms}}{407,763 \text{ liters}} = \frac{0.88 \text{ organism}}{L}$$

Answer: 88 *L. pneumophillae* bacteria per 100 liters of air. It has been estimated that a single *L. pneumophillae* bacterium in 50 liters of inhaled air is an infectious dose in a susceptible human host. A technician standing near the centrifuge would be at a greater risk since there would have been a burst of microbial aerosol concentration considerably higher than 0.88 organism/L.

380. What is the density of dry air at 300°F and 760 mm Hg?

$$\text{Air density} = \left[\frac{0.075 \, lb}{ft^3}\right]\left[\frac{530°R}{460°R + 300°R}\right] = \frac{0.052 \, lb}{ft^3}$$

Answer: 0.052 pound per cubic foot, or 69.3% that of dry air at NTP (0.075 lb/ft³). That is, as a given volume of air is heated, it expands, the gaseous air molecules become farther apart, and therefore, the density decreases. Mass is conserved, but now in a larger volume.

381. Three pounds of aluminum phosphide (AlP, molecular weight = 57.96 grams/mole) are used to fumigate a 100,000 cubic foot grain storage building by the generation of phosphine gas (PH_3, molecular weight = 34.00 grams/mole) from a quantitative reaction of the dry AlP pellets with atmospheric water vapor:

$$2AlP + 3H_2O \rightarrow 2PH_3 \uparrow + Al_2O_3$$

What is the final concentration of phosphine gas in the building assuming that the grain net void volume and headspace volume total 20% of the building volume?

$$3 \text{ pounds AlP} = 1361 \text{ grams AlP}$$

1361 grams AlP/57.96 grams/mole = 23.48 moles of AlP. Therefore, since 1 mole of aluminum phosphide generates 1 mole of phosphine, 23.48 moles of PH_3 gas are released from the solid aluminum phosphide pellets.

$$23.48 \text{ moles} \times 34.00 \text{ grams PH}_3/\text{mole} = 798.3 \text{ grams of PH}_3$$

$$20\% \times 100,000 \text{ ft}^3 = 20,000 \text{ ft}^3 = 566.3 \text{ m}^3$$

$$\text{ppm} = \frac{(\text{mg/m}^3) \times 24.45}{\text{molecular weight}} = \frac{(798,300 \text{ mg/566.3 m}^3) \times 24.45}{34.00 \text{ grams/mole}} = 1014 \text{ ppm}$$

Answer: 1014 ppm$_v$ PH$_3$. TLV-TWA = 0.3 ppm$_v$. STEL = 1 ppm$_v$. The IDLH for PH$_3$ is 200 ppm$_v$. Application of 3 pounds of aluminum phosphide is excessive, an "overkill," since the LC$_{50}$ for rats (4 hours) is 11 ppm$_v$, and the LCL$_o$s for rabbits = 2500 ppm$_v$ for 20 minutes, for mice = 273 ppm$_v$ for 2 hours, for guinea pigs = 101 ppm$_v$ for 4 hours, and 50 ppm$_v$ for 2 hours for cats. The smallest amount of phosphine gas necessary to achieve sufficient vermin control should be chosen. An EPA-licensed phosphine fumigator can prescribe the safest effective dose for rodent and insect control. One should question if there is sufficient water vapor (relative humidity) in the air space to quantitatively convert AlP into phosphine gas. If not, one must ensure sufficient mechanical (water nebulizer) humidification of the atmosphere. Careful application of phosphine is required because human deaths have been associated with its use as a fumigant for insects in sealed railroad cars and other infested locations.

382. A 4-inch, full-port floating ball valve with self-relieving seats is in an outdoor line transporting dry, liquid chlorine at 500°F and 80 psia. This valve is inadvertently left closed when the line is drained and purged for maintenance. As a result, 1 L of liquid chlorine is trapped in the valve's body cavity. Piping on downstream side had been removed for repairs. It is a hot and sunny day. Before long, the sun heats the valve up to 160°F increasing cavity's pressure to 300 psig. The valve's seats are designed to relieve between 150 and 400 psia, and one seat does at 250 psia. The pressurized liquid chlorine immediately vaporizes forming a blast of gas. How much chlorine is released? In what air volume is produced a 1000 ppm$_v$ cloud of chlorine gas at 760 mm Hg? Chlorine's liquid density at 160°F is 1237 kg/m³ is 1237 g/L. Molecular weight of chlorine is 70.906 grams gram-mole^{-1}.

$$PV = nRT$$

$$160°F = 344 \text{ K (kelvin)}$$

$$760 \text{ mm Hg} = 1 \text{ atmosphere}$$

$$V = \frac{nRT}{P} = \frac{(1237 \text{ g/L})/(70.906 \text{ g/mole})(0.0821 \text{ L-atm/K-mole})(344 \text{ K})}{1 \text{ atmosphere}}$$
$$= 492.7 \text{ L}$$

As the pressurized, liquid chlorine rapidly blasts from the valve, perhaps injuring or killing somebody nearby, 491.7 liters of gas are released. One liter of 100% Cl_2 gas remains in the valve body cavity for some time until it later slowly diffuses into the surrounding atmosphere.

$$1000\,ppm = 0.1\%$$

$$\frac{491.7\,L}{491,700\,L} \times 10^6 = 1000\,ppm\ Cl_2$$

$$\frac{491.7\,L}{0.001} = 491,700\,L$$

$$\sqrt[3]{491,700\,L} = \sqrt[3]{17,364\,ft^3} = 25.9\,feet$$

Answer: 491,700 liters of chlorine gas was released. If confined to a volume of 26 feet to a side, the cube would contain approximately an average of 1000 ppm$_v$ of gas. The IDLH for chlorine is 30 ppm$_v$. Only a few inhalations of this concentration could be fatal to unprotected persons. If the valve rupture was closer to the ground (instead of a point-source release of gas), the spread of gas cloud along the x and y axes would be greater than 26 feet; accordingly, the escape distance for unprotected workers would be more than 26 feet. If released in a corner, for example, and under a low-lying roof or cover, the lateral escape distances become still longer. For example, if the Cl_2 gas was released in a corner under an 8-foot ceiling, the lateral distances in both of the horizontal dimensions become:

$$\sqrt{\frac{17,364\,ft^3}{8\,ft}} = \sqrt{2171ft^2} = 46.6\,feet\ \text{making safe evacuation less likely.}$$

The Cl_2 gas concentration in the corner near the valve would, of course, be much greater than 1000 ppm$_v$. The blast effects might incapacitate a worker so greatly that she/he would be unable to evacuate the area.

383. Household chlorine bleach is a common, inexpensive, highly effective disinfectant often prescribed by industrial hygienists to sanitize and disinfect (but not sterilize) inanimate surfaces. Most commercial solutions as purchased are 5.25% sodium hypochlorite in water (NaOCl). It is thought that the ClO$^-$ ions destroy bacteria by oxidizing life-sustaining compounds within them. How much stock bleach solution (at 5.25% NaOCl) must be diluted with water to make 1-gallon of a 3000 ppm disinfecting solution?

Have: 5.25% solution (w/v) = 5.25 g NaOCl/100 mL = 0.0525 g/mL

$$= 52,250\,ppm$$

Want: 3000 ppm solution = 3000 mg NaOCl/L = 3 mg/mL = 0.003 g/mL

$$1 \text{ gallon} = 3785 \text{ mL}$$

$$3785 \text{ mL/gallon} \times 0.003 \text{ g NaOCl/mL} = 11.36 \text{ g NaOCl/gallon}$$

$$\frac{11.36 \text{ g NaOCl}}{0.0525 \text{ g/mL NaOCl}} = 216.4 \text{ mL} = 7.3 \text{ ounces} = 0.92 \text{ cup}$$

Answer: Dilute almost one cup of household chlorine bleach to 1 gallon with tap water to make a 3000 ppm disinfecting solution.

384. A gas cylinder contains 1260 grams of hydrogen fluoride at 17.5 atmospheres and 25°C. What mass of HF gas would be released if this cylinder was heated to 90°C, the valve was opened until the gas pressure dropped to 1 atmosphere, and the temperature was maintained at 90°C?
    The HF jet gas volume is determined from the initial physical conditions:

$$n = (1260 \text{ grams}) \left[\frac{1 \text{ mole}}{20.01 \text{ grams}} \right] = 62.97 \text{ moles of HF}$$

$$V = \frac{nRT}{P} = \frac{(62.97 \text{ moles})(0.0821 \text{ L-atm/mole-K})(273 \text{ K} + 25 \text{ K})}{17.5 \text{ atmospheres}}$$
$$= 88.0 \text{ liters}$$

The final number of moles is

$$n = \frac{PV}{RT} = \frac{(1.00 \text{ atm})(88.0 \text{ liters})}{(0.0821 \text{ L-atm/mole-K})(363 \text{ K})} = 2.95 \text{ moles of HF}$$

$$(2.95 \text{ moles HF}) \left[\frac{20.01 \text{ g}}{\text{mole}} \right] = 59.0 \text{ grams of HF}$$

$$(1260 \text{ grams HF initial}) - (59 \text{ grams final}) = 1201 \text{ grams of HF}$$

Answer: 1201 grams of HF gas escaped from the cylinder.

385. Calculate the composition of the vapor phase at 30°C that is in equilibrium with a solution of benzene (35 mole %) and toluene (65 mole %). The vapor pressures of these aromatic hydrocarbons at this temperature are 119 and 37 mm Hg, respectively.

$$\text{Partial total vapor pressure from benzene} = (0.35)(119 \text{ mm Hg})$$
$$= 41.7 \text{ mm Hg}$$

Partial total vapor pressure from toluene = (0.65) (37 mm Hg)
= 24.1 mm Hg

Total vapor pressure = 41.7 mm Hg + 24.1 mm Hg = 65.8 mm Hg

The percent vapor composition is calculated by applying Dalton's law of partial pressures:

$$\text{concentration}_{\text{benzene}} = \frac{VP_{\text{benzene}}}{VP_{\text{total}}} = \frac{41.7\,\text{mm Hg}}{65.8\,\text{mm Hg}} = 0.634 = 63.4\%\,\text{benzene}$$

$$\text{concentration}_{\text{toluene}} = \frac{VP_{\text{toluene}}}{VP_{\text{total}}} = \frac{24.1\,\text{mm Hg}}{65.8\,\text{mm Hg}} = 0.366 = 1.000 - 0.634$$

$$= 36.6\%\,\text{toluene}$$

Answers: The vapor phase at saturation is 63.4% benzene and 36.6% toluene. Note how the more volatile component (benzene with its higher vapor pressure) is enriched from 35% in the liquid phase to 63.4% in the vapor phase. Strictly speaking, the vapor pressure of each aromatic organic chemical must be added to the denominator because as each evaporates, they create back pressure in a closed system that hinders evaporation. Regardless, saturation will be reached as long as walls of the container are not colder than the interior atmosphere. Partial condensation of vapor-phase organics will occur, in such case, to the liquid phase.

386. The relative variability of the randomly distributed errors in a normal distribution is referred to as the CV. In air sampling, there are numerous opportunities for errors (analysis, sample collection, air flow rate calibration, etc.). Collectively, there can be a CV for the combined errors. For the following sampling and analytical procedures, rank their respective CVs: colorimetric detector tubes, rotameters on personal pumps, charcoal tubes, asbestos, respirable dust, gross dust.

Answers:

Rotameter on personal pumps (sampling only)	0.05 CV
Gross dust (sampling/analytical)	0.05
Respirable dust, except coal mine dust (sampling/analytical)	0.09
Charcoal tubes (sampling/analytical)	0.10
Colorimetric detector tubes	0.14
Asbestos (sampling/counting)	0.24–0.38

387. An 8.3 L/min (nominal) critical orifice was calibrated in Philadelphia at 8.9 L/min at 72°F. This critical orifice was used to collect a general area air sample for silica dust at 45°F in a foundry also in Philadelphia. What was the actual air flow rate through the critical orifice? Assume identical barometric pressures for calibration and air-sampling conditions.

$$Q_{actual} = Q_{indicated} \sqrt{\frac{T_{actual}}{T_{calibration}}}$$

$$Q_{actual} = \sqrt{\frac{460° + 45°}{460° + 72°}} = 8.67\,L/min$$

Answer: 8.67 liters of air per minute. If the barometric pressures were different between the calibration and sampling conditions, use this standard formula:

$$Q_{actual} = Q_{indicated} \sqrt{\frac{P_{calib} \times T_{actual}}{P_{actual} \times T_{calib}}}$$

Always ensure that all units are the same and that temperatures are expressed in degrees absolute. Note that the square root function is always used with variable orifice fluid flow meters.

388. 700 cubic yards of dry, densely packed cement dust are transferred approximately every hour during a 3-minute period from the calciner drying kiln into a storage silo. The procedure requires the operator to turn on a fan for the mechanical local exhaust system prior to dumping and allowing it to run as cement dust cascades into the silo. Displaced dusty air exits the silo through a 3' × 5' hole. An engineer selected a dust ventilation capture velocity of 300 fpm for the hole face. Is this sufficient to prevent escape of dust into the surrounding atmosphere? The air void space in the packed cement dust (say, 10% of the total volume) does not enter into calculations since this air, along with the cement dust, is expelled from the silo.

$$700\,yd^3 \times 27\,ft^3/yd^3 = 18,900\,ft^3$$

$18,900\,ft^3/3\,minutes = 6300\,ft^3$ of dusty air are displaced from the silo per minute.

Since the cement transfer process occurs in "burps" over the 3 minutes, allow a ventilation design contingency factor of 5: $6300\,ft^3/minute \times 5 = 31,500\,cfm$

$$Q = AV$$

$$V = \frac{Q}{A} = \frac{31,500\,cfm}{3\,ft \times 5\,ft} = 2100\,ft/minute$$

Answer: A capture velocity of 300 fpm is too low. With this velocity, 6300 cfm of dusty air are being displaced while the fan attempts to exhaust 4500 cfm ($15\,ft^2 \times 300\,fpm$). The silo emits, on average, 1800 ft³ of dusty air per minute if the 300 fpm capture velocity is selected. Choose the fan, dust collector, and appropriately sized ducts to provide silo exit air capture velocity of about 2100 feet per minute. Decreasing the exhaust opening size increases capture velocity, or, alternatively, a smaller fan and exhaust system could be used with the 2100 cfm capture velocity.

This approach can also be used when designing and selecting mechanical local exhaust ventilation systems for drum and tank-filling operations. For example, if it requires 2 minutes to fill a 55-gallon drum with volatile solvents (27.5 gallons per minute = 3.7 ft³/minute), a marginal ventilation system would be able to capture the 3.7 ft³ of air saturated with solvent vapors every minute. In typical, good industrial hygiene practice, a larger volume is normally chosen such as, in this case, a 20 or so cfm exhaust "elephant trunk" placed within 4 inches of the bung fill hole.

389. Organic vapor emissions from point sources (VOC, or volatile organic carbon) can be determined in a variety of ways depending on the source, the hydrocarbon, and other fugitive vapor release factors. What formula can be used to estimate yearly vapor losses ("breathing loss") from volatile hydrocarbon storage tanks?

Answer: A good calculation procedure is found in the Environmental Protection Agency Publication *AP-42*:

$$L_B = 2.26 \times 10^{-2} \, M_v \left[\frac{P}{P_A - P} \right]^{0.68} D^{1.73} \, H^{0.51} \, \Delta T^{0.5} \, F_p \, C \, K_c$$

where
  L_B = organic vapor breathing loss, lb/year
  M_V = molecular weight of the vapor, lb/lb-mol
  P = true vapor pressure, psia
  P_A = average atmospheric pressure, psia
  D = tank diameter, ft
  H = average vapor space height, ft
  T = average ambient diurnal temperature (Δ equals the daily change), °F
  F_P = tank's paint factor, dimensionless (e.g., shiny aluminum versus flack black)
  C = adjustment factor for small tanks, dimensionless
  K_C = product factor, dimensionless

The EPA publication provides data on common hydrocarbons and various emission factors and coefficients. *EPA-405/4-88-004* and *EPA-450/2-90-001a* are other useful publications.

The working loss of a tank is the amount of volatile vapors displaced during loading and unloading the tank. Assuming that there is no vapor recovery system, this can be estimated by the equation:

$$L_W = 2.40 \times 10^{-5} \, M_v \, PV \, N \, K_N \, K_C$$

where

L_W = working loss, lb/year
V = tank capacity, gallons
N = number of capacity turnovers during the year, dimensionless
K_N = turnover factor from *AP-42*

390. Silane (SiH_4, silicon tetrahydride) has a TLV of 5 ppm$_v$. This pyrophoric gas readily reacts with water vapor in the atmosphere according to the equation:

$$SiH_4 + 2H_2O \rightarrow SiO_2 + 4H_2$$

Assume the atmosphere in a 20,000 gallon confined space containing 10,000 ppm$_v$ SiH_4 vapor (1%). There is sufficient water vapor present to quantitatively hydrolyze the silane to silicon dioxide aerosol and hydrogen gas. Is an explosive atmosphere generated (LEL $H_2 \cong 4\%$ in air)? Molecular weight of silane is 32.12 gram/gram-mole. Assume the heat of chemical reaction is sufficient to ignite hydrogen gas. Silane is a colorless, spontaneously flammable, and pyrophoric gas. Most likely, silane vapor reacts with atmospheric water vapor to form silicic acid (hydrated silica, $SiO_2 \cdot n\text{-}H_2O$). The formation of crystalline silica in an aerosol cloud appears unlikely; amorphous silica appears to be the most likely reaction product.

$$\frac{mg}{m^3} = \frac{ppm \times molecular\,weight}{24.45\,L/gram\text{-}mole} = \frac{10,000\,ppm \times 32.12\,g/mole}{24.45\,L/g\text{-}mole}$$

$$= 13,137\,mg/m^3$$

$$20,000\,gallons = 75.71\,m^3$$

$$(13,137\,mg/m^3) \times 75.71\,m^3 = 994,602\,mg\,SiH_4$$

$$\frac{994.6\,grams\,SiH_4}{32.12\,grams\,SiH_4/mole} = 30.97\,moles\,of\,SiH_4\,are\,in\,the\,tank's\,atmosphere$$

Since 1 mole of silane generates 4 moles of hydrogen gas, 30.97 moles will produce 123.88 moles of hydrogen. 123.88 moles × 2.016 grams/mole = 247.7 grams of hydrogen gas.

$$ppm = \frac{(247,700\,mg\,H_2/75.71\,m^3) \times 24.45}{2.016\,g/mole} = 39,679\,ppm\,H_2$$

Answer: 39,679 ppm$_v$ hydrogen gas equals 3.97% or right at the LEL of 4% hydrogen in air. While one debates if there is quantitative conversion of silane into hydrogen, since silane has a "built-in match," this tank and much nearby shortly will become history. Prevention of silicosis is not a key issue here!

391. What is the average vapor concentration after one cup of benzene evaporates into a building the size of a football field with a 14-foot ceiling? Assume that there is no dilution, only mixing, ventilation as the benzene evaporates. In other words, what is the equilibrium vapor concentration within the building assuming no dilution?

$$300' \times 160' \times 14' = 672,000\,\text{ft}^3 = 19,030\,\text{m}^3$$

$$\text{one cup} = 236.6\,\text{mL}$$

$$236.6\,\text{mL} \times 0.88\,\text{g/mL} = 208.208\,\text{g} = 208,208\,\text{mg}$$

$$\text{ppm} = \frac{(\text{mg/m}^3) \times 24.45}{\text{molecular weight}} = \frac{(208,208\,\text{mg}/19,030\,\text{m}^3) \times 24.45}{78.11} = 3.4\,\text{ppm}$$

Answer: 3.4 ppm$_v$ benzene vapor, or 34 times the NIOSH REL of 0.1 ppm$_v$. Such calculations of hypothetical examples are helpful to explain low air contaminant concentrations to workers, supervisors, and to lay persons, such as jurors, our courts, your spouse and children, and to the neighbors. Using another familiar example—a standard sewing thimble of 2.2 mL volume: spill this amount of benzene in 1-$\frac{1}{2}$ car garage (14' × 20' × 8'), and the resultant vapor concentration becomes 9.5 ppm$_v$—almost 10 times the OSHA PEL (assuming, of course, the garage is unventilated).

392. What formula can be used to determine the amount of oxygen required for perfect combustion of any fuel?

Answer: The approximate theoretical volume of oxygen required to burn a fuel containing carbon, hydrogen, sulfur, and oxygen is

$$\frac{\text{volume of oxygen}}{\text{pound of fuel}} = 359\,\text{ft}^3 \ \ \text{or} \ \ 1710\,\text{ft}^3 \left[\frac{C}{12} + \frac{H_2}{4} + \frac{S}{32} - \frac{O_2}{32} \right]$$

The coefficient of 359 is the volume (in ft^3) of 1 pound-mole of O$_2$ at 32°F and 1 atmosphere of pressure. Use 1710 as the coefficient to obtain the theoretical volume of air. This formula is used for any dry fuel (e.g., oil, coal, wood, gasoline, propane, kerosene, etc.). C, H$_2$, S, and O$_2$ are decimal weights of these elements in 1 pound of fuel (e.g., for a fuel with 81% carbon, C = 0.81).

393. Flue gases (containing SO$_2$) at 280°F are fed to a power plant stack through a 3-feet diameter duct. The centerline Pitot tube measurement is 0.7 inch of water. The static pressure duct wall manometer reads—0.6 inch of water. The gas contains 0.45 mole-% SO$_2$. The barometric pressure is 725 mm Hg. What is the emission rate of SO$_2$ gas from this stack in pounds per hour?

Assume that the Pitot traverse and static pressure gauge exceed 10 duct diameters downstream of any gas flow laminarity disturbances.

$$\Delta P = hp = [0.7 \, \text{inch} - (-0.6 \, \text{inch})] \times \frac{1 \, \text{ft}}{12 \, \text{in}} \times \frac{62.4 \, \text{lb}}{\text{ft}^3} = \frac{6.76 \, \text{lb}}{\text{ft}^3}$$

$$P_{\text{air}} = \frac{1 \, \text{lb} - \text{mole}}{359 \, \text{ft}^3} \times \frac{29 \, \text{lb}}{\text{mole}} \times \frac{492°\text{R}}{740°\text{R}} \times \frac{725 \, \text{mm Hg}}{760 \, \text{mm Hg}} = \frac{0.05123 \, \text{lb}}{\text{ft}^3}$$

$$\frac{v^2_{\text{max}}}{2 \, g_c} = \frac{\Delta P}{r} = h_v$$

$$v_{\text{max}} = \sqrt{2 \, g_c \times \frac{\Delta P}{r}} = \sqrt{\frac{(2)(32.2)(6.76)}{0.05123 \, \text{lb/ft}^3}}$$

$$v_{\text{max}} = \frac{92.2 \, \text{ft}}{\text{second}}$$

$$\frac{v_{\text{avg}}}{v_{\text{max}}} = 0.81$$

$$v_{\text{avg}} = (0.\,81)(92.2 \, \text{ft/second}) = 74.7 \, \text{ft/second}$$

$$Q = A \, v_{\text{avg}} = \frac{\pi}{4}(3 \, \text{ft})^2 \times \frac{74.7 \, \text{ft}}{\text{second}} = \frac{528 \, \text{ft}^3}{\text{second}}$$

$$SO_2 \, \text{gas emission rate} = (528)(0.05123)(3600 \, \text{seconds/hour})\left(\frac{1 \, \text{lb-mole}}{29 \, \text{lb}}\right)$$

$$= \left(\frac{64 \, \text{lb} \, SO_2}{\text{lb mole}}\right)(0.0045) = \frac{967 \, \text{lb} \, SO_2}{\text{hour}}$$

Answer: 967 pounds of SO_2 gas per hour.

394. A gas mixture containing hydrogen chloride and dry air at 22 psia and 22°C is bubbled through a gas scrubber containing 180 mL of 0.01 N NaOH solution. The collection efficiency of the gas bubbler is 100%. The remaining NaOH is back-titrated requiring 15.63 mL of 0.1 N HCl. The dry air mixture has a wet test meter volume reading of 1.0 L at a pressure of 740 mm Hg and 25°C. What is the mass fraction and the mole fraction of HCl gas in the original gas mixture?

$$NaOH + HCl \rightarrow NaCl + H_2O$$

moles:

$$NaOH \text{ at start} = (0.18)(0.01) = 0.0018 \text{ mole}$$

$$NaOH \text{ at end} = (0.01563)(0.1) = 0.001563 \text{ mole}$$

$$HCl \text{ scrubbed} = \Delta NaOH_{start} - NaOH_{end} = 0.000237 \text{ mole}$$

Assume no water vapor in the air when volume was measured in wet test meter:

$$\text{moles of air} = \frac{PV}{RT} = \frac{[740 \text{ mm Hg}/760 \text{ mm Hg}](1000 \text{ mL})}{(82.06)(298)} = 0.0398 \text{ mole}$$

Assume water vapor in air:

$$P_{H_2O} = 0.4594 \text{ psia} = 23.76 \text{ mm Hg}$$

$$P_{air} = 740 \text{ mm Hg} - 23.76 \text{ mm Hg} = 716.24 \text{ mm Hg}$$

$$\text{Moles of air} = \frac{PV}{RT} = \frac{[716.24/760](1000 \text{ mL})}{(82.06)(298)} = 0.0385 \text{ mole}$$

Water vapor in air:

$$Y_{HCl} = \frac{0.000237}{0.000237 + 0.0385} = 0.0061$$

$$Y_{HCl} = \frac{y_{HCl}[M_{HCl}/M_{air}]}{1 + y_{HCl}[(M_{HCl}/M_{air}) - 1]} = \frac{(0.0061)[36.5/29]}{(1 + 0.0061)[(36.5/29) - 1]}$$
$$= 0.0295$$

No water vapor in air:

$$y = \frac{0.000237}{0.000237 + 0.0398} = 0.0059$$

$$y = \frac{(0.0059)[36.5/29]}{(1 + 0.0059)[(36.5/29) - 1]} = 0.0286$$

Answers: 0.0295 and 0.0286 mole fractions of HCl gas, respectively, in moist and in dry air (0.0295 and 0.0286 mole HCl, respectively/liter of moist or dry air). Converting, for example, 0.0295 mole of HCl/L into ppm_v:

$$0.0295 \text{ mole HCl} \times 36.46 \text{ g HCl/mole} = 1.076 \text{ g HCl} = 1,076,000 \text{ mcg HCl}$$

$$\frac{(1,076,000 \text{ mcg HCl/L}) \times 24.45 \text{ L/g-mole}}{36.46 \text{ g/g-mole}} = 721,563 \text{ ppm HCl} = 72.16\%$$

Such an obviously high concentration of HCl in this gas sample shows that it is not a workplace air sample. This is the concentration one might expect in a chemical process stream, say in the product effluent of an HCl-manufacturing plant: $Cl_2 + H_2 \rightarrow 2HCl$. A 1-liter gas sample from such a process stream would normally be sufficient.

395. The concentration of iron oxide fume (as Fe_2O_3) in the breathing zone of a scarfer of new steel ingots is 4.7 mg/m^3. The TLV for iron oxide fume is 5 mg/m^3 (as Fe). What is the exposure of this scarfer as Fe?

The molecular weights of Fe_2O_3 and Fe = 159.69 and 55.847, respectively.

$$\frac{2 \text{ Fe}}{Fe_2O_3} = \frac{2 \times 55.847}{159.69} = 0.6994 = 69.94\% \text{ Fe}$$

$$4.7 \text{ mg Fe}_2O_3/\text{m}^3 \times 0.6994 = 3.29 \text{ mg Fe/m}^3$$

396. The static pressure at a fan outlet is 0.5 inch of water. The static pressure at the fan's inlet is -6.5 inches of water. The duct velocity pressure is 0.8 inch of water. What is the fan static pressure?

$$FSP = SP_{outlet} - SP_{inlet} - VP = +0.5'' - (-6.5'') - 0.8'' = 6.2 \text{ inches of water}$$

397. Carbon dioxide gas measurements can be used to estimate the percent of outdoor air admitted to a building where, for example, outdoor air is 400 ppm_v CO_2, return air is 750 ppm_v CO_2, and mixed air is 650 ppm_v CO_2. What is the percent outdoor air supplied under these conditions?

$$\% \text{ outdoor air} = \frac{\text{ppm}_{\text{return air}} - \text{ppm}_{\text{mixed air}}}{\text{ppm}_{\text{return air}} - \text{ppm}_{\text{outdoor air}}} \times 100$$

$$= \frac{750 \text{ ppm}_v - 650 \text{ ppm}_v}{750 \text{ ppm}_v - 400 \text{ ppm}_v} \times 100 = \frac{100 \text{ ppm}_v}{350 \text{ ppm}_v} \times 100$$

$$= 28.6\%$$

Answer: Approximately 29% of the building's general ventilation air is supplied from the outside. Therefore, about 71% of the air in the building is being recirculated.

398. A 13,500 gallon liquid incinerator feed tank is filled and emptied daily with a PCB-contaminated solution of ethyl acetate (100 ppm_m PCBs). The tank is

vented to the atmosphere and does not have a vapor recovery system. Does the incinerator or the tank release more hydrocarbons to the atmosphere? The vapor pressure of ethyl acetate at NTP is 73 mm Hg. The vapor pressure of PCBs, compared to the vapor pressure of ethyl acetate, is negligible. The design combustion efficiency of this incinerator is verified at 99.99%. The specific gravity of ethyl acetate is 0.90 g/mL. The molecular weight of ethyl acetate is 88.1 grams/gram-mole.

$$\text{Partial pressure of ethyl acetate in tank} = \frac{73 \, \text{mm Hg}}{760 \, \text{mm Hg}}$$

$$= 0.096 \, \text{atmosphere at NTP}$$

$$\text{Vapor concentration in tank} = \frac{PM}{RT}$$

$$= \frac{(0.096 \, \text{atm})(88.1 \, \text{lb/lb-mole})}{0.7302 \, \text{atm} - \text{ft}^3/\text{lb-mole} - {}^\circ\text{R}(460 + 75)}$$

$$= 0.0217 \, \text{lb/ft}^3$$

The daily emission rate from the tank's open vent

$$= \frac{13{,}500 \, \text{gallons}}{\text{day}} \times \frac{0.134 \, \text{ft}^3}{\text{gallon}} \times \frac{0.0217 \, \text{lb}}{\text{ft}^3} = \frac{39.3 \, \text{lb}}{\text{day}}$$

The daily hydrocarbon emission rate from this incinerator

$$= \frac{13{,}500 \, \text{gallons}}{\text{day}} \times 0.90 \times \frac{8.34 \, \text{lb}}{\text{gallon}} \times (1 - 0.9999) = \frac{10.1 \, \text{lb}}{\text{day}}$$

Answer: The tank's vent emits almost 4 times the amount of vapors released from the incinerator's stack. Engineering controls for the tank include an inert gas blanket (nitrogen) to reduce explosion hazards. Ethyl acetate vapors will pass through a regenerative carbon adsorber as the tank is filled. Do not vent these vapors into the incinerator to prevent flashback explosive fire! Present combined daily VOC emissions for this system are 39.3 lb plus 10.1 lb = 49.4 pounds. Note that if the incinerator efficiency drops from 99.99% to 99.95%, incinerator's contribution of VOC to the environment exceeds the fugitive VOC emissions from the tank by 50.7 lb/day − 39.3 lb/day = 11.4 lb/day.

$$\frac{13{,}500 \, \text{gallons}}{\text{day}} \times 0.90 \times \frac{8.34 \, \text{lb}}{\text{gallon}} \times (1 - 0.9995) = \frac{50.7 \, \text{lb}}{\text{day}}$$

399. Leakage of highly and acutely toxic gases (e.g., arsine) from pressurized systems, especially into confined work spaces with poor ventilation, is problematic for the chemical safety engineer and workers in those spaces. Dish detergent solutions and commercial leak detection liquids can be used to determine the volumetric flow rate of gas leaks. The emission rate can be estimated with a small ruler scale and a stopwatch. The diameter of an enlarging gas bubble is measured as a function of time of growth. The assumption is made that the bubble is a sphere. Calculate leak rate of 12% phosgene gas in nitrogen if a 2-cm diameter bubble is emitted from a gas line fitting in 7.3 seconds.

$$V = \frac{(1/6) \times \pi \times d^3}{t} = \frac{(1/6) \times \pi \times (2\,\text{cm})^3}{7.3\,\text{seconds}} = \frac{0.574\,\text{cm}^3}{\text{second}}$$

$$= \frac{0.0344\,\text{L}}{\text{minute}} = \frac{0.0122\,\text{ft}^3}{\text{minute}}$$

$$\frac{0.0122\,\text{ft}^3}{\text{minute}} \times 0.12 = \frac{0.001464\,\text{ft}^3\,\text{COCl}_2}{\text{minute}}$$

where
   V = gas leak rate, volume per time
   d = diameter of gas bubble at time t, and
   t = elapsed time to form the gas bubble with diameter, d

Answer: Approximately 0.001464 ft³ of phosgene gas is emitted per minute. If the leak is not repaired immediately or mechanical local exhaust ventilation is not readily available, then the volume of dilution air required to reduce $COCl_2$ gas to a safer concentration can be calculated. In general, it is poor industrial hygiene practice to rely upon dilution ventilation for highly toxic gases. In this example, the amount and rate of gas emitted appear to be relatively tiny; however, any leak of a highly toxic gas, such as phosgene, must be staunched immediately. That is, pronto, stat.

400. Helium, a gas considerably less dense than air (0.0103 lb/ft³; air = 0.075 lb/ft³), is used, among other purposes, to inflate party balloons. Most of us know that when one inhales helium gas, our voice becomes "squeaky" due to a lower density of the helium and air gas mixture passing over vocal cords in our larynx. Fatalities have resulted when persons have deliberately inhaled helium released from cylinders of pressurized gas. Deaths have been attributed to physical asphyxiation from the gas and from increased intrapulmonary pressure. Direct inhalation of helium from a commercial balloon-filling system can pose a greater hazard than inhaling helium from a party balloon. How can pressure of the inhaled helium gas dispensed from a commercial system kill one essentially instantly?

   A difference of 30 mm Hg pressure between the surrounding lung pressure and the intrapulmonary pressure can be fatal if the exposure is prolonged. As the helium (or any inhaled gas) pressure differential increases to 80–100 mm Hg, death is rapid from rupture of alveoli. Prompt hemorrhage results in drowning asphyxiation fatalities.

Commercial helium balloon-filling systems typically deliver gas at 5 cubic feet per minute (2.36 liters/minute). Lung volumes vary from person to person. Total lung capacity for an average adult man is 5.6 liters, and, for an average adult woman, it is 4.4 liters. For this evaluation, assume a 10-year-old boy with a total lung volume of 3 liters directly inhaled helium from a commercial gas-dispensing system by placing his mouth directly over the gas dispensing spigot while cracking the valve.

Lung bursting pressure = 80–100 mm Hg = 1.55–1.93 lb/in^2 (psi)

As added measure of safety, use the lower limit of 1.55 psi for lung rupture. The questions become: (1) How much gas volume must be added to the total lung volume to cause a pressure increase of 1.55 psi? and (2) How quickly could this occur?

Boyle's gas law applies: $P_1/P_2 = V_2/V_1 + v$, where P_1 and P_2 = original and final gas pressures, respectively; and V_1 and V_2 = the original and final gas volumes, respectively. v = the additional gas volume to be added to a 3 liter lung volume to cause a pressure increase of 1.55 psi. Atmospheric pressure is assumed to be 14.7 psi. Pressure at other altitudes can be substituted for this sea-level pressure.

$$\frac{14.7\,\text{psi}}{14.7\,\text{psi} + 1.55\,\text{psi}} = \frac{3.0\,\text{L}}{3.0\,\text{L} + v}(3.0\,\text{L})(16.25\,\text{psi}) = (14.7\,\text{psi})(3.0\,\text{L} + v)$$

$v = 0.316$ L, the additional lung volume required.

The minimum time required to produce this additional lung volume at the rupture pressure (1.55 psi) is calculated by dividing the additional lung volume of 0.316 L by the maximum helium gas flow rate, or

$$\frac{0.316\,\text{L}}{2.36\,\text{L/second}} = 0.134\ \text{second}$$

Answer: The increased inhaled gas volume in a person with this lung capacity is only 0.316 L, and the pressure increase time is no less than 0.134 second. But, since lungs are compliant and are not a rigid structure, time required to reach this increased pressure may be a tad longer. Regardless, these calculations show that inhalation of helium gas directly from portable balloon-filling tanks is extremely hazardous. Since the portable tanks have initial pressures above 200 psi, there is ample pressure to cause fatal injuries in persons who place their mouth directly over the helium discharge gas valve.

401. A chemical reaction is predicted to release 590 cubic meters of process air that will contain 4390 ppm$_v$ of H_2S. A combination of sodium hydroxide (NaOH) solution followed by a sodium hypochlorite (NaOCl) solution oxidation is used to remove the H_2S from the process gas stream. The H_2S neutralization and oxidation reactions are

$$H_2S + 2NaOH \leftrightarrow Na_2S + 2H_2O$$

$$Na_2S + 4NaOCl \rightarrow Na_2SO_4 + 4NaCl$$

The H_2S gas will be scrubbed through a column containing 55 gallons of 1% NaOH solution in water. It is assumed that, if there is sufficient NaOH present, the H_2S gas absorption will be 100% efficient if the gas scrubbing rate is sufficiently slow. The resultant sodium sulfide solution is then reacted with the sodium hypochlorite solution for final oxidation. Is there sufficient NaOH present?

$$\frac{mg}{m^3} = \frac{ppm \times molecular\,weight}{24.45} = \frac{4390\,ppm \times 34.08}{24.45} = \frac{6119\,mg\,H_2S}{m^3}$$

$$\frac{6119\,mg\,H_2S}{m^3} \times 590\,m^3 = 3{,}610{,}210\,mg\,H_2S = 3610\,grams\,of\,H_2S$$

$$\frac{3610\,g\,H_2S}{34.08\,g\,H_2S/mole} = 105.9\,moles\,of\,H_2S\,are\,to\,be\,scrubbed$$

Therefore, $105.9\,moles \times 2 = 211.8\,moles$ of NaOH are needed for 100% reaction.

$$55\,gallons = 208.2\,L$$

$$1\%\,(w/v)\,solution = 10\,grams\,NaOH/liter$$

$$208.2\,L \times 10\,g\,NaOH/L = 2082\,grams\,of\,sodium\,hydroxide$$

$$\frac{2082\,g\,NaOH}{40.00\,g\,NaOH/mole} = 52.05\,moles\,of\,NaOH \ll 211.8\,moles$$
$$of\,NaOH\,required$$

Answer: There is insufficient NaOH present in the gas scrubber solution for the removal of H_2S from the process stream. Increase the concentration to at least 5% to ensure stoichiometric reaction.

A 5% solution contains 260.25 moles of NaOH, or 48.45 moles in excess. Consideration should be given to scrubbing this gas process stream through two absorption columns connected in series with the first containing 55 gallons of a 4% solution of NaOH, and with the second containing 55 gallons of a 2% solution of NaOH. After the reaction is completed, combined scrubber solutions are oxidized with sodium hypochlorite.

Using a flare to oxidize the H_2S or injection through a high-level stack to dilute the gas to safer ground-level concentrations is risky, possibly much more costly, and likely prohibited. Issues of inhalation toxicity and community odor would remain.

In 1950, in Poza Rica, Mexico, a flare failed to oxidize H_2S to SO_2 gas at a sulfur recovery unit in a refinery, and, during the night, 22 persons in the community died with another 320 hospitalized. H_2S is a potent chemical asphyxiant with an IDLH of $100\,ppm_v$. CO and HCN, two other chemical asphyxiant gases, have IDLHs of 1200 and $50\,ppm_v$, respectively.

402. The concentration of radon-222 in the basement of a newly constructed house is 1.85×10^{-6} mole per liter. Assuming that basement air remains static, and there is no further release of Rn-222, what is the concentration after 2.3 days? The half-life of Rn-222 is 3.8 days.

$$\text{Amount of Rn-222 remaining} = (1.85 \times 10^{-6} \text{ mole/L})\left[\frac{1}{2}\right]^{2.3\,days/3.8\,days}$$

$$= (1.85 \times 10^{-6} \text{ mole/L})\left[\frac{1}{2}\right]^{0.605}$$

$$= 1.22 \times 10^{-6} \text{ mole/L}$$

Answer: 1.22×10^{-6} mole per liter, or $(1.22 \times 10^{-6}\text{mole/L}) \times (222\,g$ Rn/mole$) = 270.84 \times 10^{-6}$ g/L $= 271.84\,mcg/L$.

403. 10.9 liters of ambient, community air at $740\,mm$ Hg and $19.0°C$ were bubbled through lime water [an aqueous suspension of $Ca(OH)_2$]. All carbon dioxide gas in air sample was precipitated as calcium carbonate. If the precipitate weighed $0.058\,g$, what was the percent by volume of CO_2 in this air sample? Molecular weights of carbon dioxide and calcium carbonate are, respectively, 44.01 and 100.09 grams per gram-mole. Disregard the low solubility of calcium carbonate in water ($0.00153\,g/100\,cc$ at $25°C$).

$$CO_2 + Ca(OH)_2 \rightarrow CaCO_3 \downarrow + H_2O$$

$$10.9\,L = 0.0109\,m^3$$

$$\frac{0.058\,g\,CaCO_3}{100.09\,g\,CaCO_3/mole} = 0.00058 \text{ mole of } CaCO_3$$

Therefore, 0.00058 mole of CO_2 was in the 10.9 L air sample.

$$0.00058 \text{ mole of } CO_2 \times \left[\frac{44.01 \text{ g } CO_2}{\text{mole}} \right] = 0.0255 \text{ g of } CO_2 = 25.5 \text{ mg } CO_2$$

$$
\begin{aligned}
\text{ppm} &= \frac{\text{mg}}{\text{m}^3} \times \frac{22.4}{\text{molecular weight}} \times \frac{\text{absol. temp.}}{273.15 \text{ K}} \times \frac{760 \text{ mm Hg}}{\text{pressure, mm Hg}} \\
&= \frac{25.5 \text{ mg } CO_2}{0.0109 \text{ m}^3} \times \frac{22.4}{44.01} \times \frac{273.15°C + 19°C}{273.15} \times \frac{760 \text{ mm Hg}}{740 \text{ mm Hg}} \\
&= 1308 \text{ ppm } CO_2
\end{aligned}
$$

Answer: 1308 ppm$_v$ CO_2 = 0.1308$_{\%v}$. This is obviously polluted air because the typical ambient concentration of CO_2 in relatively clean urban air is below 400 ppm$_v$. In such an atmosphere, there would most likely be several other air contaminants, for example, CO, soot, PNAHs, particulates, HCHO, O_3, NO_x, SO_2, and so on.

404. Inert gases, such as nitrogen, are used to exclude oxygen from systems where an explosion hazard exists and/or the product or materials are subject to air oxidation. Ventilation formulae are presented in this book to enable the engineer to calculate how much nitrogen is needed to dilute an initial 21% ambient oxygen concentration down to some predetermined level, say <1% O_2. Devise some simple guidelines to enable a chemical process attendant to do this without resorting to complex first-order ventilation air contaminant decay equations.

Answer: The concentration of oxygen in the ambient air is 20.95$_v$%, so after seven inert gas volume exchanges in a system, the oxygen concentration decays to less than 0.1%. Thus, total nitrogen required for sweep purging would be 7 times the headspace volume. For example, a 10,000 gallon tank containing 2000 gallons of liquid would require 7 × 8000 gallons = 56,000 gallons of nitrogen with good mixing of the sweep gas with the oxygen-containing air.

Since fans are rated in cubic feet of air per minute, convert gallons to cubic feet by multiplying by 0.134 (56,000 gallons × 0.134 = 7504 ft³). Divide rated fan capacity (in cfm) into total headspace volume to obtain fan operating time, for example, with a 500 cfm fan blowing nitrogen into the space: 7504 ft³/500 cfm = 15 minutes. Apply a safety factor if there is poor mixing of the nitrogen with the air. Verify the oxygen concentration by obtaining representative gas samples.

In closed systems, a mass gas flow meter or a critical flow orifice can measure the nitrogen delivery. People have died from asphyxiation when they unknowingly entered confined spaces that contained inert atmospheres. Such spaces must be clearly posted to warn of the inhalation hazard. All elements of a confined space entry program must be in place to conserve the health and safety of entrants (see 29 *CFR* 1910.146). Padlocking inert atmosphere spaces to impede unauthorized entry is a tremendous idea; hazard warning signs are never enough.

405. An ambient air sample is collected in a partially evacuated 3.2 L glass flask for gas analysis. The barometric pressure at the time of sampling is 728 mm Hg, and the air temperature is 25°C. The relative humidity is 45%. The pressure remaining in the flask after evacuation (but before air sampling) is 480 mm Hg. What volume of air is sampled? The vapor pressure of water at 25°C (77°F) is 23.76 mm Hg.

The partial pressure of the air sample due to water vapor = 0.45 × 23.76 mm Hg = 10.7 mm Hg.

$$V = V_a \times \frac{P_{bar} - P_{H_2O} - P_{partial}}{760\,\text{mm Hg}} \times \frac{273°C}{273°C + 25°C}$$

$$= 3.2\,\text{L} \times \frac{728\,\text{mm Hg} - (10.7\,\text{mm Hg} + 480\,\text{mm Hg})}{760\,\text{mm Hg}} \times \frac{273}{273 + 25}$$

$$= 3.2\,\text{L} \times \frac{237.3}{760} \times \frac{273}{298} = 0.915\text{L},$$

where

P_{bar} = barometric pressure

P_{H_2O} = partial pressure of water vapor

$P_{partial}$ = pressure remaining in the flask after evacuation

Answer: 0.915 liter note that these calculations correct gas-sampling conditions to STP (0°C and 760 mm Hg).

406. Air was sampled for sodium nitrate dust and nitric acid gas at an average flow rate of 1.73 liters per minute through a membrane filter followed by an impinger that contained 9.7 mL of a dilute NaOH. Assuming 100% collection efficiency and an air-sampling duration of 29.5 minutes, how much dust and gas were present if the filter contained 161 mcg of NO_3^-, and the impinger contained 2.7 mcg of NO_3^- per mL? The respective molecular weights of $NaNO_3$ and HNO_3 are 84.99 and 63.01. The molecular weight of nitrate anion, NO_3^-, is 62.01. Assume that this air sample was obtained at 25°C and 760 mm Hg.

It is reasonable to assume HNO_3 gas passed through the filter and was collected by the impinger and that essentially no sodium nitrate dust passed the filter.

$$1.73\,\text{L/min} \times 29.5\,\text{minutes} = 51.035\,\text{L}$$

$$161\,\text{mcg}\,NO_3^- \times \frac{84.99}{62.01} = 220.7\,\text{mcg}\,NaNO_3$$

$$\frac{220.7 \, \text{mcg NaNO}_3}{51.035 \, \text{L}} = 4.32 \, \text{mg NaNO}_3/\text{m}^3$$

$$\text{ppm} = \frac{(26.6 \, \text{mcg HNO}_3/51.035 \, \text{L}) \times 63.01}{24.45 \, \text{L/g-mole}} = 1.34 \, \text{ppm HNO}_3$$

Answers: 4.33 mg $NaNO_3$ dust/m³ and 1.34 ppm$_v$ HNO_3 gas—a dusty, irri-tating atmosphere. The ACGIH TLVs for HNO_3 gas, total inhalable particu-lates (NOC), and respirable particulates are 2 ppm$_v$, 10 mg/m³, and 3 mg/m³, respectively. If we assume one-third of total airborne dust by mass was respi-rable, what was the additive mixture exposure?

$$\frac{1.44 \, \text{mg/m}^3}{3 \, \text{mg/m}^3} + \frac{2.89 \, \text{mg/m}^3}{10 \, \text{mg/m}^3} + \frac{1.34 \, \text{ppm}}{2 \, \text{ppm}} = 1.44$$

This clearly is an overexposure if this air sample represented a worker's breathing zone and the exposure was prolonged. A short-term air sample, such as this, is perfectly acceptable to demonstrate the magnitude of expo-sures and if engineering and work practice controls are clearly required.

407. A worker exposed to a steady-state concentration of an air contaminant is required to wear an air-purifying respirator with a protection factor of 100. What is effective protection factor (P_{effect}) if he removes his respirator for only a 30-second period in his 8-hour work day?

$$480 \, \text{minutes} \times (60 \, \text{seconds/minute}) = 28{,}800 \, \text{seconds}$$

$$\frac{30 \, \text{seconds}}{28{,}800 \, \text{seconds}} \times 100 = 0.104\% \text{ of the exposure time}$$

$$1 - 0.104 = 99.986 = \% \text{ of exposure time when protection is provided}$$

$$P_{\text{effect}} = \frac{100}{[99.986/100] + 0.104} = 90.59$$

$$100 - 90.59 = 9.41$$

Answer: Removing respirator for only 30 seconds during an 8-hour work day reduces protection factor by 9.4%.

408. In the preceding example, what is P_{effect} if his respirator is worn 90% of the time?

$$480 \, \text{minutes} \times 0.1 = 48 \, \text{minutes}$$

$$P_{effect} = \frac{100}{[90/100] + 10} = 9.2$$

$$100 - 9.2 = 90.8$$

Answer: If worker's respirator is removed for any combinations of time that total 48 minutes during an 8-hour work day, respirators with protection factors of 1000 and 100 provide nearly the same level of respiratory protection. This respirator's protection factor, in this example, is reduced by an astounding 91%.

409. A worker's exposure to an air contaminant released from a point emission source can be estimated using a dispersion calculation method. If an industrial process, for example, emits 33 milligrams of hydrogen fluoride gas per second from a point source (such as a leaking valve in a pressurized line) in a 10 mph horizontal wind flow, what is a worker's most likely exposure to this gas if he is 50 feet away from the point emission source (i.e., downwind of the source)? What would the gas concentration be approximately if he was 10 feet away?

$$C = \frac{Q}{k u x^n},$$

where
   C = concentration of air contaminant expected in the breathing zone, mg/m^3
   k = a constant, 0.136
   u = wind speed, meters/second (with 0.5 m/sec being a minimum velocity)
   x = distance between the worker and the source, meters
   n = a constant, 1.84
   Q = emission rate, milligrams/second

$$10 \text{ mph} = 4.47 \text{ meters/second}$$

$$50 \text{ feet} = 15.24 \text{ meters}$$

$$C = \frac{33 \text{ mg HF/s}}{(0.136)(4.47 \text{ m/s})(15.24 \text{ meters})^{1.84}} = 0.36 \text{ mg/m}^3$$

$$10 \text{ feet} = 3.048 \text{ meters}$$

$$C = \frac{33 \text{ mg HF/s}}{(0.136)(4.47 \text{ m/s})(3.048 \text{ meters})^{1.84}} = 6.98 \text{ mg HF/m}^3$$

Answer: 0.36 mg HF/m^3 at 50 feet away. The TLV (ceiling concentration) for HF is 2.3 mg/m^3. At 10 feet, the predicted exposure concentration is 7 mg

HF/m³, or 3 × TLV (C). This illustrates a fundamental and primary industrial hygiene control method: reduce exposure by increasing the distance
between the worker and environmental stressors (air contaminant, heat,
noise, radiation, etc.). Move exposed workers away, and repair leak immediately. Arrange medical evaluation for those exposed.

410. Dry air at sea level has a total atmospheric pressure of 760 mm Hg. Oxygen
comprises 20.95%$_v$ (159.22 mm Hg), and 79.05%$_v$ (600.78 mm Hg) is comprised of nitrogen, argon, other inert gases, and carbon dioxide. If air is
humidified to 70% at 77°F (25°C), what is the change in the partial pressure
of the gases of composition? Vapor pressure of water at 77°F is 23.76 mm Hg.
0.7 × 23.76 mm Hg = 16.63 mm Hg (the partial pressure due to water vapor)

For oxygen: 20.95% × (760 − 16.63) mm Hg = 155.73 mm Hg

For nitrogen, argon, and so on: 79.05% × (760 − 16.63) mm Hg = 587.63 mm Hg

Answer: 155.7 mm Hg for oxygen and 587.6 mm Hg for the remaining gases.

411. The maximum allowable concentration of calcium carbonate dust (limestone)
in a worker's breathing zone is selected to be 10% of the TLV = 1 mg/m³
(TLV = 10 mg/m³). The total general dilution ventilation through his workspace is 10,000 cfm with 20% of the air (2000 cfm) recirculated. The ventilation mixing factor, K, is estimated at 3. What is the maximum allowable
concentration of calcium carbonate dust in the discharge air from the dust
collector before mixing with room air to meet the 1 mg/m³ maximum exposure limit?

$$C_R = \frac{1}{2}(\text{TLV} - C_o) \times \frac{Q_T}{Q_R} \times \frac{1}{K},$$

where
   C_R = concentration of contaminant in exit air from the collector before
      mixing
   Q_T = total ventilation through the affected workspace (ft³/minute)
   Q_R = recirculated air flow (ft³/minute)
   K = ventilation mixing factor, usually varying between 3 and 10 with one = excel-
      lent conditions, three = good mixing conditions, and 10 = extremely poor
      mixing conditions.
   TLV = Threshold limit value of the air contaminant. Only relatively non-
      toxic airborne contaminants may be recirculated—and then with exqui-
      site engineering controls. Recirculation of carcinogens and reproductive
      health toxicants must not be permitted.
   C_o = concentration of contaminant in worker's breathing zone with local
      exhaust discharged outside

$$C_R = \frac{1}{2} \times (10\,\text{mg/m}^3 - 1\,\text{mg/m}^3) \times \frac{10{,}000\,\text{cfm}}{2000\,\text{cfm}} \times \frac{1}{3} = 7.5\,\text{mg/m}^3$$

Answer: Maximum concentration in the discharge air of the collector may be no more than 7.5 mg/m³ (before mixing with dilution air) to ensure breathing zone concentrations do not exceed 1 mg/m³. Limestone typically contains small, but toxicologically significant amounts of crystalline silica. Such exposures must be quantified.

412. A coal-burning power plant emits 0.2% SO_2 by volume at a temperature of 260°F from its stack. The total volume of gases emitted is 5×10^5 cubic feet per minute. What is the SO_2 emission rate in units of mass/time (seconds)? The atmospheric pressure is 980 millibars (i.e., 0.98 bar).

$$\text{Molecular weight of } SO_2 = 64$$

$$260°F = 399.8 \text{ kelvin}$$

$$500{,}000\,\text{cfm} = 236\,\text{m}^3/\text{second}$$

$$\frac{\text{mass}}{\text{time}} = \frac{V}{t} \times \frac{P M}{R T} = \frac{(236\,\text{m}^3/\text{second})(0.002)(64)(980)}{(0.0832)(399.8\,\text{K})}$$

$$= 890\,\text{grams}\,SO_2/\text{second}$$

Answer: 890 grams of SO_2 are emitted per second.

413. A boiler maker removing fly ash from an oil-fired furnace air pollution electrostatic precipitator is exposed to an average total dust concentration of 3 mg/m³ over his 8-hour work shift. He does not use a respirator. Without an analysis of constituents of this fine dust, is this a potential significant inhalation hazard? In other words, should one not implement industrial hygiene controls because some information is lacking? After all, if this is only a "nuisance" dust (TLVs: inhalable particulates = 10 mg/m³; respirable particulates = 3 mg/m³ if NOC), an over exposure does not appear indicated. The author abhors the term "nuisance dust."

Typical Fuel Oil Ash Metal Analysis (As Elemental, but Most Likely Exist as Oxides)	
Iron	22.90 weight%
Aluminum	21.90
Vanadium	19.60*
Silicon	16.42
Nickel	11.86*
Magnesium	1.78
Chromium	1.37*
Calcium	1.14

Sodium	1.00
Cobalt	0.91*
Titanium	0.55
Molybdenum	0.23
Lead	0.17*
Copper	0.05
Silver	0.03
Miscellaneous	0.09
Total	100%

Source: Air Pollution Engineering Manual, U.S. Department of Health, Education, and Welfare 1967.

Applying the above typical percentages to just the * highlighted higher toxicity metals:

$3\,mg/m^3 \times 0.196$ (vanadium) $= 0.59\,mg\ V/m^3$	TLV $= 0.05\,mg/m^3$ (as respirable V_2O_5 dust or fume)
$3\,mg/m^3 \times 0.1186$ (nickel) $= 0.36\,mg\ Ni/m^3$	TLV $= 0.05\,mg/m^3$ (soluble Ni) and $0.1\,mg/m^3$ for (insoluble Ni)
$3\,mg/m^3 \times 0.0137$ (chromium) $= 0.04\,mg$ total chromium/m^3	TLV $=$ various: $0.01–0.5\,mg/m^3$
$3\,mg/m^3 \times 0.0091$ (cobalt) $= 0.03\,mg\ Co/m^3$	TLV $= 0.02\,mg/m^3$
$3\,mg/m^3 \times 0.0017$ (lead) $= 0.005\,mg\ Pb/m^3$	PEL $= 0.05\,mg/m^3$

Answer: Excessive exposures to several metals appear likely—particularly vanadium and nickel. Control methods are indicated. There might be crystalline silica (SiO_2) in the total silicon analysis. For this type of work, the author maintains that only an air-supplied full face respirator, or in some cases, a powered air-purifying mask, be used. Regular medical surveillance for those doing this work is suggested.

414. Inert gas blanketing techniques are often used to help prevent explosions in closed systems. Nitrogen or steam is often used for this. Vacuum purging with nitrogen gas is done by evacuating gases and vapors from the system and then increasing system pressure back to atmospheric pressure with N_2. This is done repeatedly until the desired minimum oxygen concentration is achieved. What would the oxygen concentration be after three purges in a system where attainable vacuum pressure is 0.6 atmosphere?

$$O_n = 21P^n,$$

where
$O_n =$ oxygen concentration after n purges
$P =$ vacuum pressure, bar absolute

$$0.6 \text{ atmosphere} = 0.608 \text{ bar}$$

$$O_n = 21(0.608)^3 = 4.72\%$$

To determine the number of purges to achieve a desired oxygen concentration:

$$n = \frac{\log[O_n/21]}{\log P} = \text{(from the preceding example)},$$

$$\frac{\log[4.72\%/21\%]}{\log 0.608 \text{ bar}} = \frac{-0.648}{-0.216} = 3$$

Answer: 4.72% oxygen by volume. This assumes inert gas does not contain any oxygen, and that the initial concentration of oxygen is 21%. Oxygen concentration should be regularly verified by reliable gas analysis. These are highly oxygen-deficient atmospheres without warning properties. Such systems must be posted, secured, and treated per OSHA's confined space entry procedure. If steam is used, special care must be exercised because many do not regard water vapor as an inert gas. The author investigated an asphyxiation fatality from steam because nitrogen was unavailable to create an inert head space blanket. The tank entrant, presumably (and the site safety engineer!) assumed that steam contained sufficient oxygen to support life. The tank atmosphere, tested after the worker's body was extracted, contained 2% oxygen by volume. Refer to Problem 417.

415. Secondary dust explosions can occur in buildings when an initial explosion from any cause entrains settled dust into the atmosphere, and the dust cloud is ignited. Assume that a small building 20 meters long, 10 meters wide, and 5 meters high has 2 millimeters of settled explosive dust dispersed evenly upon the floor. The bulk density of this dust is 450 kg/m³. If a small explosion in this building entrains 50% of the settled dust uniformly into the building's air, and there is a sufficiently hot ignition source, will there be a secondary explosion after the first blast wave?

50% of 2 mm bulk dust layer = 1 mm of settled dust suspended in the air

$$1 \text{ mm} = 0.001 \text{ meter}$$

$$\text{Floor area} = 20 \text{ meters} \times 10 \text{ meters} = 200 \text{ meters}^2$$

$$\text{Volume of dust suspended in air} = 200 \text{ meters}^2 \times 0.001 \text{ meters} = 0.2 \text{ m}^3$$

$$\text{Mass of dust suspended in air } 0.2 \text{ m}^3 \times 450 \text{ kg/m}^3 = 90 \text{ kg} = 90,000 \text{ g}$$

$$\text{Building volume} = 20 \text{ m} \times 10 \text{ m} \times 5 \text{ m} = 1000 \text{ m}^3$$

$$\text{Concentration of explosive dust in atmosphere} = \frac{90{,}000\,g}{1000\,m^3} = \frac{90\,\text{grams}}{m^3}$$

Answer: 90 grams of dust per cubic meter of air. Ranges of LELs for explosive dusts in air are approximately 10–2000 g/m³. So, depending upon this particular dust, the blast wave from the initial explosion could entrain enough dust to result in a normally more substantial secondary dust explosion. For example, some minimum dust concentrations for explosion in air (grams/m³) are: cornstarch (40), sugar (35), wheat flour (50), liver protein (45), aluminum stearate (15), coal (50), soap (45), rubber (25), aluminum (40), iron (250), zinc (480), and magnesium (10). Particle size of the dust (i.e., total surface area) is a very important variable. Refer to Problem 377.

416. Calculate the vapor/hazard ratio numbers for toluene and ethyl acetate. The vapor pressure and TLV for toluene are, respectively, 21 mm Hg and 20 ppm$_v$, and for ethyl acetate, respectively, are 73 mm Hg and 400 ppm$_v$.

First, calculate the equilibrium saturation concentrations for each organic vapor:

$$\text{For toluene: } \frac{21\,mm\,Hg}{760\,mm\,Hg} \times 10^6 = 27{,}632\,ppm$$

$$\text{For ethyl acetate: } \frac{73\,mm\,Hg}{760\,mm\,Hg} \times 10^6 = 96{,}053\,ppm$$

Next, divide the equilibrium saturation concentrations for each organic vapor by its TLV:

$$\text{For toluene: } \frac{27{,}632\,ppm}{20\,ppm} = 1381.6\ (\text{dimensionless})$$

$$\text{For ethyl acetate: } \frac{96{,}053\,ppm}{400\,ppm} = 240\ (\text{dimensionless})$$

Answer: The vapor/hazard ratio for ethyl acetate is 1381.6/240 = 5.76 times below that of toluene indicating that, although it is more volatile than toluene and reaches higher concentrations more quickly, its lower inhalation toxicity makes it perhaps a better choice in selecting a solvent if fire issues are effectively controlled. Toluene, unlike ethyl acetate, is skin-absorbed, greater systemic toxicant, and teratogen. While the vapor/hazard ratio is a reasonably good approach to consider when selecting solvents, other industrial hygiene and chemical safety engineering factors must obviously be considered, for example, flash point, ignition temperature, systemic versus local toxic effects, carcinogenicity, teratogenicity, solvent mixtures, consequences and degrees of exposures, and so on. This approach to risk management demonstrates that TLVs and PELs, alone, must not be used solely in arriving at a solvent selection.

417. Instead of vacuum purging to reduce the concentration of oxygen in a flammable atmosphere vessel (see Problem 414), pressure purging can be performed. An inert gas, such as nitrogen, is used to pressurize the vessel. The pressure is then relieved in a safe area, and the process is repeated until a desirable minimum oxygen concentration is achieved. What is oxygen concentration in the head space of a vessel after three pressure purges if the oxygen concentration of the purge gas is 3%, and the purge pressure is 2.3 bar?

$$O_n = O_p + (O_i - O_p)\left[\frac{1}{P^n}\right],$$

where

O_n = oxygen concentration after n purges
O_i = initial oxygen concentration, normally taken as 21%
O_p = oxygen concentration of the purge gas
P = purge pressure, bar absolute, and
n = number of inert gas pressurizations

$$O_n = 3\%(21\% - 3\%)\left[\frac{1}{2.3^3}\right] = 4.44\%$$

Answer: 4.44% oxygen by volume. The number of pressure purges necessary to achieve a preselected minimum oxygen concentration can be calculated by rearranging the above equation:

$$n = \frac{\log\left[(O_i - O_p)/(O_n - O_p)\right]}{\log P} = \text{(using the example)},$$

$$\frac{\log\left[(21\% - 3\%)/(4.44\% - 3\%)\right]}{\log 2.3} = \frac{\log\left[18/1.44\right]}{\log 2.3} = 3.03$$

Pressure purging uses more inert gas than vacuum purging to achieve the same reduced oxygen concentration. An advantage of pressure purging over vacuum purging are the briefer cycle times; that is, it generally takes longer to develop vacuum than it does to pressurize a vessel. Cost and performance determine the best method that, in some cases, can be a combination of the two procedures. The vessel must be rated for the maximum pressure and/or vacuum before either method may be employed, and allow a safety design factor before proceeding.

The volume of inert gas required to reduce the oxygen level from O_1 to O_2 is

$$Qt = V \ln \left[\frac{O_1 - O_0}{O_2 - O_0} \right],$$

where
     Q = volumetric flow rate
     t = time
     V = vessel volume
     O_0 = inlet oxygen concentration by volume
     O_1 = initial oxygen concentration by volume in vessel
     O_2 = final oxygen concentration by volume in vessel

418. Flow-through ventilation, or dilution purging, can be performed to create an inert (or reduced oxygen) atmosphere in a space with explosive vapors, gases, or dusts. The inert gas, usually nitrogen, flows longitudinally through the vessel and sweeps oxygen from the system. Choice of such a procedure requires careful placement of the inert gas entry point(s) and flow gas exit point(s). What is the concentration of oxygen in the exit gas after 10 minutes if the initial concentration of oxygen in a tank was 21%, the inert gas flow rate is 2 cubic meters per second, the volume of the tank head space is 700 m³, and the oxygen concentration of the purge gas is 2%?

$$O_f = O_p + [(O_i - O_p)e^{[-Qt/V]}],$$

where
     O_f = oxygen concentration after time, t,
     O_p = oxygen concentration of the inert purge gas,
     O_i = initial oxygen concentration (usually taken as 21% O_2, or ambient),
     Q = purge gas flow rate, and
     V = head space volume of vessel.

$$10 \text{ minutes} = 600 \text{ seconds}$$

$$O_f = 2\% + [(21\% - 2\%)\, e^{[(-2\,\text{m}^3/\text{second} \times 600 \text{ seconds})/700 \text{ m}^3]}] = 5.42\% \, O_2$$

This equation can be rearranged to determine purge time to achieve the desired minimum oxygen concentration, in this case, 5.42%:

$$t = -\frac{V}{Q} \ln \left[\frac{O_f - O_p}{O_i - O_p} \right] = -\frac{700 \, \text{m}^3}{2 \, \text{m}^3/\text{second}} \ln \left[\frac{5.42\% - 2\%}{21\% - 2\%} \right]$$

$$= -(350 \, \text{seconds}) \ln \frac{3.24}{18} = 600 \text{ seconds} = 10 \text{ minutes}$$

419. What is the final relative humidity if air at 95°F dry bulb and 40% relative humidity is cooled to 70°F dry bulb without the addition or removal of moisture?

This can be determined from a good psychrometric chart or calculated as follows:

Since a fixed air mass is cooled, its volume becomes smaller according to Charles' law: volume of a mass of gas is directly proportional to its absolute temperature when the pressure is held constant, or, using $1000\,ft^3$ as the original volume:

$$\frac{V_i}{T_i} = \frac{V_f}{T_f}, \quad \text{rearranging: } V_f = \frac{V_i\,T_f}{T_i} = \frac{(1000\,ft^3)(294.26\,K)}{308.15\,K} = 954.92\,ft^3$$

At 95°F, the vapor pressure of water is 42.18 mm Hg, and at 70°F, the vapor pressure of water is approximately 18.68 mm Hg.

$$\text{A. ppm }H_2O\text{ vapor at } 95\,°F = 0.4 \times \frac{42.18\,mm\,Hg}{760\,mm\,Hg} \times 10^6 = 22,200\,ppm$$

(100% relative humidity would be $55,500\,ppm_v = 5.55\%$ water vapor.)

The gram molar gas volumes at 760 mm Hg and 70 and 95°F are, respectively, 24.13 and 25.27 L.

$$\frac{mg\,H_2O}{m^3} = \frac{ppm \times \text{molecular weight}}{25.27} = \frac{22,200 \times 18}{25.27} = \frac{15,813\,mg}{m^3}$$

$15,813\,mg/m^3 = 447.776\,mg/ft^3$ at 95°F $= 447,776\,mg$ of water vapor in $1000\,ft^3$ at 70°F: $447,776\,mg/954.92\,ft^3 = 469\,mg/ft^3 = 16,563\,mg/m^3$

$$ppm = \frac{(16,563\,mg/m^3) \times 24.13}{18} = 22,204\,ppm$$

Note how ppm_v are constant between the two conditions of temperature. Equation A (above) can be rearranged to solve for the relative humidity at the new condition:

$$\%\text{ relative humidity} = \frac{ppm\,H_2O\text{ vapor (at elevated DB temperature)}}{[(\text{vapor pressure, mm Hg (at reduced DB temperature)})/}$$
$$760\,mm\,Hg) \times 10^4]$$

$$= \frac{22,200\,ppm}{(18.68\,mm\,Hg/760\,mm\,Hg) \times 10^4} = 90.3\%$$

Answer: 90.3% relative humidity at 70°F—in other words, cooler, moister air. Significant cooling (air conditioning) can be achieved if much of the moisture is removed from this air before it is supplied to occupied areas. If using psychometric charts, note that the dew point temperature does not change between these two conditions.

420. A rectangular ventilation duct is 14" × 20". What is the equivalent diameter size for a round duct?

$$D_{equiv} = (1.3)\left[\frac{(A \times B)^{0.625}}{(A + B)^{0.25}}\right],$$

where

D_{equiv} = equivalent diameter of the round duct size for a rectangular duct, inches

A = one side of the rectangular duct, inches

B = adjacent size of the rectangular duct, inches

$$D_{equiv} = (1.3)\left[\frac{(14 \times 20)^{0.625}}{(14 + 20)^{0.25}}\right] = (1.3)\left[\frac{(280)^{0.625}}{(34)^{0.25}}\right] = 18.2 \text{ inches}$$

Actually, select an 18-inch diameter duct since this size is essentially an "off-the-shelf" item. Selection of a smaller diameter increases duct transport velocity. In general, round ducts are stronger than un-braced rectangular ducts.

421. In a 6-foot diameter duct, a 10-point Pitot tube traverse in each of two directions gave the following readings (in inches of water) and corresponding duct velocities calculated as shown below:

0.70" (4180 fpm)	0.62" (3930 fpm)	Gas temperature = 300°F
0.79 (4440)	0.65 (4030)	
0.83 (4550)	0.67 (4080)	Altitude = 100 feet above sea level
0.89 (4710)	0.75 (4330)	
0.91 (4760)	0.90 (4740)	Dew point of gas = 140°F
0.90 (4740)	0.89 (4730)	
0.93 (4820)	0.89 (4730)	
0.85 (4620)	0.89 (4730)	
0.80 (4470)	0.70 (4180)	
0.78 (4420)	0.70 (4180)	
Σ = 45,710	Σ = 43,660	

What is the actual air flow rate and velocity? What is the standard air flow rate?

$$V = 174 \sqrt{P_v \frac{(t + 460)}{K \times d}},$$

where

V = duct velocity, fpm

P_v = velocity pressure, inches of water

t = air stream temperature, °F

K = relative density for altitude (= 1.0 for altitudes <1000 feet)

d = relative density correction factor for moisture. (Note that increased amounts of water vapor reduce the gas density since water vapor molecules weigh less than air molecules: 18 vs. 29.)

The moisture content for 140°F dew point air is 0.17 pound of water/pound of dry air. From standard psychrometric tables and charts, d is 0.918.

Taking the first velocity pressure reading of 0.70" above:

$$V = 174 \sqrt{(0.70)\frac{(300 + 460)}{(1)(0.918)}}$$

$$= 4180\,\text{fpm. Repeat this for the other 19 readings!}$$

therefore,

$$V_{avg} = \frac{45,710\,\text{fpm} + 43,660\,\text{fpm}}{20} = 4470\,\text{fpm}$$

$$\text{area of six-foot diameter duct} = \frac{\pi}{4} \times (6\,\text{ft})^2 = 28.3\,\text{ft}^2$$

Air flow rate = 28.3 ft² × 4470 fpm = 126,500 actual cfm

because the temperature is 300°F, the standard air flow rate is

$$126,500\,\text{cfm} \times \frac{460 + 70}{460 + 300} = 88,200\,\text{standard cfm}$$

Answer: 126,500 actual cfm. 88,200 standard cfm. 4470 fpm duct velocity. If the gas was cooled to standard conditions, water condensation will occur. These calculations are tedious if not daunting. But there is relief for those who do these often in the form of high-quality computer software. Contact ASHRAE for a list of suppliers.

422. A ventilation system branch duct with a 10,000 cfm design volume has a static pressure of −2.1 inches of water at the branch entry. The main, carrying a volume of 50,000 cfm, has a static pressure of −2.40 inches of water where the branch enters. What volume will be drawn through the branch at a balanced condition?

$$Q_b = Q_o \sqrt{\frac{P_b}{P_o}},$$

where

Q_b = air flow volume required for balance
Q_o = design air flow volume
P_b = static pressure required for balance
P_o = static pressure originally calculated for balance

$$Q_b = 10,000\,\text{cfm} \times \sqrt{\frac{2.40}{2.10}} = 10,700\,\text{cfm}$$

423. An exhaust ventilation hood is required for gas tungsten-arc welding of non-ferrous metals at an assembly line production operation. The hood face can be no closer than 9 inches from the arc. A 6-inch flange can be installed on the 8" × 12" hood face. Because of variable cross-drafts that are difficult to control, a capture velocity of 300 fpm is desired. What exhaust volume is required?

$$Q = K(10x^2 + A)V_x,$$

where

Q = exhaust volume, cfm
X = the distance from the center of the hood face to the farthest point of welding fume, smoke, and gases release, feet
A = hood face area (not including the flange), ft^2
V_x = minimum capture velocity, fpm
K = 1.0 for an unflanged hood; 0.75 for a large-flanged hood

$$Q = 0.75\left[10(0.75\,\text{ft})^2 + \left(\frac{8\,\text{inches} \times 12\,\text{inches}}{144\,\text{in}^2/\text{ft}^2}\right)\right]300\,\text{fpm} = 1416\,\text{cfm}$$

Be cognizant of the generation of ozone gas a considerable distance from the arc, such as occurs with TIG welding of aluminum or other "white" metals.

424. In preceding problem (423), select a round duct that will provide a minimum duct transport velocity of 3000 fpm.

$$A = \frac{Q}{V} = \frac{1416\,\text{cfm}}{3000\,\text{fpm}} = 0.472\,\text{ft}^2$$

A 9-inch internal diameter duct has a cross-sectional area of 0.4418 ft^2.

$$V = \frac{Q}{A} = \frac{1416\,\text{cfm}}{0.4418\,\text{fpm}} = 3205\,\text{fpm}$$

Answer: Nine-inch ID duct. When selecting a duct to achieve minimum transport velocity, typically one should select the next smaller size so pollutant transport minimum velocity is achieved. For welding fume, in general, the minimum duct transport velocity is typically 2000 fpm. In some cases, 3000 fpm would be a better design choice especially if large airborne particles are generated (e.g., welding slag).

425. Refer to Problem 419. If there was 60% condensation of the water vapor in 70°F and 90.3% relative humidity air, how much water vapor (in gallons) would be condensed per hour from a single-pass ventilation system delivering 20,000 cfm?

$$\text{Vapor pressure of water at } 70°F = 18.68\,\text{mm Hg}$$

$$0.903 \times 18.68\,\text{mm Hg} = 16.87\,\text{mg Hg}$$

$$\text{ppm}\,H_2O\,\text{vapor} = \frac{16.87\,\text{mm Hg}}{760\,\text{mm Hg}} \times 10^6 = 22,197\,\text{ppm}$$

$$\frac{\text{mg}\,H_2O}{m^3} = \frac{\text{ppm} \times \text{molecular weight}}{24.13\,\text{L/gram-mole}} = \frac{22,197\,\text{ppm} \times 18}{24.13}$$

$$= \frac{16,558\,\text{mg}\,H_2O}{m^3}$$

$$16,558\,\text{mg}\,H_2O/m^3 = 468.9\,\text{mg/ft}^3$$

$$\frac{468.9\,\text{mg}}{ft^3} \times \frac{20,000\,ft^3}{\text{minute}} \times \frac{60\,\text{minutes}}{\text{hour}} = \frac{5.63 \times 10^8\,\text{mg}\,H_2O}{\text{hour}}$$

$$= \frac{1241\,\text{pounds}\,H_2O}{\text{hour}}$$

Since 60% of the water vapor condenses: 0.6 (1241 lb/hour) = 744.6 lb/ hour

$$\frac{744.6\,\text{lb/hour}}{8.33\,\text{lb/gallon}} = \frac{89.4\,\text{gallons}}{\text{hour}}$$

Answer: 89.4 gallons of water vapor will condense into liquid water every hour.

426. Community air was sampled for 30 days and 17 hours at 7.3 cfm through an 8" × 10" high-efficiency particulate filter. A 2 cm × 2 cm section of filter contained 89 micrograms of lead. What was the average airborne lead concentration during the sampling period (assuming lead mass was evenly distributed across the filter face)?

$$[30\,\text{days} \times (24\,\text{hours/day})] + 17\,\text{hours} = 737\,\text{hours} = 44,220\,\text{minutes}$$

$$44,220\,\text{minutes} \times 7.3\,\text{cfm} = 322,806\,ft^3 = 9140.8\,m^3$$

$$4\,cm^2 = 0.62\,in^2$$

$$(80\,in^2/0.62\,in^2) = 129.03 = \text{filter area multiplication factor}$$

$$89\,mcg\,Pb \times 129.03 = 11{,}484\,mcg\,Pb$$

$$\frac{11{,}484\,mcg\,Pb}{9140.8\,m^3} = \frac{1.26\,mcg\,Pb}{m^3}$$

Los Angeles, for example, before tetraethyl lead was eliminated from motor fuels, typically had an arithmetic mean and a geometric mean of 1.7 and 1.08 mcg Pb/m³, respectively.

427. A 120-liter cylinder of nitric oxide at 5.2 atmosphere and 20°C developed a leak. When the leak was repaired, 2.1 atmosphere of nitric oxide gas remained in the cylinder that was still at 20°C. How many moles and grams of NO gas escaped?

Moles of gas escaped = original moles of gas − final moles of gas

$$= \frac{(5.2\,atm)(120\,L)}{(0.0821)(293\,K)} - \frac{(2.1\,atm)(120\,L)}{(0.0821)(293\,K)} = 25.94\,moles - 10.48\,moles$$

$$= 15.46\,moles$$

$$15.46\,moles\,NO \times 30.01\,grams/mole = 463.95\,grams\,of\,NO$$

Answer: 15.46 moles and 463.95 grams of NO gas escaped from this gas cylinder.

428. If, in the preceding problem, NO gas escaped into an empty 30' × 80' × 18' room without ventilation, what NO gas concentration could be predicted after thorough mixing?

$$15.46\,moles\,NO \times 30.01\,grams/mole = 463.95\,grams$$

$$30' \times 80' \times 18' = 43{,}200\,ft^3 = 1223.3\,m^3$$

$$ppm = \frac{(463{,}950\,mg\,/\,1223.3\,m^3) \times 24.45}{30.01} = 309\,ppm\,NO$$

Answer: 309 ppm$_v$ NO (TLV is 25 ppm$_v$). NO converts into NO_2 at a steady rate following first-order decay reaction kinetics, so the atmosphere in the room contains a mixture of noxious gases: $2NO + O_2 \rightarrow 2NO_2 \leftrightarrow N_2O_4$. Beware all ye who enter without respiratory protection!

429. A chemical operator inadvertently poured 890 grams of potassium iodide into a 70-gallon vat of sulfuric acid. There was excess sulfuric acid to oxidize the reducing agent (KI) to hydrogen sulfide. What volume of H_2S was produced and released to the atmosphere assuming negligible solubility of the H_2S in the sulfuric acid?

$$8KI + xs\ 5H_2SO_4 \rightarrow 4K_2SO_4 + 4I_2 \uparrow + H_2S \uparrow + 4H_2O$$

Eight moles of KI generate 1 mol of H_2S, or

$$(890\,g\,KI)\left[\frac{1\,mole\,KI}{166\,g\,KI}\right]\left[\frac{1\,mole\,H_2S}{8\,moles\,KI}\right]\left[\frac{24.45\,L\,at\,NTP}{mole\,H_2S}\right]$$

$$= 16.39\,liters\,of\,H_2S.$$

16,390,000 liters of air would be necessary to dilute this rotten egg odor gas to 1 ppm. Iodine vapor would also be generated. Arguably, iodine, as a weak oxidizing vapor, could react with H_2S, a weak reducing agent, in the atmosphere to produce elemental sulfur and hydriodic acid:

$$I_2 + H_2S \longrightarrow 2\,HI + S_o$$

The iodine is dissolved in a dilute solution of potassium iodide. When the purple of I_2 disappears, the H_2S has been oxidized, and one can then calculate the amount of H_2S that was present in the "titrated" air sample.

The rate of such reaction would depend, in part, on the intermolecular collisions of the molecules. That is, higher concentrations, temperature, and pressure will drive a faster rate of reaction. See Problem 703.

430. Calculate LCL and UCL for a time-weighted air-sampling result of 28 ppm$_v$ for a worker. The PEL/TLV for this organic vapor is 25 ppm$_v$. The combined analytical, sampling, and other cumulative errors are ±19.5% (sampling analytical error, SAE; see Problem 5 for the procedure to calculate SAE).

$$LCL = \frac{EC}{PEL\,or\,TLV} - SAE \ \ and \ \ UCL = \frac{EC}{PEL\,or\,TLV} + SAE,$$

where
   LCL and UCL = lower and upper confidence limits, respectively. The UCL is rarely used in normal industrial hygiene compliance determinations.
   EC = exposure concentration (ppm, f/cc, or mg/m^3)
   PEL or TLV = permissible exposure limit or threshold limit value
   SAE = sampling and analytical error as a decimal (absolute—disregard algebraic signs)

$$LCL = (28/25) - 0.195 = 0.925$$
$$UCL = (28/25) + 0.195 = 1.315$$

Answer: LCL is 0.925. UCL is 1.315. Since the LCL is less than 1, and assuming a normal distribution for exposure levels, a violation of the PEL and TLV has not occurred. However, since the exposure is well above the action level, industrial hygiene controls are warranted. The level of confidence in this statistical parameter is 95% which means that, to a 95% degree of certainty, the true value lies between 92.5% and 131.5% of the reported value, or 25.9–36.8 ppm$_v$.

431. Automobiles are multicoat spray painted with an average of 1.1 gallons of high-solids paint (43% solids) per vehicle. The paint solvents are primarily a mixture of aromatic hydrocarbons, ketones, alcohols, and naphtha. The automobile bodies are baked to dryness in an oven at 350°F for 13 minutes. An air pollution control engineer wants to know what the solvent evaporation rates are in pints per vehicle and the total solvent vapor emissions if this assembly plant produces 42 vehicles per hour. Can you help him?

   Essentially 57% of the paint mass evaporates in the baking oven (100–43%). A small amount evaporates in the booth vestibules and spray booths leading to oven. Regardless, all solvent sources contribute to volatile organic carbon atmospheric emissions (VOCs).

$$(0.57) \, (1.1 \text{ gallons/vehicle}) \, (8 \text{ pints/gallon}) = 5 \text{ pints/vehicle}$$

$$5 \text{ pints/vehicle} \times 42 \text{ vehicles/hour} = 210 \text{ pints/hour} = 3.5 \text{ pints/minute}$$

5 pints/vehicle. 3.5 pints/minute. 210 pints/hour. Emissions in baking ovens lend themselves to engineering controls, whereas vapors released outside of the ovens are fugitive emissions that are more difficult to control.

432. How much water vapor is in a building 15 meters × 30 meters × 5 meters if relative humidity is 40%, and dry bulb temperature is 86°F (30°C)? The vapor pressure of water at this temperature is 31.82 Torr (31.82 mm Hg). Assume an empty building.

$$15 \, \text{m} \times 30 \, \text{m} \times 5 \, \text{m} = 2250 \, \text{m}^3 = 2{,}250{,}000 \, \text{L}$$

$$P = (0.40) \, (P_{vapor}) = (0.40) \, (31.82 \text{ Torr}) = 12.73 \text{ Torr (partial pressure due to water vapor)}$$

$$n = \frac{PV}{RT} = \frac{[(12.73/760)\text{atm}](2{,}250{,}000\,\text{L})}{(0.0821\,\text{L-atm}/\text{mole-K})(303.15\,\text{K})}$$
$$= 1514.2 \text{ moles of water vapor}$$

$$(1514.2\,\text{moles}) \left[\frac{18.0\,\text{g}}{\text{mole}} \right] \left[\frac{1\,\text{kg}}{10^3\,\text{g}} \right] = 27.26\,\text{kg of water vapor}$$

Answer: 27.26 kilograms of water vapor (60.1 pounds).

433. Confined, occupied spaces without outside fresh air ventilation (e.g., sub-
marines, space capsules, diving bells) can accumulate significant levels of
carbon dioxide gas from human respiration. Lithium oxide, Li_2O, is the
most efficient CO_2 gas scavenger. What is the CO_2 absorption efficiency of
Li_2O in liters of CO_2 (at STP) per kilogram?

$$Li_2O + CO_2 \rightarrow Li_2CO_3 \text{ (lithium carbonate)}$$

$$(1\,\text{kg}\,Li_2O) \left[\frac{1000\,\text{g}}{\text{kg}} \right] \left[\frac{1\,\text{mole}\,Li_2O}{29.88\,\text{g}\,Li_2O} \right] \left[\frac{22.4\,\text{L at STP}}{\text{mole of }CO_2} \right] \left[\frac{1\,\text{mol}\,CO_2}{1\,\text{mol}\,Li_2O} \right]$$

$$= 749.7\,\text{liters}$$

Answer: Each kilogram of lithium oxide theoretically absorbs 749.7 liters
of carbon dioxide gas at STP. Knowing the maximum CO_2 gas generation
rate for a number of people present (body mass, metabolic rates, etc.), the
amount of lithium oxide required for an air-cleansing system can be calculated. At STP, CO_2 weighs $1977\,\text{g/m}^3 = 1.977$ grams per liter. Exhaled air
contains up to 5.6% CO_2 by volume.

434. A brine-ammonia refrigeration plant loses an average of $35\,\text{ft}^3$ of ammonia
every day from leaking valves and process equipment as determined from
the purchase records. What is the dilution ventilation air volume necessary
to control to no more than 10% of the TLV of 25 ppm for NH_3?

$$\frac{35\,\text{ft}^3/\text{day}}{24\,\text{hours/day}} = \frac{1.458\,\text{ft}^3}{\text{hour}} \times \frac{\text{hour}}{60\,\text{minutes}} = 0.0243\,\text{ft}^3/\text{minute}$$

$$25\,\text{ppm}_v \times 0.1 = 2.5\,\text{ppm}_v$$

$$\frac{0.0243\,\text{ft}^3/\text{minute}}{2.5\,\text{ppm}} \times 10^6 = 9720\,\text{cfm}$$

Answer: 9720 cubic feet of fresh dilution air per minute assuming good
mixing. Repair the leaking equipment first with local exhaust ventilation
placed at the source during the repairs; then consider increasing the dilution
ventilation only if needed.

435. An elderly, senile man, in an attempt to cool his top-floor, nonair condi-
tioned and nonventilated room, placed a 50-lb block of dry ice next to his

bed. When his body was recovered from the $12' \times 16' \times 8'$ room, the dry ice block weighed about 22 pounds. What was the most likely average oxygen concentration in the air in the room just before the man's body was discovered? What was the most likely average CO_2 gas concentration?

Assume a worst-case scenario in which there was no ventilation or air exchange in the room.

$$\frac{50\,lb\,CO_2 - 22\,lb\,CO_2}{12\,ft \times 16\,ft \times 8\,ft} = \frac{28\,lb}{1536\,ft^3} = \frac{0.0182\,lb\,CO_2}{ft^3}$$

$$\frac{0.0182\,lb}{ft^3} \times \frac{35.315\,ft^3}{m^3} \times \frac{kg}{2.205\,lb} \times \frac{1000\,g}{kg} \times \frac{1000\,mg}{g}$$
$$= \frac{291,489\,mg\,CO_2}{m^3}$$

$$ppm = \frac{(291,489\,mg/m^3) \times 24.45}{44.01} = 161,938\,ppm\,CO_2 = 16.19\%\,CO_2$$

$$100\%\,air - 16.19\%\,CO_2 = 83.81\%\,air\,remaining$$

$$83.81\%\,air \times 0.2095 = 17.55\%\,oxygen$$

Answer: About 17.6% oxygen and 16.2% CO_2. OSHA regards an atmosphere containing less than 19.5% oxygen as oxygen-deficient. The NIOSH IDLH for CO_2 is 40,000 ppm (4%). Protracted exposure of healthy adults to air containing 17.5% oxygen, while physiologically taxing, would probably not cause significant adverse effects. In this case, however, CO_2 contributes toxic effects as a respiratory and cardiac stimulant, prolonged CO_2 inhalation can produce metabolic acidosis, the gentleman was old (with, possibly, poor cardiovascular status), and heat strain was, most likely, an added myocardial stressor. Moreover, since the block of dry ice was located next to the bed where his body was located, pockets of carbon dioxide gas (e.g., say, 50%?) could have surrounded his head for times sufficiently long to produce anoxia, asphyxiation, and a heart attack.

436. The carbon dioxide gas exhaled by laboratory rats into the test chambers used in experimental toxicology will be absorbed so that the chamber atmosphere can be recirculated. How much carbon dioxide gas can be absorbed into a solution that contains 900 grams of potassium hydroxide? Assume 100% CO_2 gas absorption. The molecular weights of potassium hydroxide and carbon dioxide are 56.11 and 44.01, respectively.

$$CO_2 + 2KOH \rightarrow K_2CO_3 + H_2O$$

K_2CO_3, by the way, is also known as potash and was originally obtained from the po̲tassium-rich a̲shes of wood fires. Potash and lard, or animal fat, were mixed to produce the original soap.

Therefore, 2 moles of KOH (or NaOH) will absorb 1 mole of CO_2 gas.

$$\frac{900 \text{ g KOH}}{56.11 \text{ g/mole}} = 16.04 \text{ moles of KOH}$$

16.04 moles of KOH will absorb 8.02 moles of CO_2 gas.

$$8.02 \text{ moles CO}_2 \times \frac{44.01 \text{ grams}}{\text{mole}} = 352.96 \text{ grams of CO}_2$$

Refer to Problem 433 and see how lithium oxide generally is more efficient (kilogram per kilogram) as a CO_2 gas scavenger. The higher cost of lithium oxide, however, might argue for the purchase of the less expensive KOH or NaOH.

437. What is the density of a mixture of 500 ppm methyl alcohol vapor in air?

The "apparent" molecular weight of dry air (based on the volume% composition of nitrogen, oxygen, argon, CO_2, etc.) is nearly 29 (exactly = 28.941 grams/gram-mole of air). The density of dry air is 1.2 mg/mL at 25°C and 760 mm Hg. The density of air is defined as 1.00 (no units, a relative number). Since methanol has a molecular weight of 32.01 grams/gram-mole, the density of 100% MeOH vapor (i.e., 10^6 ppm MeOH) compared to air is (since a mole of MeOH has greater mass than a "mole" of air):

$$\frac{32.01}{28.941} = 1.106$$

$$500 \text{ ppm}_v \text{ MeOH} = 0.05 \text{ vol}\%$$

For MeOH vapor:	0.0005×1.106	= 0.00055
For air:	$\underline{0.9995 \times 1.00}$	= $\underline{0.9995}$
Total	1.0000	1.00005

Answer: 1.00005 (no units) is essentially the same as air since (1) MeOH has a molecular weight near air, and (2) 500 ppm$_v$ MeOH is relatively dilute concentration of vapor, that is, 500/1,000,000.

438. The United States Nuclear Regulatory Commission established in their document NUREG-1391 (*Chemical Toxicity of Uranium Hexafluoride Compared to the Acute Effects of Radiation*) the values for uranium and hydrogen fluoride gas that should be used for accident analysis purposes at facilities that process large quantities of UF_6. UF_6 reacts exothermically

with moisture in air to form uranyl fluoride (UO_2F_2) and hydrogen fluoride gas (HF):

$$UF_6 + 2H_2O \rightarrow UO_2F_2 + 4HF + heat$$

The human inhalation exposure value for hydrogen fluoride gas is

$$HF\, concentration = \left[\frac{25\,mg}{m^3}\right]\sqrt{\frac{30\,minutes}{t}},$$

where t is the time in minutes of the duration of exposure. If the exposure duration is 15 minutes, what is the maximum acceptable HF gas concentration in ppm_v using these NRC criteria?

$$HF\, concentration = \left[\frac{25\,mg}{m^3}\right]\sqrt{\frac{30\,minutes}{15\,minutes}} = \frac{35.4\,mg}{m^3}$$

$$ppm = \frac{(mg/m^3) \times 24.45}{molecular\, weight\, of\, HF} = \frac{(35.4\,mg/m^3) \times 24.45}{20.01}$$

$$= 43.3\,ppm\, HF$$

This exceeds the ACGIH "ceiling" TLV of 3 ppm_v. AIHA Emergency Response Planning Guideline for HF gas for up to 1-hour exposure is 2 ppm_v (with most "not experiencing other than mild transient adverse health effects or perceiving a clearly defined objectionable odor"); for an ERPG-2 ("Nearly all individuals could be exposed for up to one hour without experiencing or developing irreversible or serious health effects or symptoms that could impair abilities to take protective action."), the value for HF is 20 ppm_v; for an ERPG-3 ("nearly all individuals could be exposed for up to one hour without having life-threatening health effects"), the value for HF gas is 50 ppm_v.

439. If the air temperature (dry bulb) is 86°F, and the dew point is 68°F, what is relative humidity? The pressure of saturated water vapor at 86°F is 31.82 mm Hg. At 68°F, the pressure of saturated water vapor is 17.54 mm Hg.

$$Relative\, humidity = \frac{saturated\, pressure\, of\, water\, vapor\, in\, air\, at\, dew\, point}{pressure\, of\, water\, vapor\, in\, saturated\, air\, at\, 86°F}$$

$$= \frac{17.54\,mm\, Hg}{31.82\,mm\, Hg} \times 100 = 55.1\%$$

The dew point is the temperature at which the atmosphere would be saturated with contained water vapor. We observe dew point phenomena on a humid summer day with a glass of ice tea. The dew point is the tempera-ture of the cold glass upon which condensation of atmospheric moisture accumulates.

440. Fifty pounds of sodium hydroxide pellets are rapidly dumped into an open top steel tank containing 40 gallons of water. The solution becomes very hot, and there is evolution of an irritating caustic mist aerosol. After returning to room temperature, 42.9 gallons of solution remain in the tank. Correcting for blank sodium content, the solution contained 80,070 mg Na/L. If the building in which the NaOH aerosol was released was 35' × 120' × 18', how much sodium hydroxide was released into the air? What was the average mist concentration if the room did not have ventilation?

Molecular weights of Na and NaOH, respectively, are 23 and 40.

$$42.9 \, \text{gallons} \times \frac{80.07 \, \text{g Na}}{\text{L}} \times \frac{1 \, \text{L}}{0.264 \, \text{gallon}} \times \frac{40}{23} \times \frac{1 \, \text{lb}}{453.59 \, \text{g}}$$

$$= 49.88 \, \text{lb NaOH}$$

$$\frac{(50 \, \text{lb NaOH} - 49.88 \, \text{lb NaOH})}{35 \, \text{ft} \times 120 \, \text{ft} \times 18 \, \text{ft}} = \frac{0.12 \, \text{lb NaOH}}{75,600 \, \text{ft}^3}$$

$$\frac{0.12 \, \text{lb NaOH}}{75,600 \, \text{ft}^3} \times \frac{35.314 \, \text{ft}^3}{\text{m}^3} \times \frac{459.59 \, \text{g}}{\text{lb}} \times \frac{1000 \, \text{mg}}{\text{g}} = \frac{25.8 \, \text{mg NaOH}}{\text{m}^3}$$

Answer: 0.12 lb of NaOH was released into the air. Average concentration of NaOH mist in the air was 25.8 mg/m³ (ACGIH ceiling TLV = 2 mg/m³). The mist concentration near the tank, obviously, initially would have been much higher.

441. A grave explosion hazard presents in the molten materials industry when materials above 100°C are charged in vessels that contain liquid water. Huge explosions resulting in fatalities, injuries, and substantial property damages have occurred when, for example, molten metal is poured into receiving vessels, or onto a floor where water—even a tiny amount—accumulated. The higher the molten material temperature, the greater the amount of water, and the higher the rate of addition all contributively add to explosion's magnitude. Calculate volume of steam produced when 1 ton of molten iron is carelessly added to a 4-feet internal diameter pouring ladle containing water 3-inches deep. Atmospheric pressure is 720 mm Hg.

Water volume $= \pi \, r^2 h = \pi \, (2 \, \text{ft})^2 \times 0.25 \, \text{ft} = 3.1416 \, \text{ft}^3 = 88,960 \, \text{mL}$ of liquid water

$$\text{Density of water} = 1.00 \, \text{g/mL}$$

$$PV = nRT, \text{ the universal gas law}$$

$$\frac{88,960\,g\,H_2O}{18.0\,g/mole} = 4942\,moles\,of\,water$$

$$T = 100°C + 273\,K = 373\,K$$

$$V = \frac{nRT}{P} = \frac{(4942\,moles)(0.0821\,atm\text{-}liter/mole\text{-}K)(373\,K)}{(720\,mm\,Hg/760\,mm\,Hg) = 0.947\,atmosphere} = 159,810\,L$$

Answer: The liquid volume of 88.96 liters of water explosively expands into a 159,810 liter steam cloud. For this reason, torches or other means are applied to keep the interiors of receiving vessels bone dry before addition of molten metals, glass, oil, plastics, and so on. Besides the plume of steam, the molten metal spews from the ladle with dire consequences if workers are nearby.

442. A reasonable, reliable microscopic quantification of fibers on air filters using phase-contrast microscopy (PCM) is 10 fibers per 100 fields. What are the equivalent, reliable PCM quantifications for sampling volumes of 3000, 5000, and 7500 L of air?

   855 mm² is the collection area of a 37 mm diameter air filter. Obviously, a smaller diameter filter (e.g., 25 mm) could improve detection limits.

   0.003 mm² is the size of a typical field of view for a PC microscope that varies between 0.003 and 0.006 mm² for different microscopes. Larger fields of view will improve (decrease) the limit of reliable quantification.

$$Quantification\,limit = \frac{10\,fibers/100\,fields}{3000\,liters} \times \frac{855\,mm^2}{0.003\,mm^2} \times \frac{1\,liter}{1000\,cc}$$

$$= \frac{0.01\,fiber}{cc}$$

$$\frac{0.01\,f/cc}{x\,f/cc} = \frac{5000\,L}{3000\,L}\quad x = 0.006\,f/cc\qquad \frac{0.01\,f/cc}{x\,f/cc} = \frac{7500\,L}{3000\,L}\quad x = 0.004\,f/cc$$

Answers: 0.01 f/cc, 0.006 f/cc, and 0.004 f/cc, respectively for air-sampling volumes of 3000 L, 5000 L, and 7500 L. NIOSH PCM Method 740 will improve the reliable limit of quantification for the same air-sampling volumes.

443. Johnny yanks the thermometer from his mouth and throws it against the wall. The thermometer smashes releasing 0.5 mL of mercury that spills on the carpet of his bedroom. Johnny lives in a 1500 square feet bungalow without a basement. The ceilings are 8-feet high. Assuming absolutely no

ventilation of his house, what maximum average mercury vapor concentration can be, in time, achieved within this house? Assume no adsorption on surfaces.

This problem can be approached in no less than three ways: collect air samples until the equilibrium Hg vapor concentrations are reached; perform calculations assuming all mercury evaporates into the house atmosphere; and calculate the maximum vapor concentration based upon mercury's vapor pressure. Mercury's molecular weight, density, and vapor pressure at normal temperature are 200.6, 13.6 g/mL, and 0.0012 mm Hg, respectively.

Mass/volume method:

$$0.5\,mL \times 13.6\,g/mL = 6.8\,g = 6800\,mg \text{ of mercury evaporate}$$

$$1500\,ft^2 \times 8\,ft = 12,000\,ft^3 = 339.8\,m^3$$

$$\frac{6800\,mg}{339.8\,m^3} = 20.01\,mg/m^3$$

Vapor pressure method:

$$\frac{0.0012\,mm\,Hg}{760\,mm\,Hg} \times 10^6 = 1.579\,ppm\,Hg \text{ vapor at saturation}$$

$$mg\,Hg/m^3 = \frac{ppm \times mol.wt.}{24.45\,L/g\text{-mole}} = \frac{1.579\,ppm \times 200.6}{24.45} = 12.95\,mg/m^3$$

Answer: Note large variance between calculated concentrations. Because air cannot contain more than saturated concentrations, the correct answer is 12.95 mg/m³. This is considerably greater than ACGIH's TLV for mercury of 0.025 mg/m³ for workplaces that, in turn, is greater than what should be present inside a residence. Without recovery of the spilled mercury, encapsulation, or chemical binding of the mercury, and/or good ventilation, Johnny risks mercury poisoning. In summary, there is a sufficient amount of liquid mercury in a standard household thermometer to saturate air in 1500 square feet house with poisonous vapors.

444. Estimate the volumetric air flow rate above a slab of recently rolled steel that is 30-feet long and 8-feet wide. Estimated rate of heat loss from the 10-inch thick slab is an average of 5,500,000 BTU/hour for the first 4 hours out of the rolling mill. Assume the slab is surrounded on four sides by 40-feet high aluminum sheet radiation panels so that most of the radiant portion of heat loss is converted into sensible heat and hot air convection currents. The height of the steel rolling mill is 100 feet above the floor.

The air motion may be estimated by: $Q_z = 1.9 Z^{1.5} \sqrt[3]{q}$

where

Q_z = air flow rate at effective height Z (in feet), in cfm. The Z factor accounts for an envelope of air that expands as it rises from the hot body.

q = convection heat loss from the hot body, BTU/hour

$$Z = Y + 2B$$

Y = actual height above hot body, feet (in this case, at the top of the heat shields)

B = largest horizontal dimension of hot body, feet (Note how the shorter dimension of width does not enter into the calculations.)

Z = 40 ft + (2 × 30 ft) = 100 feet (coincident with the height of the steel rolling mill)

$$Q_z = (1.9)(100)^{1.5} \left[\sqrt[3]{5,500,000} \right] = 335,000 \, \text{cfm}$$

445. Assume a branch in an exhaust ventilation system has a design volume of 10,000 cfm with a calculated static pressure, P_s, of 2.10 inches water gauge. The main exhaust duct, carrying an exhaust volume of 50,000 cfm, has a P_s of 2.40 inches water gauge where the branch enters. What volume of air will be drawn through the branch at balanced conditions?

$$Q_b = 10,000 \, \text{cfm} \sqrt{\frac{2.40 \, \text{inches}}{2.10 \, \text{inches}}} = 10,700 \text{ cubic feet of air per minute}$$

446. What is the worst-case peak blood methyl alcohol level in a 70-kg man exposed for 8 hours to the ACGIH TLV of 200 ppm$_v$ for methanol vapor?

"Standard 70-kg man" has a 60% water content and inhales 10 cubic meters of air during an 8-hour workday.

$$\frac{\text{mg}}{\text{m}^3} = \frac{\text{ppm} \times \text{molecular weight}}{24.45} = \frac{200 \times 32.04}{24.45} = 262 \, \text{mg/m}^3$$

$$262 \, \text{mg/m}^3 \times 10 \, \text{m}^3 = 2620 \, \text{mg MeOH}$$

$$70 \, \text{kg} \times 0.6 = 42 \, \text{kg H}_2\text{O} = 42 \, \text{L H}_2\text{O}$$

Peak blood level of MeOH = 2620 mg/42 L = 62 mg/L = 6.2 mg/dL.

This is close to 5% of the dose reported to cause acute irreversible toxic effects. Furthermore, since the blood half-life of methanol is 3 hours, the body burden would be negligible when the worker returns to his job the next day. It appears unlikely that this dose would cause ocular toxicity. However, MeOH is absorbed through intact healthy skin, and industrial hygienists, workers, and the workers' supervisors must be mindful of this potential route of exposure. The preceding calculations assume most MeOH distributes equally in aqueous compartments of the body; a very little dissolves in lipid compartments as well (fatty tissues, CNS, PNS, marrow). The calculations are "worst case" and do not consider the on-going metabolism and excretion of MeOH throughout the work day.

447. How many tons of air are exhausted yearly from a laboratory exhaust hood with full open face dimensions of 24" × 48"? The face capture velocities are 130, 140, 120, 120, 140, 130, 110, 120, and 110 cfm/ft². The hood operates 8 hours per day, 5 days a week, for 50 weeks a year.

$$\left[\frac{130 + 140 + 120 + 120 + 140 + 130 + 110 + 120 + 110}{9}\right] \text{cfm/ft}^2$$

$$= 124.4 \text{ cfm/ft}^2$$

$$24" \times 48" = 2' \times 4' = 8 \text{ ft}^2$$

$$8 \text{ ft}^2 \times 124.4 \text{ cfm/ft}^2 = 995.2 \text{ cfm}$$

$$\frac{995.2 \text{ ft}^3}{\text{minute}} \times \frac{60 \text{ minutes}}{\text{hour}} \times \frac{8 \text{ hours}}{\text{day}} \times \frac{5 \text{ days}}{\text{week}} \times \frac{50 \text{ weeks}}{\text{year}} \times \frac{0.075 \text{ lb}}{\text{ft}^3}$$

$$\times \frac{\text{ton}}{2000 \text{ lb}} = 4478.4 \text{ tons/year}$$

448. What is the solvent vapor emission rate from an open surface tank that is 2.5' × 4' containing TCE at room temperature? The molecular weight and the vapor pressure of TCE are 131.4 grams/gram-mole and 58 mm Hg, respectively. The barometric pressure is 748 mm Hg. The air velocity passing over the surface of the tank is 200 feet per minute. The air temperature is 78°F.

The following EPA formula can be used to estimate the emission rate, q, in grams per second:

$$q = \frac{8.24 \times 10^{-8} \times M^{0.835} \times P \left[(1/29) + (1/M)\right]^{0.25} \times U^{0.5} \times A}{T^{0.05} \times L^{0.5} \times P_t^{0.5}}$$

M is the molecular weight. P is the vapor pressure, mm Hg. U is the air velocity, fpm. A is the surface area, cm². T is the air temperature, kelvin. L is the length of the liquid surface, cm. P_t is the total pressure, atmospheres.

$$A = 2.5' \times 4' = 10.0 \text{ ft}^2 = 9290.3 \text{ cm}^2$$

$$T = 78°F = 298.7 \text{ K}$$

$$L = 4' = 121.9\,cm$$

$$P_t = (748\,mm\ Hg/760\,mm\ Hg) = 0.984\ atm$$

$$q = \frac{8.24 \times 10^{-8} \times 131.4^{0.835} \times 58\left[(1/29) + (1/131.4)\right]^{0.25} \times 200^{0.5} \times 9290.3}{298.7^{0.05} \times 121.9^{0.5} \times 0.984^{0.5}}$$

$$= 1.15\,g/sec$$

Answer: 1.15 grams of TCE evaporates every second. Refer to Problem 449.

449. Referring to Problem 448, the depth of the TCE in the tank was 23-$\frac{3}{4}$" on Monday at 8:00 am. On Friday, at 4:00 pm, the solvent depth was 21-$\frac{1}{2}$". If the system was not used for de-greasing (i.e., there was no physical carry-out of the liquid solvent, and there were no additions), what was the vapor emission rate assuming system values identical to those specified in Problem 448? Assume that the exhaust ventilation system for the tank operated 8 h/day. The specific gravity of TCE is 1.46 g/mL.

$$23.75" - 21.5" = 2.25"\ \text{evaporative loss}$$

$$2.5' = 30"$$

$$4' = 48"$$

$$2.25" \times 30" \times 48" = 3240\ in^3 = 3234.7\,cm^3$$

$$117{,}987\,mL \times 1.46\,g/mL = 172{,}261\,g$$

$$8\,hours/day \times 5\,days \times 60\,minutes/hour \times 60\,seconds/minute$$
$$= 144{,}000\,seconds$$

$$172{,}261\,g/144{,}000\,seconds = 1.20\,g/sec$$

Answer: 1.20 grams of TCE evaporated every second. This semi-empirical approach used in this problem provides a result essentially identical to calculated theoretical amount applying the equation in Problem 448 (1.15 g/second).

450. Pure helium gas passes through a rotameter at an indicated flow rate of 1.7 L/min. The rotameter was calibrated using air at 29.9 inches of water and 70°F. Helium has a specific gravity of 0.138 at 70°F and 29.9 inches of water (air = 1.00). If the helium temperature is 100°F, and its pressure is 33 inches of water, what is the actual helium gas flow rate, k?

$$k = \sqrt{\frac{460 + 100}{460 + 70} \times \frac{29.9}{33} \times \frac{1.00}{0.138}} = 2.63$$

$$1.7\,L/m \times 2.63 = 4.47\,L/m$$

Answer: 4.47 liters of helium per minute, or 2.63 times greater than the rate indicated on the rotameter.

451. Workplace air contains 7.89 mg of total particulates per cubic meter. What is the airborne dust concentration on a mass/mass basis expressed in ppm_m? Disregard water vapor content of air; that is, assume dry air (0% relative humidity). Assume NTP (760 mm Hg and 25°C).

$$1\,m^3 = 35.3\,ft^3$$

$$35.3\,ft^3 \times 0.075\,lb/ft^3 = 2.6475\,lb \text{ of air per cubic meter}$$

$$1\,lb = 454,000\,mg$$

$$2.6475\,lb \times (454,000\,mg/lb) = 1,201,965\,mg \text{ per cubic meter}$$

$$(7.89\,mg/1,201,965\,mg) \times 10^6 = 6.56\,ppm_m$$

It is unusual to express airborne dust concentrations on this mass-to-mass basis. However, mechanical and chemical engineers often do so in selecting fans for air pollution control devices, that is, "How many pounds of dust plus pounds of air must this fan handle per minute?" This, in part, determines the fan brake horsepower requirements. System flow resistance pressures plus pressure loss across the collector are also needed to select the proper type and to determine the fan size and fan rpm.

452. Carbon dioxide gas accumulation in our atmosphere is a major contributor to the global warming of our planet. An ambient air sample detects 452 ppm_v CO_2 in the mid-Pacific Ocean remote from anthropogenic and geologic sources. Convert this to mg/m³ to determine how much pure carbon must be stoichiometrically burned (oxidized) to achieve this amount of CO_2 in 1 m³.

$$mg/m^3 = \frac{ppm_v \times molecular\ weight}{24.45\,L/gram\text{-}mole} = \frac{452 \times 44}{24.45} = 813.4\,mg\,CO_2/m^3$$

$$C + O_2 \longrightarrow CO_2$$

That is, 1 mole of carbon produces 1 mole of CO_2.

$$\frac{moleculare\ weight\ C}{molecular\ weight\ CO_2} \times 100 = \%\,C = \frac{12}{44} \times 100 = 27.27\%$$

$$813.4\,\text{mg CO}_2 \times 0.2727 = 221.8\,\text{mg C}$$

Answer: 221.8 mg of carbon stoichiometrically oxidizes to 813.4 mg of carbon dioxide. 591.6 milligrams of oxygen are required.

Note: 221.8 mg C + 591.6 mg O_2 produces 813.4 mg CO_2.

453. Air pollutant concentrations expressed as mass per unit volume of atmospheric air (e.g., mg/m³, µg/m³, lb/SCF, grains/SCF) at sea level decrease with an increase in altitude according to the equation $C_a = (C)(0.9877^a)$, where a = altitude, in 100s of meters; C = concentration at sea level, in mass per unit volume; and C_a = concentration at altitude a, in mass/volume. It is very easy to overlook this correction. For example, consider Albuquerque where an average altitude = 1779 meters, and the sea-level concentration of TSP = 0.027 mg/m³. What is the corrected concentration?

$$C_a = (0.027)(0.9887^{17.79}) = 0.022\,\text{mg/m}^3$$

454. There are situations when air sampling is obviously not required. The following are examples of some. Assume you were just hired as the industrial hygienist for a multiple product, mid-size chemical manufacturing plant in Houston. You report to work, meet your coworkers, find the coffee pot, stow your books and other materials, pick up your PPE, adjust your chair, and boot your computer. You spend the rest of the morning studying your company's organization chart. After lunch, you conduct a brief walk-through inspection of several nearby plants and processes to learn your way around the complex. You encounter the following situations. What would you do? Rank and justify your risk management plans.
   a. A worker without a respirator is enveloped in clouds of silica flour dust as she slits 17 bags and dumps the powder into a slurry mixing tank.
   b. A pipe fitter is exposed to cough-producing concentrations of chlorine gas while making adjustments and repairs to a chemical plant process line.
   c. Refrigeration plant process workers are exposed to ammonia gas that regularly causes throat irritation and lacrimation.
   d. Employees are periodically exposed to offensive odors intruding from a nearby meat scrap rendering plant. It smells offal.
   e. Extensive green mold is amplifying on the ceiling of the plant manager's office.
   f. An unprotected worker is about to enter a process tank to help another worker who collapsed from unknown causes.

Answer: f > b > c > a > d > e Obvious problems require prompt intervention—not air sampling! Focus on immediate, unknown conditions in this confined space and the immediate, acute toxicants (Cl_2, NH_3, and chloramines if the ammonia and chlorine leaks are nearby). Next, control silica dust exposures (prompt respiratory protection while you design a mechanical local exhaust ventilation system). Later prevail on management of the meat rendering plant to control their air emissions. In one sense, this approach follows "Pareto's Principle": concentrate on important few first and, for the moment, disregard the relatively unimportant many. A former student—with tongue in cheek—suggested control of the mold first by cozying up to the plant manager for political bargaining purposes later!

455. A side-draft bench hood has three slots on 12-inch centers each $2" \times 3'\text{--}4"$ and all with slot velocities of 2000 fpm. How deep must the exhaust plenum be to ensure uniform capture velocity across the booth's hood face?

$$3\,(2" \times 40")/144 \text{ in}^2/\text{ft}^2 = 1.667\,\text{ft}^2$$

$$2.667\,\text{ft}^2 \times 2000\,\text{fpm} = 3334\,\text{cfm}$$

$$Q = AV$$

$$A = Q/V = 3334\,\text{cfm}/1000\,\text{fpm} = 3.334\,\text{ft}^2$$

The maximum plenum velocity must not exceed $\frac{1}{2}$ the slot velocity to ensure equal air distribution across all three slots (in this case, 1000 fpm). As a precaution, the plenum could be, say, 20%, deeper to provide added assurance. Let us apply 800 fpm as the maximum plenum velocity. Thus, the plenum's cross sectional area is

$$3334\,\text{cfm}/800\,\text{fpm} = 4.168\,\text{ft}^2$$

$$\text{Plenum area, } A = L \times W$$

$$W = A/L = 4.168\,\text{ft}^2/3.333\,\text{ft} = 1.25\,\text{ft deep} = 1'\text{--}3"$$

Narrow plenums do not provide uniform air distribution across all slots. With such, more air will crowd its way into slots closest to the fan while those farther from the fan will be starved for air. Deep plenums promote even exhaust distribution for all slots.

456. Many toxic chemicals can be dissolved or suspended in water or solvents and then dispersed and dispensed as a spray, fog, or mist aerosol. Concentration might be low as dispensed, but the carrier phase quickly evaporates into the surrounding air (assuming <100% relative humidity) producing smaller particles enriched with the toxicants. The particles assume spherical shape. Consider 10-μm diameter particles with 0.05% (vol/vol) concentration quaternary ammonium chloride disinfectant. What particle diameters develop after the aqueous phase evaporates?

$$\text{Volume of sphere, } V = (4/3)\,\pi r^3$$

$$\frac{4}{3}\,\pi\,(5\mu)^3 \; = V = 523.6\,\mu^3$$

$$0.05\% \text{ of } 523.6\,\mu^3 = 0.0005 \times 523.6\,\mu^3 = 0.2618\,\mu^3.$$

That is, 99.95% of particle was water, and, after evaporation, $0.2618\,\mu^3$ active, toxic ingredient remains solely as 100% quaternary ammonium chloride.

Rearrange the sphere volume equation to calculate diameter of new particles:

$$r^3 \; = \; \frac{V}{(4/3)\,\pi} \; = \; \frac{0.2618\,\mu^3}{1.333\,\pi} \; = 0.0625\,\mu^3$$

$r = 0.0625^{0.333}$
$\mu = 0.397\,\mu$
$2r = \varnothing = 0.794\,\mu$

10 micron particles became 0.794 micron in $\varnothing$. Concentration increased from 0.05% to 100%. Particles are more respirable into bronchioles and alveoli of the respiratory tract, whereas the 10-micron particles were non-respirable. This is toxicologically profound. Be mindful of this concept if one, for example, asserts, "Chemical is present in trace amounts. These particles are too big to inhale. You need not worry." Worry. Moreover, even if inhaled at 0.05%, it is virtually 100%, because respiratory tract is coated with water on the mucociliary escalator and the nasopharyngeal areas. A good industrial hygiene control method would be the barring of any dispensing of toxicants as mists, sprays, fogs, or nebulized aerosols. Alternatively, dispense with a brush, roller, tube, wipe pads, or dauber. Refer to Problems 482 and 487.

457. An atmosphere's oxygen concentration increased from 21 vol.% to 70 vol.%, say from a leaking oxygen cylinder in a gas bottle storage room. One lights a cigarette in the now oxygen-enriched atmosphere. Will an explosion and ensuing fire occur?

An explosion will not occur. However, the cigarette will burn much more rapidly. You might recall oxygen, by itself, is not flammable, combustible, or explosive. It simply is an oxidant, and with more present, flammability and explosibility of any oxidizable materials are promoted. Oxygen adheres to many materials including hair and clothing creating a fire trap for those so contaminated. Yes, oxygen gas can be an atmospheric contaminant.

The author investigated a major fire where one worker sustained significant facial burn injuries. The man worked in an oxygen gas bottling plant where smoking and other ignition sources were prohibited at the gas charging rack. He left the work area on a break, lit his cigarette, and his bushy beard almost instantaneously burst into flames. This led to a law suit naming cigarette lighter manufacturer defendant of a faulty product. Indeed, the lighter was not flawed whatsoever as expert engineers demonstrated to the Court. The employer failed to require its bottling employees to "de-oxygenate" in a fresh air curtain or an air booth before engaging in smoking. Such a booth is easily engineered. Residence time in the air booth and quantity of fresh air flow can be determined to design such a booth to strip and flush adherent oxygen gas from a worker's hair and clothing before she/he exits to a safe haven.

458. 350 grams of polonium-210 are securely sealed in a properly labeled container for shipment to a remote radioactive waste storage facility. ^{210}Po has a half-life of 138.376 days. The container will reside on a secure shipping dock for 7 days before an 8-day transfer to the storage site. ^{210}Po decays almost exclusively into alpha particles with rare decay into a gamma ray ($\approx$ 1/100,000 disintegrations). How much ^{210}Po remains when this storage cask arrives at radionuclide storage facility? ^{210}Po decay kinetics is second order.

This problem can be solved two ways:

$$\text{Storage and transit time} = 7 \text{ days} + 8 \text{ days} = 15 \text{ days}$$

The radioactive decay rate coefficient is

$$k = \frac{0.693}{t_{1/2}} = \frac{0.693}{138.376 \text{ days}} = 0.005008/\text{day}.$$

The amount of ^{210}Po remaining after 15 days is

$$C = C_o\, e^{-kT} = (350\,\text{g})\, e^{-(0.005008)\,(15 \text{ days})} = 324.67\,\text{grams}$$

The second method is to calculate the number of half-lives occurring over 15 days:

$$\text{Elapsed time}/T_{1/2} = 15 \text{ days}/138.376 \text{ days} = 0.1084 \text{ half-life elapsed}$$

$$(0.5)^{0.1084}\,(350\,\text{grams}) = 324.67\,\text{grams remain}$$

$$100\% - (324.67\,\text{grams}/350\,\text{grams}) = 7.237\% \text{ decayed over 15 days}$$

There is excellent agreement between both calculation methods.

459. The OSHA *Hazard Communication Standard* (29 *CFR* 1910.1200) stipulates that manufacturers, importers, and others list every hazardous material

greater than 1% concentration and carcinogens at greater than 0.1% in their Material Safety Data Sheet (MSDS). Many, however, do not appear to understand that hazardous materials below these concentrations must also be listed if it is foreseeable lower concentrations (e.g., < 0.1%) can present a health or safety risk to workers. *To wit*:

> "Information must also be included on the MSDS for ingredients of a mixture present in concentrations less than 1% (or 0.1% for carcinogens) when the hazardous substance may be released in a concentration which exceeds a PEL or TLV <u>or</u> may present a health risk to exposed employees. An example of the latter may be TDI [toluene 2, 4 – (or 2, 6 –) diisocyanate] because it is a sensitizer in very small concentrations, thereby presenting a health risk that must be noted on MSDSs." (emphasis added, *Occupational Safety and Health Administration Directives Pertaining to the Hazard Communication Standard*, CPL 2-2-38A)

As an example: What is the benzene vapor concentration in the air of an empty, unventilated garage where an epoxy resin paint was used to cover the floor? The floor area is 20' × 30'. Ceiling height is 9 feet. Three liters of solvent ("toluene") evaporated, and the benzene concentration of mixture is 0.0007 mL/mL (volume/volume) or in%, (0.0007 mL/1 mL) × 100 = 0.07% (= 700 ppm$_v$).

$$3 \text{ liters} = 3000 \text{ mL}$$

$$3000 \text{ mL} \times 0.0007 \text{ mL/mL} = 2.1 \text{ mL of benzene evaporated}$$

$$2.1 \text{ mL} \times 0.8787 \text{ g/mL} = 1.845 \text{ grams} = 1845 \text{ mg}$$

$$20' \times 30' \times 9' = 5400 \text{ ft}^3 = 152.9 \text{ m}^3$$

$$\text{ppm}_v = \frac{(\text{mg/m}^3) \times 24.45}{\text{molecular weight}} = \frac{(1845 \text{ mg/152.9 m}^3) \times 24.45}{78.11 \text{ grams/gram-mole}} = 3.78 \text{ ppm}_v$$

> OSHA's PEL = 1 ppm$_v$, ACGIH TLV = 0.5 ppm$_v$, and NIOSH REL = 0.1 ppm$_v$.

The benzene vapor concentration greatly exceeds exposure limits and guidelines. Moreover, benzene is absorbed through intact, healthy skin, and, once absorbed into perfusing blood, is distributed systemically to every organ and tissue including the bone marrow. Benzene is a genotoxic human carcinogen with no limit below which exposure is "safe." The benzene concentration of 0.0007 mL/mL must be reported on MSDS. There is no "safe" level of exposure to benzene vapor. One can only say that lower levels, all other factors being equal, are "safer." Work with your product suppliers to ensure that the benzene concentration of products they manufacture are below 1 ppm$_v$ (0.0001%) or, better, analytically nondetectable below 1 ppm$_v$. The cost for the safer solvent blend may be more than offset by the high costs of complying with OSHA's *Benzene Standard* (29 *CFR* 1910.1028) (air sampling and analyses, medical surveillance, biological monitoring, education and training, engineering, record keeping, and so on). See Problem 710.

460. In the previous problem (459), one must also consider toluene vapor concentration. What hazards does toluene vapor pose? Think fire, inhalation, skin absorption, CNS toxicant, and reproductive health hazard.

$$(3000\,\text{mL toluene and benzene}) - 2.1\,\text{mL benzene} = 2997.9\,\text{mL toluene}$$

$$\text{Density of toluene} = 0.8669\,\text{g/mL}$$

$$\text{Molecular weight of toluene} = 92.14\,\text{grams/gram-mole}$$

$$2997.9\,\text{mL} \times 0.8669\,\text{g/mL} = 2598.88\,\text{g} = 2{,}598{,}880\,\text{g}$$

$$\text{ppm} = \frac{(\text{mg/m}^3) \times 24.45}{\text{molecular weight}} = \frac{(2{,}598{,}880\,\text{mg}/152.9\,\text{m}^3) \times 24.45}{92.14\,\text{grams/mole}}$$
$$= 689{,}631\,\text{ppm}_\text{v}$$

This toluene vapor concentration greatly exceeds the OSHA PEL of $100\,\text{ppm}_\text{v}$ and ACGIH TLV of $50\,\text{ppm}_\text{v}$ and NIOSH IDLH of $2000\,\text{ppm}_\text{v}$. It greatly exceeds UEL of 6.75% ($67{,}500\,\text{ppm}_\text{v}$). Be wary, however, because $689{,}631\,\text{ppm}_\text{v}$ concentration is the average vapor concentration throughout the garage (68.96%$_\text{v}$). Although the toluene vapor concentration exceeds its flammability limit, dilution with fresh air will lower the vapor concentration so that, when in the LEL–UEL range and a source of sufficient ignition potential is present, there will be a huge explosion. Furthermore, even though the amount of benzene present is tiny from a flammability perspective, it adds to toluene's fire hazard.

461. A dairy cow kicks a bucket containing $2\text{-}\frac{1}{2}$ gallons of gasoline onto the floor of a metal pole barn. The barn, tightly sealed with no ventilation, has dimensions of $85' \times 44' \times 18'$. After extended time, does the barn's atmosphere present a gasoline vapor explosion risk? Average molecular weight of gasoline is $108\,\text{g/g-mole}$, and its specific gravity is $0.8\,\text{g/mL}$.

$$\text{Barn's gross volume} = 85' \times 44' \times 18' = 67{,}320\,\text{ft}^3 = 1906.3\,\text{m}^3$$

Subtract 10% of the barn volume (cattle, stalls, bedding, equipment, etc.) to obtain the approximate net volume:

$$1906.3\,\text{m}^3 - 190.6\,\text{m}^3 = 1715.7\,\text{m}^3$$

$$2\text{-}\tfrac{1}{2} \text{ gallons} = 9463.3\,\text{mL}$$

$$9463.3\,\text{mL} \times 0.8\,\text{g/mL} = 7570.64\,\text{g} = 7{,}570{,}640\,\text{mg}$$

$$\text{ppm}_v = \frac{(7{,}570{,}640\,\text{mg}/1715.7\,\text{m}^3) \times 24.45}{108\,\text{g/g-mole}} = 999\,\text{ppm}_v \cong 0.1\%$$

The average concentration of gasoline vapor is below lower explosive of 1.4%, or 14,000 ppm$_v$. Averages, as are well known, can be very misleading. Until gasoline vapors diffuse throughout this barn, there is a gasoline vapor layer hovering just above the floor with pockets of vapor above LEL. With a sufficiently strong ignition source present long enough to ignite vapors, there will be a flash fire. Indeed, until mixed with air, there is a vapor layer exceeding UEL. This layer then dilutes into the dangerous range between the LEL and the UEL.

462. A 6' × 6' × 6' steel tote bin containing extremely fine aluminum powder is dumped into a large steel mixing tank. The transfer normally requires about 40 seconds, but after 30 seconds there is a massive explosion with burn injuries, fractures, and lacerations to the operator and his helper. There are significant building structural damages. What happened?

Aluminum powder is explosive in air in the presence of a strong source of ignition. The most likely ignition source was a streaming current of electrons as the powder was dumped. An electrostatic charge accumulates in tote bin and in the receiving vessel. Bonding by attaching conductive copper wire and clips between both of the containers and grounding by attaching conductive copper wire from tote bin and receiving vessel to a substantial ground dissipates the accumulated electrical charge before it can reach a spark ignition temperature. Transferring the powder very slowly helps to hinder charge accumulation. The use of inert gases such as nitrogen, argon, or CO_2 in the head spaces also helps; however, the best controls are bonding and grounding coupled with a slow transfer generally exceed inert gas because pockets of air (O_2) might exist in which inert gas cannot reach. Many other materials (inorganic metals: e.g., zinc, iron, copper; organic: e.g., flour, rice, plastics) behave similarly. Work with your material suppliers and their fire safety product stewards to develop a robust risk management plan.

463. A surfboard manufacturer employs five people who, on average, make one board/every hour and 20 minutes. This southern California plant is in a climate that permits open doors and windows on most days. Boards are assembled in jigs on open benches. The five workers' assembly tables are in an open area 30' × 50'. Toluene is the sole solvent for the resin used to glue the laminates together. The company owner hires an industrial hygienist to

determine the best and least costly vapor control methods. How would you proceed?

First, meet with the purchasing agent to see how much resin is bought per month or year—or even by season because surfing might even be seasonal in California, and manufacturing might be by seasonal demand. Next determine the amount of solvent in resin. It is reasonable to assume toluene losses are by evaporation into room air because boards are not cured in ovens. Let us say by calculations, each board evaporates 0.21 pounds of toluene on average. So, for all workers:

$$\frac{0.21\,lb}{board} \times \frac{one\,board}{80\,minutes} \times 5\,workers = 0.013125\,lb/minute$$

We cannot rely on natural ventilation because it unreliably varies by direction, time of day, temperature, and velocity. Portable pedestal fans are out of the question because they could compete with each other, and one worker will claim it's too drafty, and another will demand more dilution air. The net result is that fans will be positioned at whims of the workers. The inhalation toxicity of toluene is too great so that, in this situation, entirely dilution ventilation is unacceptable. Let us then consider mechanical dilution ventilation and mechanical local exhaust ventilation. The general dilution ventilation equation is

$$Q_D = K \times \frac{387 \times 10^6 \times W}{MW \times TLV}$$

where
  Q_d = fresh dilution air required, cfm
  K = an air mixing factor where 1 = outstanding, and 10 = dismal
  W = pounds of solvent (or gas) generated and dispersed per minute
  MW = molecular weight of solvent or gas
  TLV = fraction of TLV, PEL, or REL that you decide reasonably assures
    health protection.

We select $K = 1$. We have W. MW for toluene = 92.13 g/g-mole. TLV = 20 ppm$_v$.

We select 0.05 TLV = 1 ppm$_v$ because the occupational reproductive guide is 2.5 ppm$_v$ to prevent hearing loss in the developing embryo/fetus. Moreover, using a low TLV fraction provides additional protection from benzene. Now, let us plug and chug:

$$Q_D = 1 \times \frac{387 \times 10^6 \times 0.013125}{92.13 \times 0.05 \times 20} = 55,132.7\,cfm$$

The work area is 30' × 50' = 1500 ft², 55,132.7 cfm/1500 ft² = an average air flow rate of 36.75 fpm. This is low because most workers claim "still air" when the air flow over their skin is less than 50 fpm. Perhaps the work benches can be placed closer to each other, say 20' × 30' = 600 ft², 55,132.7 cfm/600 ft² = 91.9 fpm. The workers have a work space of [(20' × 30')/5] = 120 ft², or an area approximately 11 feet square. This work area appears adequate for surfboard jig construction table, tools, and materials.

Now, let us compare this general dilution ventilation with mechanical local exhaust ventilation. The longest surfboard (a "*FunBoard*") made by this company is 8'-4". Most are about 7 feet. A local exhaust side-draft bench hood 10-feet wide and 22 inches high (1.83') is satisfactory for workers. A face capture velocity of 100 fpm is satisfactory if cross drafts are carefully controlled. Therefore, the total exhaust volume for five hoods is

$$\text{cfm} = 5 \times [(10' \times 1.83') \times 100\,\text{fpm}] = 9150\,\text{cfm}$$

There appears to be little question that the far superior local exhaust ventilation prevails over general dilution ventilation. Because of production demands, one more worker was hired. This permitted construction of three back-to-back hoods with a team of two at each. As each worker finished one side of a board, they simultaneously passed it across to their teammate. This eliminated removing a board while still emitting toluene vapors to turn it around outside of the booth.

Other elements of the industrial hygiene program included hazard communication training, training in proper work practices, respiratory protection program for those desiring additional protection, protective gloves and aprons, hearing conservation, workplace lighting, medical surveillance, supervisory surveillance, fire protection, ensuring benzene-free solvent from supplier, and ventilation system preventative maintenance.

464. A petroleum refinery's tank farm has a 250,000 gallon gasoline storage tank that is bottom-filled at a steady 16,350 gallons per hour. Naturally, for each gallon added to this tank, 1 gallon of air rich in gasoline vapor is displaced to the atmosphere. We wish to prevent community air pollution by designing and installing an excellent vapor recovery system. Establish design engineering specifications.

$$\frac{16,350\,\text{gallons}}{\text{hour}} \times \frac{\text{hour}}{60\,\text{minutes}} = 272.5\,\text{gallons/minute}$$

$$\frac{272.5\,\text{gallons}}{\text{minute}} \times \frac{0.1337\,\text{ft}^3}{\text{gallon}} = 36.4\,\text{cfm}$$

This is such a modest amount of ventilation that one might be tempted to increase it 10-fold (364 cfm) because of gasoline's flammability and "just to be safe." We should be mindful, however, that increasing air flow over the gasoline will promote its evaporation. Therefore, a moderate increase to only 50 cfm should suffice. The vapors must pass through an explosion-proof air-vapor handling system that will include an intrinsically safe explosion proof fan, bonding and grounding, spark arrestors, and introduction of inert gas (N_2) to ensure gasoline is 10% or less of its LEL. The fan must be sized to include the air + gasoline vapor + N_2 gas.

465. In the previous problem, a large organic carbon column adsorber with a capacity of 930 pounds of gasoline will be used to scavenge the gasoline vapors. Calculate the column packing change-out time if nitrogen gas is not injected into the vapor stream. This is also achievable because the gasoline vapors exceed their UEL. Base calculations on a 50 cfm fan as described in the previous problem (464).

Vapor pressure of gasoline is 7–11 psi at 0°C. This equals 362–569 mm Hg. Ethanol-free gasoline has a broad range of vapor pressures because it contains numerous aliphatic, aromatic, and naphthenic hydrocarbons all with very different volatility and evaporation rates. Moreover, the season of the year and the climate determine gasoline's composition. Considering the above variables, let us select a conservative vapor pressure during warm weather of 600 mm Hg allowing for a safety factor. This refinery is located in Houston (i.e., 1 atmosphere = 760 mm Hg).

$$\text{Gasoline vapor concentration in ppm}_v = (600 \text{ mm Hg}/760 \text{ mm Hg})$$
$$\times 10^6 = 789,474 \text{ ppm}_v$$

Average molecular weight of gasoline = 108 g/gram-mole

$$\text{Saturation vapor concentration in mg/m}^3 = \frac{789,474 \text{ ppm}_v \times 108 \text{ g/g-mole}}{24.45 \text{ L/g-mole}}$$
$$= 3,487,247 \text{ mg/m}^3$$
$$= 3487.2 \text{ grams/m}^3$$

$$50 \text{ cfm} = 1.416 \text{ m}^3/\text{minute}$$

$$1.416 \text{ m}^3/\text{minute} \times 3487.2 \text{ grams/m}^3 = 4937.9 \text{ grams/minute}$$

$$4937.9 \text{ grams/minute} = 10.89 \text{ pounds/minute}$$

$$930 \text{ lbs}/10.89 \text{ lbs/minute} = 85.4 \text{ minutes per change-out time} = 1.4 \text{ hours}$$

This is highly unreasonable and would be essentially a fulltime job for one or more refinery employees. Alternate engineering controls must be investigated. Perhaps three, for example, of these vapor scavengers could be assembled in parallel so that two are being automatically desorbed of gasoline while one operates. Maybe larger adsorption columns are available. Since headspace vapors are saturated at almost 79% by volume, perhaps parallel refrigeration tubes could be located near the top of the tank to provide vapor condensation surfaces. Combinations of simple engineering controls appear necessary. In any procedure, robust oversight and engineered redundancy must be in place.

466. A fan was selected to supply 35,530 cfm at 8" SP. The fan operates at 1230 rpm and requires 61.0 bhp. After installation, it is desired to increase fan output 20%. At what rpm must the fan operate? What SP will develop? What bhp is required?

$$\text{cfm varies as rpm: } 1230 \text{ cfm} \times 1.20 = 1476 \text{ rpm}$$

$$\text{SP varies as (rpm)}^2 = (1476 \text{ cfm}/1230 \text{ cfm})^2 \times 8" = 11.52" \text{ SP}$$

$$\text{bhp varies as (rpm)}^3 = (1476 \text{ cfm}/1230 \text{ cfm})^3 \times 61.0 = 105.4 \text{ bhp}$$

Note while fan output went up 20%, the brake horsepower requirements increased a substantial (105.4 bhp/61.0 bhp) = 1.728, or 173%. In these days of energy = money, very careful system design is necessary. If only a 10% "safety factor" is added to system volume, the horsepower increase will be 33% per the third fan law. Evaluation should be made weighing necessity of this "safety factor" against the substantial energy cost penalty incurred.

467. A fan was initially installed to deliver 10,300 cfm at 2-$\frac{1}{4}$" SP and to operate at 877 rpm requiring 5.20 bhp. After installation and performance testing, it was found that the system only delivered 9150 cfm at 2-$\frac{1}{2}$" SP and used 4.70 BHP. This suggested the original calculations were in error, or the system was not installed according to the plan's specifications. What fan rpm and bhp would be necessary to develop the required 10,300 cfm? What SP should have been calculated?

$$\text{cfm varies as rpm: } (10{,}300 \text{ cfm}/9150 \text{ cfm}) \, (877 \text{ rpm}) = 987 \text{ rpm}$$

$$\text{SP varies as (rpm)}^2 = (987/877)^2 \, (2.50") = 3.17" \text{ SP}$$

$$\text{bhp varies as (rpm)}^3 = (987 \text{ rpm}/877 \text{ rpm})^3 \times 4.70 = 6.70 \text{ bhp}$$

468. An exhaust ventilation system was calculated to require a static pressure equal to 2" water gauge at an air flow rate of 1000 cfm. Without any physical changes to the system, what is the required SP if it is desired to increase the air flow to 1500 cfm?

$$(1500 \text{ cfm}/1000 \text{ cfm})^2 \times 2" \text{ SP} = 4.5" \text{ SP}$$

469. A train derails. Two tank cars—one containing 7400 gallons anhydrous NH_3, and the other with 8950 gallons of 30% (mass/volume) of HCl connected together—rupture and react chemically and immediately in a large pool formed in the right-of-way ditch. What is the maximum amount of ammonium chloride fume that could be formed? What is the reality check?

$$NH_3 + HCl \rightarrow NH_4Cl + heat$$

That is, 1 mole of ammonia reacts with 1 mole of hydrogen chloride to yield 1 mole of ammonium chloride. A question then becomes: What is the limiting chemical reactant? How many moles of NH_3 and HCl are available to react?

Liquid NH_3 density $= 42.57$ pounds per cubic foot @ $-28°F$

7400 gallons $NH_3 \times 0.1337\,ft^3/gallon = 989.38\,ft^3$

$989.38\,ft^3 \times 42.57\,lb/ft^3 = 42,117.9\,lb$

$42,117.9\,lb \times 453.59\,lb/g = 19,356,965.7\,g\ NH_3$

$19,356,965.7\,g/17\,g/mole = 1,138,645.0\,moles\ of\ NH_3$

density of 30% HCl in water is $1.1443\,g/mL$

8950 gallons $\times 3785.4\,mL/gallon = 33,879,330\,mL$

$33,879,330\,mL \times 1.1443\,g/mL = 38,768,117.3\,g\ HCl$

$38,768,117.3\,g/36.5\,g/mole = 1,062,140\,moles\ of\ HCl$

Therefore, because of the limiting amount of 1,062,140 moles of HCl, the number of moles of NH_4Cl can be no larger. That is, the amount of HCl limits the amount of NH_4Cl that can be formed under perfect reaction conditions. Such calculations are essential to emergency responders in developing a community protection risk management and spill response plan.

Molecular weight $NH_4Cl = 53.49\,g/g$-mole

$53.49\,g/g$-mole $\times 1,062,140\,moles\ NH_4Cl = 56,813,868.6\,g = 56,813.9\,kg.$

In reality, however, there will never be a stoichiometric reaction. The great heat of reaction will volatilize much liquid ammonia and its gas and hydrogen chloride gas before they react *in situ* and in the atmosphere. Perfect reaction conditions would not occur in a railroad ditch. Would it not be prudent to separate these tank cars in the freight train marshaling yard with, for example, boxcars and hopper cars of wheat, rice, coal, marsh-mallows, carrots, stoves, and crushed granite—all virtually are nonreactive with NH_3 and HCl and each other?

470. Industrial hygienists are seldom requested to determine combustion air needs for fuel-fired furnaces, boilers, and other equipment. This is nor-mally the province of chemical, mechanical, and combustion engineers. Once in awhile, they are asked to ensure the volume of combustion air is high enough to hinder CO gas formation. They might be requested, in a forensic sense, to determine if the amount of combustion air was sufficient to prevent deaths and poisoning from CO gas inhalation. The following table is helpful in these calculations:

Combustible Substance	Pounds of Air/Pound of Combustible
Carbon	11.5
Hydrogen	34.3
Sulfur	4.3

Consider a fuel oil with 86.1% carbon, 13.8% hydrogen, and 0.1% sulfur. This fuel oil weighs 6.8 pounds per gallon. It will be burned nearly stoichio-metrically at the rate of 5 gallons per hour. How much combustion air is required at NTP with an excess of 10%?

$$0.861 \text{ C} \times 11.5 \times 6.8 = 67.3 \text{ lbs of air}$$

$$0.138 \text{ H} \times 34.3 \times 6.8 = 32.2 \text{ lbs of air}$$

$$0.001 \text{ S} \times 4.3 \times 6.8 = 0.03 \text{ lbs of air}$$

$$\text{Total} = 99.53 \text{ lbs of air required for pound of this fuel}$$

$$99.53 \text{ lbs of air} \times 1.10 \text{ excess} \times 5 \text{ gallons/hour} = 547.4 \text{ lb air/hour}$$

$$(547.4 \text{ lbs/hour})/(60 \text{ minutes/hour}) = 9.12 \text{ lbs of air/minute}$$

$$\text{At NTP: } (9.12 \text{ lbs/minute})/(0.075 \text{ lbs/ft}^3) = 121.6 \text{ cfm}$$

How much excess air is supplied as a safety factor depends on several variables: fuel type, combustion efficiency, CO gas alarms, and so forth. For fuel oils 1–6, excess air ranges between 5% and 20%, for natural gas: 5–15%, for wood: 10–25%, for coal: 10–60%, and for metallurgical coke: 10–30%.

471. A 10-year-old sister finds her $2\text{-}\frac{1}{2}$-year-old brother dead inside an empty toy chest. Apparently, the chest's hinged wood lid was too heavy for him to open. His death was ruled as physical asphyxiation from oxygen insufficiency;

however, the county medical examiner, during forensic evaluations, was uncertain if this boy's death was unintentional or homicidal. She requests services of an industrial hygienist to determine oxygen depletion profile over time inside the toy chest. The chest was tightly sealed, so there was no air exchange while the boy was inside. The interior dimensions were $35" \times 18" \times 15"$. The boy weighed 34 pounds naked.

Assumptions:

Ambient concentration of oxygen was 20.95% (v/v).

Boy's body density was 1.07 g/mL.

Oxygen consumption by boy was 0.13 L O_2/minute.

Exposure was at, or near, sea level, say <1000 feet.

Nobody opened the lid while the boy was inside.

He had good health with no CO poisoning risk factors.

Toy chest coffin: $35" \times 18" \times 15" = 9450$ in^3

9450 in^3/1728 in^3/ft^3 = 5.47 ft^3

5.47 ft^3 $\times$ 28.312 L/ft^3 = 154.87 L

34 lbs $\times$ 453.592 g/lb = 15,422 g

Boy's volume = 15,422 g $\times$ mL/1.07 g/mL = 14,413 mL = 14.413 L

Net volume of toy chest with boy inside = 154.87 L – 14.413 L = 140.46 L

140.46 L air $\times$ 0.2095 = 29.43 L of O_2 in toy chest when the boy entered

After 30 minutes: 30 minutes $\times$ 0.13 L/minute = 3.9 L O_2 consumed

After 60 minutes: 60 minutes $\times$ 0.13 L/minute = 7.8 L O_2 consumed

After 120 minutes: 120 minutes $\times$ 0.13 L/minute = 15.6 L O_2 consumed

After 120 minutes: (existing O_2 concentration/initial O_2 concentration) $\times$ 100 = (15.6 L/29.43 L) 100 = 53.0% O_2 remains after 2 hours.

20.95% O_2 in ambient air $\times$ 0.53 = 11.1% O_2 in toy chest after 2 hours.

At near 12% O_2, unconsciousness is approaching, and death occurs at nearly 6% O_2 or less.

As we see, this boy's asphyxiation death was slow—perhaps over 3 hours. Of course, he's exhaling CO_2 which stimulates the heart's carotid body CO_2 receptor to increase both respiration rate and heart rate. With diminishing oxygen, he gets weaker, strength goes down, and his ability to continue screaming for help will decrease. His most likely profound fatigue from escaping attempts accelerate his oxygen consumption and CO_2 generation into the surrounding "air." If this boy's death was ultimately ruled unintentional, this certainly argues for toy chest makers to foresee asphyxiation fatalities. An opening, say $1" \times 12"$ at the top front under the lid's edge with equal area openings at the back side near the bottom will promote natural ventilation. Body warmth produces a "chimney effect," air rises and exits at top front, with make-up air naturally entering through the back bottom area. Cries for help could be more easily heard with the openings. Of course, manufacturers must install hazard warnings with pictographs similar to what we see on 5-gallon pails. A lighter lid would have allowed this boy to extricate himself or at least to open lid partially to scream for help. A police whistle attached by a short chain to the inside, with parental training, could augment good safety engineering design. Forensically, of course, if the boy could not open the lid from the inside, could he open it from the outside? Was it open when he approached the chest? Or, was he in the toy chest while another closed the lid unintentionally without noticing he was inside, or was there intentional joking or teasing; or was it deliberate murder?

472. A direct-reading, real-time airborne particle analyzer measured the following in the air of a newly constructed house during dry wall contractor clean-up:

Particle Diameter (μ)	Particle Count/Liter of Air
0.3	47,907
0.5	2153
1	222
2	96
5	17
10	7

Assume all particles are spherical with a density of 1.82 g/cm^3. What is the weight fraction for each particle size group?

The best approach is to determine volume of one particle in each group, multiply this by number of particles in the group, and then multiply by density of particles. Add total masses from each group of particles to calculate the percent weight distribution by mass fraction.

$$\text{Sphere volume} = V = \frac{4}{3}\pi r^3 = 1.333 \times \pi (0.15\mu)^3 = 0.014137\mu^3$$

$$47{,}907 \text{ particles} \times 0.014137\,\mu^3/\text{particle} = 677.26\,\mu^3$$

$$1\,\text{cm} = 10^4\,\mu$$

$$1\,\text{cm}^3 = 10^4\,\mu \times 10^4\,\mu \times 10^4\,\mu = 10^{12}\,\mu^3$$

$$677.26\,\mu^3/1 \times 10^{12}\,\mu^3/\text{cm}^3 = 0.00000000067726\,\text{cm}^3$$

$$0.00000000067726\,\text{cm}^3 \times 1.82\,\text{g/cm}^3 = 0.0000000012326\,\text{g}$$

$$= 0.0000012326\,\text{mg} = 0.00123 \text{ micrograms}$$

That is, 47,907 particles weigh a mere 1.23 nanograms!

$$V = \frac{4}{3}\pi r^3 = 1.333 \times \pi (0.25\mu)^3 = 0.06545\mu^3$$

$$2153 \text{ particles} \times 0.06545\,\mu^3/\text{particle} = 140.91\,\mu^3$$

$$140.91\,\mu^3/1 \times 10^{12}\,\mu^3/\text{cm}^3 = 0.00000000014091\,\text{cm}^3$$

0.00000000014091 cm³ × 1.82 g/cm³ = 0.00000000025646 g
= 0.0000256 μg = 0.0256 nanograms

$$V = \frac{4}{3}\pi r^3 = 1.333 \times \pi (0.5\mu)^3 = 0.52359\mu^3$$

222 particles × 0.52359 μ³/particle = 116.24 μ³

116.24 μ³/1 × 10¹² μ³/cm³ = 0.00000000011624 cm³

0.00000000011624 cm³ × 1.82 g/cm³ = 0.00000000021156 g
= 0.000224 μg = 0.224 nanograms

$$V = \frac{4}{3}\pi r^3 = 1.333 \times \pi (1.0\mu)^3 = 4.1887\mu^3$$

96 particles × 4.1887 μ³/particle = 402.12 μ³

402.12 μ³/1 × 10¹² μ³/cm³ = 0.0000000006402 cm³

0.0000000006402 cm³ × 1.82 g/cm³ = 0.0000000011652 g = 0.001165 μg
= 1.165 nanograms

$$V = \frac{4}{3}\pi r^3 = 1.333 \times \pi (2.5\mu)^3 = 65.448\mu^3$$

17 particles × 65.448 μ³/particle = 1113 μ³

1112.6 μ³/1 × 10¹² μ³/cm³ = 0.000000001113 cm³

0.000000001113 cm³ × 1.82 g/cm³ = 0.0000000020257 g = 0.002026 μg
= 2.026 nanograms
How about all these zeroes!

$$V = \frac{4}{3}\pi r^3 = 1.333 \times \pi (5.0\mu)^3 = 523.59\mu^3$$

7 particles × 523.59 μ³/particle = 3665.13 μ³

3665.13 μ³/1 × 10¹² μ³/cm³ = 0.000000003665 cm³

0.000000003665 cm³ × 1.82 g/cm³ = 0.0000000066703 g = 0.006670 μg
= 6.670 nanograms
Man, what a chore to plug and chug these numbers!

473. From data in the preceding problem, create a table listing the average weight per particle versus particle size. What conclusions can you draw from this?

For 0.3μ particles: 1.23 nanograms/47,907 particles = 0.000025675 ng/particle

For 0.5μ particles: 0.0256 nanograms/2153 particles = 0.0000119 ng/particle

For 1.0μ particles: 0.224 nanograms/222 particles = 0.001 ng/particle

For 2.0μ particles: 1.165 nanograms/96 particles = 0.012 ng/particle

For 5.0μ particles: 2.026 nanograms/17 particles = 0.119 ng/particle

For 10.0μ particles: 6.670 nanograms/7 particles = 0.953 ng/particle

Particle Diameter (μ)	Average Particle Mass (nanograms)
0.3	0.0000257
0.5	0.0000119
1	0.001
2	0.012
5	0.119
10	0.953

It goes without saying that a small particle weighs less than a large particle of equal density. With every approximate doubling of particle size, there is nearly a 10-fold increase in particle weight. These data plot very nicely on semi-logarithmic graph paper. For whatever reasons, the data for 0.5 and 0.3 micron particles are anomalous. These particles appear to have a bimodal distribution suggesting an internal combustion engine source might contribute to these smaller particles.

474. From Problem 473, what is the airborne dust concentration in mg/m^3?

1.23 ng/L + 0.0256 ng/L + 0.224 ng/L + 1.165 ng/L + 2.026 ng/L

+ 6.67 ng/L = 11.3406 ng/L = 11,340.6 $\mu g/m^3$ = 11.34 mg/m^3

This very dusty air is, most likely, primarily gypsum ($CaSO_4 \cdot 2H_2O$), sanding dust from spackling compound, wood dust, insulation dust, and paper dust. Check the spackling compound for asbestos fibers (e.g., chrysotile, amosite, and others).

475. From Problem 474, what is the mass percent for the 10-micron particles?

(6.670 ng/11.3406 ng) × 100 = 58.8% (m/v)

476. From Problem 472, what is the count percent for the 10-micron particles?
[7/(47,907 + 2,153 + 222 + 96 + 17 + 7)] × 100 = (7/50,306) × 100 = 0.00139%. Notice 0.00139% of the airborne particles contribute 58.8% of the total mass of particles. This is an important distinction easily overlooked by

those responsible for ensuring high air quality in building interiors. One must carefully read the specifications and understand the performance ratings for all filters. Ask if the percent efficiency is by weight percent or count percent.

477. Consider Problems 472 through 476. A favorite watering hole for the author is *Bailey's Pub and Grille*, a sports bar in Dearborn, Michigan where this sign is posted:

> In consideration of your health and comfort, the sophisticated air filtration system of Bailey's provides an air purity factor of 99%. (For scientists among us, this is accomplished with an air in duct, electronic carbon precip-itator along with a complete air change eight times an hour – and boy, is it expensive!)

Such air quality engineering outreach is commendable. However, another cocktail lounge in a different part of town posts a sign that boasts: "Our ventilation system filters 99% of all airborne particles." This might be to console the patrons who are nonsmokers. Could this sign be misleading, untrue?

Yes, because if the air system filters are rated in weight-percent efficiency, instead of particle-percent efficiency, the sign's claim is misleading because tiny particles are more difficult and pricier to capture. Let us say, again referring to Problem 472, if only the big 2-, 5-, and 10-micron particles are filtered, this air system would be:

$[(47{,}907 + 2{,}153 + 222)/50{,}306] \times 100 = 99.95\%$ *in*efficient for the more dangerous tiny particles that comprise tobacco smoke. Perhaps their sign should state: "Our ventilation system filter is only 0.05% effective against tobacco smoke. Beware!" Now, however, Michigan has become a "No Smoking" state for public gathering places.

478. A worker has an 8-hour time-weighted average exposure to benzene vapor exactly at the NIOSH REL of 0.1 ppm$_v$. He does not have direct skin contact with solvents containing liquid benzene. His average alveolar ventilation rate is 7 liters/minute. Assume 50% of inhaled benzene vapor enters the blood perfusing his lungs. How many benzene molecules enter his body through his lungs during his work shift? Assume exposure conditions at NTP. Solving this problem helps one see through the eyes of inhalation molecular toxicologists.

$$\frac{mg}{m^3} = \frac{0.1\,ppm_v \times 78.11\,g/g\text{-mole}}{24.45\,L/g\text{-mole}} = 0.3195\,mg/m^3$$

$7\,L/minute \times 60\,minutes/hour \times 8\,hours/workday = 3360\,L$
$$= 3.36\,m^3/workday$$

$0.5 \times 0.3195\,mg/m^3 \times 3.36\,m^3 = 0.5368\,mg$ benzene absorbed into the worker's body

$0.5368\,mg$ benzene $= 0.0005368\,g$ benzene

One mole (78.11 g) of benzene contains 6.023×10^{23} molecules of benzene.

6.023×10^{23} molecules/78.11 g/g-mol = 7.71092×10^{21} molecules/g

7.71092×10^{21} molecules/g $\times 0.0005368$ g $= 4.1392 \times 10^{18}$ molecules

The number of cells in an adult human body has been estimated to be between 10 and 100 trillion. Assume an average of 55,000,000,000,000 cells.

4.1392×10^{18} molecules/55×10^{12} = 75,258 benzene molecules/cell.

The result is an average overall cell types. Considering benzene is only slightly soluble in water and much more soluble in lipids, we see that fatty portions of the body, such as the bone marrow, become enriched in benzene. This calculation demonstrates how some might claim, "0.1 ppm is so tiny as to be insignificant." To inhalation toxicologists, we see how, on molecular levels, it is not. Benzene is a genotoxic human carcinogen operative at a molecular level without a threshold of safety. This calculation was prepared to give a frame of reference, a perspective. Of course, the benzene entering the body is metabolized and excreted only after entering the blood and perfusing the liver and kidneys—and after passing through the bone marrow. A physically fit man has an average of 15.5% fat tissue. One could reasonably argue that, for such a man as our worker, his fatty tissues would enrich by $(100/15.5) \times 75,258$ molecules/cell to 485,535 benzene molecules/cell. These calculations are somewhat oversimplified because the assumption is that benzene is only intracellular and not in the interstitial fluid and mineralized tissues such as teeth and bones. One could also, as well, calculate the number of benzene molecules per gram of total body mass assuming that the worker weighed 70 kg (154 lb): 4.1392×10^{18} molecules/70,000 g = 5.913×10^{13} benzene molecules/g (59 trillion).

The human hematopoietic system is a tremendous blood cell production line. An average adult produces 2,400,000 red blood cells per second. No, "second" is not a typographical error. Bone marrow, in health, is truly a biochemical phenomenon.

479. 34 pounds *n*-hexane evaporates in an empty, small, unventilated tool shed $12' \times 18' \times 8'$. What is the NTP vapor concentration in ppm_v?

$$34 \text{ pounds} = 15,422.1 \text{ grams} = 15,422,100 \text{ milligrams}$$

$$12' \times 18' \times 8' = 1728 \text{ ft}^3 = 48.93 \text{ m}^3$$

$$ppm_v = \frac{(15,422,100 \text{ mg}/48.93 \text{ m}^3) \times 24.45}{92.13 \text{ grams/gram-mole}} = 83,646 \text{ ppm}_v$$

$$83,646 \text{ ppm}_v/(10,000 \text{ ppm}_v/1\%) = 8.3646\%_v$$

The LEL–UEL range for *n*-hexane is from 1.1–1.25 to 7.0–7.5 vol. %. So, when one opens the door and introduces air, the vapor concentration dilutes into the explosive range. Knowing the vapor concentration, one must not enter. Through a small opening (no sources of ignition such as a hot drill bit or hot saw blade), slowly introduces an inert gas (e.g., N_2, CO_2) or air until concentration is well below $10 \, ppm_v$. Remotely test shed's atmosphere for oxygen and hexane before allowing entry.

480. Consider a building with the length and width of a standard high school, college, and NFL football field excluding the end zones (300 feet long and 160 feet wide). This building has a ceiling height of 10 feet. Also assume that the building is tightly sealed: no windows, doors, or mechanical or natural ventilation. The internal volume of this building is $300' \times 160' \times 10' = 480,000 \, ft^3 = 13,592 \, m^3$.

Mary, my wife, has a standard sewing thimble with a volume of 3.4 milliliters. The density of liquid benzene is 0.88 gram/milliliter. So, if the benzene in the thimble spills inside the building and evaporates (perhaps in only a few minutes), the amount of benzene vapor in the air is $3.4 \, mL \times 0.88 \, g/mL = 2.992 \, grams = 2992 \, milligrams$.

Molecules move about. So, over a period of time, 2992 milligrams of benzene exists as vapor in a building with a volume of $13,592 \, m^3$. This mass of benzene can be converted to parts of benzene vapor per million parts of air in the building by the following equation:

$$\text{ppm} = \frac{(mg/m^3) \times 24.45 \, L/gram\text{-mole}}{\text{molecular weight of benzene}} = \frac{(2992/13,592) \times 24.45}{78.1}$$

$$= 0.069 \, \text{ppm} = 69 \, \text{ppb (v/v)}$$

This is 69% of the recommended exposure level for workers by NIOSH provided the worker does not have skin contact and does not have preexisting hematopoietic disorder. This REL is the worst acceptable concentration. Moreover, benzene is a known, recognized human carcinogen. It is genotoxic. Therefore, there is no safe level of exposure. There is no threshold for mutation.

481. Natural gas contains 87–96% methane, CH_4, the simplest organic molecule and hydrocarbon. Methane is highly combustible (LEL = 4.4–5% by volume, UEL = 15–17% by volume) with the right conditions: sufficient methane gas present, sufficient oxygen, and an ignition source. The ignition source must have sufficient energy to cause ignition (0.21 mJ at 8.5% by volume, near the mid-point between the LEL and UEL), and ignition source must be present long enough to cause ignition and combustion. When these conditions are present, chemists can write a combustion equation for this chemical reaction:

$$CH_4 + 2O_2 \xrightarrow{\Delta} 2H_2O + CO_2 + \text{heat}$$

That is, one part (a mole, or mol) of methane gas mixed with two parts of oxygen gas, O_2, in the presence of a sufficiently hot ignition source (Δ) sufficiently long will combust (oxidize) to produce (yield) two parts of water vapor and one part of carbon dioxide gas. This equation represents a perfect (stoichiometric) chemical reaction that rarely occurs outside of the laboratory inside a bomb calorimeter. If the high oxygen demand to combust methane is not met or sustained, incomplete combustion occurs resulting in additional gaseous by-products such as carbon monoxide, formaldehyde, acetic acid, and soot. Calculate how much oxygen is needed to perfectly combust, oxidize 1 kilogram of methane. Calculate mass of carbon dioxide gas and water vapor produced.

Molecular weights of CH_4, O_2, H_2O, and CO_2 are, respectively, 16, 32, 18, and 44 grams/mole.

$$1 \text{ kg } CH_4/16\,\text{g/mole} = 1000\,\text{g } CH_4/16\,\text{g/mole} = 62.5\,\text{moles of } CH_4$$

Therefore, 2×62.5 moles of O_2 are required $= 2 \times 62.5$ moles $\times 32\,$g/mole $= 4000\,$g.

Since mass is conserved, $1000\,$g $CH_4 + 4000\,$g $O_2 = 5000\,$g of H_2O and CO_2 are produced.

For H_2O, 2 moles $\times 18\,$g/mole $\times 62.5$ moles $= 2250$ grams of H_2O vapor are produced.

For CO_2, 1 mole $\times 44\,$g/mole $\times 62.5$ moles $= 2750$ grams CO_2 are produced.

$$1 \text{ kg } CH_4 + 4 \text{ kg } O_2 \xrightarrow{\Delta} 2.25 \text{ kg } H_2O + 2.75 \text{ kg } CO_2 + \text{heat}$$

482. Quaternary ammonium compounds ("quats") such as benzylkonium chloride have been used for over 50 years to preserve the sterility of inanimate objects such as medical and dental surgical instruments and endoscopes after steam or chemical sterilization and to protect eye drops and nasal sprays. Hemostats and forceps used in medical clinics, for example, are immersed in highly diluted solutions until needed. One part "quat" in 40,000 sterile normal saline solution is claimed to be bactericidal. 1/200,000 is bacteriostatic. During recent years, the use of "quats" has expanded to help sanitize surfaces such as floors, walls, and hard objects in medical facilities and nursing homes. Several quats are now commercially available through retail outlets.

Quats are potent respiratory irritants and, for some, respiratory, dermal, and ocular allergens and sensitizers. Asthma has been reported in those inhaling nebulized quat aerosols. Unfortunately, over past several years, some have indiscriminately dispersed aerosols of quats in occupied areas with a misguided notion of sterilizing air or hard surfaces. Careless dispensing as fogs, mists, aerosols, and nebulized solutions have been done in commercial/public buildings including schools, food processing plants such as dairies, residences, health care facilities, poultry farms, and offices.

Reactive airways disease syndrome (RADS), persistent bronchial asthma, and hypersensitivity pneumonitis have resulted from these dangerous practices. Commercial outlets market quats—often combined with "air fresheners"—to retail customers to spray homes. Organizations that offer air duct cleaning services wrongly fog quats in ventilation air supply distribution systems to "prevent mold growth."

Those spraying quats claim that low-pressure foggers cannot generate respirable particles; that is, they assert, there is no inhalation hazard. Consider 100-μm diameter airborne quat particles (1/40,000 concentration benzylkonium chloride) in less than a 100% relative humidity atmosphere. After the water phase evaporates (fast in low humidity air), particles are now 100% quat. What is the diameter of these new particles? Small enough to penetrate terminal airways?

$$(1/40,000) \times 100 = 0.0025\% \text{ benzylkonium chloride}$$

$$100\% - 0.0025\% = 99.9975\% \, H_2O$$

volume, V, of spherical aerosol particle $= (4/3) \, \pi \, r^3$

$$\text{Particle volume} = V = \frac{4}{3}\pi r^3 = \frac{4}{3}\pi(50\mu)^3 = 523{,}599\mu^3$$

$$H_2O \text{ volume} = 523{,}599\mu^3 \times 0.999975 = 523{,}586\mu^3 \, H_2O$$

$$\text{Toxicant volume} = 523{,}599\mu^3 - 523{,}586 \, \mu^3$$
$$= 13 \, \mu^3 \text{ benzylkonium chloride}$$

$$r^3 = \frac{V}{(4/3)\pi} = \frac{13}{(4/3)\pi} = \frac{13\mu^3}{4.1888} = 3.1035\mu^3$$

$$r = \sqrt[3]{3.1035} \, \mu^3 = 1.46\mu$$

diameter, $\varnothing$, $= 2 \, r = 2.92 \, \mu$.

100 micron diameter nonrespirable 0.0025% benzylkonium chloride particles are now 100% concentration particles with respirable diameter of 2.92 microns—much smaller and more toxic to the delicate tissues in the lower respiratory tract.

The author maintains that spray dispersion of quats in atmospheres must not be practiced because inhalation health risks far outweigh any dubious benefits of temporary "sterilization" of air or surfaces. Quats rapidly lose their activity inside ventilation systems because they degrade in the presence of organic materials by chemically bonding. Moreover, the dried particles are easily resuspended in air inhaled by those in treated areas. These, in turn, can act as allergic sensitizers. Refer to Problems 456 and 487.

Those who intend to dispense aerosols of biologically reactive agents into the air of facilities that people and animals could occupy, or air handling systems, should:

- Consult a board-certified industrial hygienist, an inhalation toxicologist, or a physician trained in occupational and environmental medicine.
- Carefully consider safer alternatives, safer application methods, and apply the principles of integrated pest management.
- Critically review activity of safest aerosol and residual potency over time.
- Discuss potential adverse health effects with building occupants. Obtain their written permission to proceed.
- Inquire about occupants' preexisting respiratory health conditions. The young, elderly, those with history of allergy, the atopic, those with chronic obstructive (chronic bronchitis, emphysema) and restrictive lung disease, and cardiovascular impairment are at greatest risk.
- Secure treated areas and post with bold warning signs before, during, and after application. Use negative pressure isolation, containment practices.
- Bar access during and after application and until robust ventilation, bake-out, ventilation protocol has been completed.
- Obtain ample air samples inside and outside the facility to ensure air quality equals or exceeds outdoor air quality before allowing occupancy.
- Establish a rigorous industrial hygiene program to conserve health of aerosol, fogging applicators including carefully crafted written work practices, selection and using high-quality personal protection equipment, careful supervisory oversight, and health and safety education and training per OSHA's *Hazard Communication Standard* (29 *CFR* 1910.1200).
- Not utter fraudulent, scientifically unsupported statements (e.g., "This will prevent mold growth for up to three years," or "Your air supply ducts will be sterilized," and "Your home and your family will be healthier").

483. A sheet metal worker constructing an exhaust ventilation system to control the exposures of workers to wood sanding dust has a limited supply of materials. He connects a 24-feet section of 16-inch diameter sheet metal duct to duct 12 inches in diameter. The 12-inch diameter duct was engineered to convey wood dust at 3500 feet per minute. What is wood dust transport velocity in 16-inch diameter section?

12-inch duct cross section area $= \pi r^2 = 3.1416 \, (6'')^2 = 113.1 \text{ in}^2 = 0.7854 \text{ ft}^2$

$$0.7854 \text{ ft}^2 \times 3500 \text{ fpm} = 2749 \text{ ft}^3/\text{minute}$$

16-inch duct cross section area $= \pi r^2 = 3.1416 \, (8'')^2 = 201.1 \text{ in}^2 = 1.397 \text{ ft}^2$

$$(2749 \text{ cfm})/1.397 \text{ ft}^2 = 1968 \text{ fpm}$$

The wood dust duct transport velocity was reduced from 3500 fpm to a dangerous 1968 fpm. This sheet metal worker does not realize the hazards that he created. Lower duct transport velocity permits wood sanding dust to settle and accumulate in wider diameter duct. This presents explosion and fire risk. A minimum duct transport velocity for fine wood dust is 3500 fpm, the engineering design velocity. Ventilation system design engineers should field test the systems ("turn key") so dangerous short-cuts are prevented by strict adherence to engineering design specifications. A duct velocity of 1968 fpm is acceptable for gases and vapors, but is far too low for particulates—especially if they are combustible, toxic, and/or explosive.

484. One liter of a volatile solvent (vapor pressure = 58 mm Hg; molecular weight = 79.6 grams/gram-mole) is spilled in a hyperbaric chamber with operating pressure of 2.6 atmospheres. After the air is saturated with solvent vapor, 845 mL of liquid solvent remains. What is saturation solvent vapor concentration?

$$1 \text{ atmosphere} = 760 \text{ mm Hg}$$

$$2.3 \text{ atmospheres} \times 760 \text{ mm Hg/atmosphere} = 1748 \text{ mm Hg}$$

$$(58 \text{ mm Hg}/1748 \text{ mm Hg}) \times 10^6 = 33{,}181 \text{ ppm}_v = 3.318\% \text{ by volume}$$

Clearly, the atmosphere in the hyperbaric chamber is toxic and might present an explosion hazard if 3.318% is above LEL and below the UEL. Entry must not be permitted until high-pressure atmosphere is safely vented through an activated charcoal trap or by some other method that does not expose people to this organic vapor.

485. Referring to Problem 484, what is vapor concentration after chamber's pressure is reduced to the standard sea-level atmospheric pressure (760 mm Hg)?

$$(58 \text{ mm Hg}/760 \text{ mm Hg}) \times 10^6 = 76{,}316 \text{ ppm}_v = 7.6316\%$$

Answer: 76,316 ppm$_v$. Note how higher pressure in the chamber suppressed evaporation of solvent. Not surprisingly, inhalation and explosion hazards more than doubled as the high pressure was reduced to atmospheric pressure.

486. Refer to Problem 483. If the liquid solvent has a density of 0.81 g/mL, what is the net volume of the hyperbaric chamber? Assume NTP.

$$1000 \text{ mL} - 845 \text{ mL} = 155 \text{ mL}, \text{ the liquid volume that evaporated}$$

$$155 \text{ mL} \times 0.81 \text{ g/mL} = 125.55 \text{ grams of solvent-saturated}$$
$$\text{chamber atmosphere}$$

$$mg/m^3 = (ppm \times molecular\ weight)/24.45\ L/g\text{-}mole$$

$$= (33,181 \times 79.6)/24.45 = 108,025\ mg/m^3$$

$$125.55\ g = 125,500\ mg$$

$$125,500\ mg/108,025\ mg/m^3 = 1.162\ m^3$$

Note that liquids are virtually incompressible, but gases are not. This method can be used to approximate the interior volume of any structure that has an irregularly shaped interior and various shaped objects within. Of course, a more accurate way—assuming contents are not harmed by water—is to fill the chamber with water and determine difference in weight loss or gain, or measure the water volume added or emptied.

487. What factors determine loss of water from tiny airborne particles? Disregard particles with molecules that are strongly hydrated (e.g., $2H_2O \cdot CaSO_4$).

The rate at which water evaporates from airborne particles depends upon the air temperature, particle temperature, turbulence of particle, atmospheric pressure, relative humidity, particle diameter, vapor pressure of water, and intermolecular forces between the solute or suspended chemical and with water (i.e., hydrates).

Small airborne particles rapidly assume air temperature even if emitted from very hot or cold processes. Normal air currents and Brownian motion provide the necessary turbulence. Relative humidity, perhaps more than any other factor, drives the rate of water loss from the aerosol particles— quickly in low humidity atmospheres and slowly in highly humid air to zero in 100% relative humidity air. See Problems 456 and 482.

488. Industrial hygienists and toxicologists often explain concepts of toxicants and toxins to workers, other health professionals, and to our courts. Many find it difficult to grasp the concepts of "very tiny." We can do this by comparing the familiar with the unfamiliar. Some helpful ways are to determine the amount of solvent vapor evaporated from a sewing thimble (3.4 mL) into a building of recognized size, say a football field excluding the end zones (300' × 165') with ceiling height of a standard house room (8'). For any airborne particulate (dust, mist, smoke, fume), we can consider mass (1 gram) inside the little packets of sucrose and artificial sweeteners we find in restaurants. Let us use the OSHA PEL for inorganic lead for workers (50 mcg/m³). What building volume, using sugar example assuming it is finely ground lead powder, would result?

$$1\ gram = 1,000,000\ micrograms$$

$$1,000,000\ micrograms/50\ mcg/m^3 = 20,000\ m^3$$

$$20,000 \, m^3 = 706,293 \, ft^3$$

$$706,293 \, ft^3/(165' \times 300') = 14.3 \, feet \, high \, ceiling$$

A picogram is a very tiny unit of mass. That is, 1 pcg = 10^{-12} gram, or 1 trillionth of a gram. Toxicologists and analytical chemists can detect numerous chemicals in about any matrix at picogram level, for example, 3 picograms of dimethyl awful stuff per deciliter whole blood. These units are virtually meaningless to most people. For reference, a deciliter = 100 mL = 3.38 ounces = 0.42 cup, or said another way, slightly less than ½ cup.

Let us put this into common sense perspective by calculating how many picograms are in a single crystal of common table salt, sodium chloride (NaCl).

The density of table salt is 2.165 grams/milliliter. Table salt crystals are nearly perfect cubes with a uniform dimension of 0.03 centimeter on each side. The volume of a single table salt crystal is, therefore 0.03 cm³ = 0.000027 cm³ = 0.000027 mL. So, the mass of a single table salt crystal is 2.165 g/mL × 0.000027 mL = 0.0000585 gram. This equals 0.0585 mg = 58.5 micrograms = 58,500 nanograms = 58,500,000 picograms.

Now, let us see how many molecules of NaCl are in a single grain of table salt.

$$Molecular \, weight \, of \, NaCl = 58.43 \, g/mole$$

$$58.43 \, g \, NaCl \, contains \, 6.022 \times 10^{23} \, molecules$$

$$(6.022 \times 10^{23} \, molecules/mole) \times (mole/58.43 \, g) = 1.0306 \times 10^{22} \, molecules/gram$$

$$(1.0306 \times 10^{22} \, molecules/gram) \times 0.0000585 \, g$$

$$= 6.0606 \times 10^{17} \, molecules \, of \, sodium \, chloride \, in \, each \, grain \, of \, table \, salt.$$

The Earth's population is 7 billion in 2012. If each person equals a picogram, how many Earth's are needed for 1 trillion people? 143 Earths.

31,688 years have 1 trillion seconds. That is, 1 second in 31,688 years is one part per trillion.

489. An investigator wants to determine particle emission rate in, for example, milligrams per cubic meter per minute of an automobile's exhaust system without a catalytic converter. What factors must she consider in her measurements and research?

Engine revolutions per minute, absolute temperature of exhaust gas, isokinetic sampling, exhaust gas pressure, engine displacement, dynamometer loading, internal diameter of exhaust pipe, Pitot tube or other measuring device of exhaust gas velocity, mass of fuel and air injected, volumetric flow rate of exhaust gas corrected for temperature and pressure, collecting media filtration efficiency, calibration of measuring instruments, and standard analytical laboratory methods.

490. A 4760 cfm air stream contains 420 ppm$_v$ chlorine gas. Local environmental stack permits no more than 1 ppm$_v$ in vented gas streams. A chemist and a mechanical engineer design a chlorine gas device that will scavenge up to 320 pounds before the device requires changing or regeneration. How long will the device operate before change-out is required?

$$\frac{4760\,ft^3}{minute} \times \frac{1\,m^3}{35.315\,ft^3} = 134.79\,m^3/minute$$

$$\frac{mg}{m^3} = \frac{ppm \times molecular weight}{24.45\ L/g\text{-}mole} = \frac{420 \times 70.91}{24.45} = 1218.1\,mg/m^3$$

$$\frac{134.79\,m^3}{minute} \times \frac{1218.1\ mg}{m^3} = 164,187.7\,mg/minute$$

$$\frac{1\ lb}{453,592.37\,mg} \times \frac{164,187.7\,mg}{minute} = \frac{0.362\ pound}{minute}$$

$$\frac{320\ pounds}{0.362\ pound/minute} = 884\,minutes = 14.7\,hours$$

Change-out or regeneration is required every 14.7 hours. However, a safety factor should be considered. For example, consider a change-out every 12 hours, and evaluate residual functional capacity by laboratory tests. These tests might reveal design capacity of 14.7 hours is acceptable, or a briefer change-out is required. Alternatively, consider operating for 16 hours (at start of the work shift), but verify Cl_2 gas concentrations downstream of collector by stack testing. Double split-stream parallel identical chlorine gas scavengers require change-out every 29.4 hours. Material costs will not increase, but labor costs should decrease because change-out crew makes fewer visits to service air pollution control device. One can also consider scaling the chlorine gas collector larger to operate longer before gas saturation. This promotes system safety because of the scavenger's higher residual capacity before Cl_2 gas breakthrough.

491. A fan was selected to supply 35,530 cfm at 8" SP. The fan operates at 1230 rpm and requires 61.0 bhp. After installation, an un-sophisticated HVAC "engineer" elects to increase fan's output by 20% for "safety." At what rpm must the fan now operate? What SP will develop? What bhp will be required?
    cfm varies as rpm: 1230 cfm × 1.20 = 1476 rpm. Do not exceed fan's rated tip speed! Blades can shatter hurling shards and projectiles of sharp metal in every direction.

sp varies as $(rpm)^2 = (1476 \, cfm/1230 \, cfm)^2 \times 8" = 11.5" \, sp$

bhp varies as $(rpm)^3 = (1476 \, cfm/1230 \, cfm)^3 \times 61.0 = 105.4 \, bhp$

Note that as the fan output increased by 20%, the brake horsepower went up substantially ($105.4 \, bhp/61.0 \, bhp = 1.728$), or by a whopping 173%. In these days of energy = money, very careful system designs are required. If only a 10% "safety factor" is added to the system volume, the horsepower increase will be 33% per the third fan law. Evaluation should be made weighing necessity of a "safety factor" against the substantial energy cost penalty incurred. Good engineering does not rely on expensive "safety factors." Design for the reasonably foreseeable, not any more.

492. The air in a tightly sealed empty building $60' \times 30' \times 18'$ contains $17 \, ppm_v$ HBr. A service door connects this building to another tightly sealed building $130' \times 80' \times 18'$. After the connecting service door is opened, contaminated air mixes with the clean air in the adjacent building. What is the HBr gas concentration after gas is homogenous throughout both rooms?

$$mg/m^3 = \frac{ppm_v \times mol.\,wt.}{24.45 \, L/g\text{-}mole} = \frac{17 \times 80.92 \, g/g\text{-}mole}{24.45} = 56.26 \, mg/m^3$$

$$60' \times 30' \times 18' = 32{,}400 \, ft^3 = 917.\,5 \, m^3$$

$56.26 \, mg/m^3 \times 917.5 \, m^3 = 51{,}618.6 \, mg$ HBr is in the air of the smaller building. Mass is conserved, but now distributes into the larger room until equilibrium is reached. At such low concentrations, this might require a long time without the aid of mechanical and/or thermal turbulence forces.

Volume of larger building $= 130' \times 18' \times 80' = 187{,}200 \, ft^3 = 5301.3 \, m^3$

Volume of both buildings $= 917.5 \, m^3 + 5310.3 \, m^3 = 6227.8 \, m^3$

$$ppm_v = \frac{(mg/m^3) \times 24.45}{mol.\,wt.} = \frac{(51{,}618.6 \, mg/6227.8 \, m^3) \times 24.45}{80.92}$$

$$= 2.50 \, ppm_v$$

$2.5 \, ppm_v$ HBr. Sometimes, simply allowing the high concentrations of a toxic gas to dilute over time to a lower concentration is a good first start to protecting workers required to enter such buildings. Of course, in the absence of air-sampling data, workers must diligently wear appropriate respiratory protection. As an aside, hydrogen bromide is reactive acid gas, and, over time, could be expected to react with surfaces in two buildings further reducing concentration—again to be verified by robust air sampling. I would select detector tubes for this type of air sampling because results are prompt, pockets of gas can easily be identified, and costs are negligible when compared to wet chemical laboratory methods.

493. A sawyer's Monday to Friday 8-hour time-weighted averages to respirable hard wood dust (oak, beech, maple) were, in mg/m³: 2.3, 0.1, 4.2, 0.9, 1.7. For the week, what was his cumulative dose to these respirable carcinogens? The cumulative exposure for T (the selected time interval) is calculated by

$$\text{cumulative}\,(T) = \sum_{i=1}^{n} C_i\, t_i$$

In this case, we select days as the time units, where $T =$ total elapsed time, $t =$ incremental time units as components of T.

$$\text{cumulative (5 days)} = (2.3 \times 1) + (0.1 \times 1) + (4.2 \times 1)$$
$$+ (0.9 \times 1) + (1.7 \times 1) = 9.2\,\text{g/m}^3$$

Cumulative dose for the week was 9.2 mg/m³-days (or 9.2 mg/m³-5-day work week). Also note that hard wood dusts are carcinogenic to the upper airways. Therefore, respirable dust samplers typically select against the larger particles. The TLVs for carcinogenic wood dusts are expressed in total wood dust per cubic meter of breathing zone air. To the unwary, reported test results are lower than the true or actual value for total airborne wood dust.

494. Refer to previous Problem (493). Calculate the sawyer's cumulative dose if his consecutive dust exposures were:

2.3 mg/m³ for 3.3 weeks
0.1 mg/m³ for 0.4 weeks (2 workdays/7 days)
4.2 mg/m³ for 2.2 weeks
0.9 mg/m³ for 1.2 weeks
1.7 mg/m³ for 2.6 weeks

$$(2.3 \times 3.3) + (0.1 \times 0.4) + (4.2 \times 2.2) + (0.9 \times 1.2) + (1.7 \times 2.6)$$
$$= 22.37\,\text{mg/m}^3\text{-weeks}$$

22.37 mg/m³-weeks. Note how longer exposures to higher dust levels contribute, not surprisingly, largest contributions to his dose. For example, the 2.3 and 4.2 mg/m³ exposures comprise $[(7.59 + 9.24)/22.37] \times 100 = 75.2\%$ of his cumulative dose for the week. The other 3 days account for less than 25% of his dose. Again, respirable dust mass concentrations are usually much lower than total airborne dust concentrations.

495. We know that 1 gram-mole of a perfect, ideal gas or vapor occupies 22.414 L at 0°C and 760 mm Hg sea-level atmospheric pressure. Practical industrial hygiene conditions assume that gases and vapors behave as if they are "ideal." Unless in a strange realm of chemical engineering where pressures and temperatures can be extreme, we normally do not need to concern ourselves with nonideal vapors and gases. A few problems in this book cover some applications of nonideal gas behavior. However, most workers do not find themselves in 0°C situations, and, for this reason, it was once decided that 77°F be selected as the reference temperature. Mathematically describe this standard mg/m³ to ppm$_v$ conversion equation using 77°F and 760 mm Hg pressure.

0°C = 32°F, 25°C = 77°F, and from the universal gas laws, we know that

$$\frac{P_1 V_1}{T_1} = \frac{P_2 V_2}{T_2}$$

Rearranging

$$V_2 = V_1 \left(\frac{P_1}{P_2}\right) \times \left(\frac{T_2}{T_1}\right) = 22.414 \text{ L} \left(\frac{760 \text{ mm}}{760 \text{ mm}}\right) \times \left(\frac{298°C}{273°C}\right)$$

= 24.467 L/gram-mole, accurately, or 24.45 liters
when rounding to 22.4 L.

There, now you know. Also, of course, when one obtains air samples at greatly different temperatures and/or pressures than 298°C and 760 mm Hg, the above equation must be used for the proper and accurate expression of results. When sampling high-temperature organic material combustion processes or calcining processes, water vapor typically accounts for a significant portion of exhaust gas volume. Corrections to dry air should be made. See Problem 47 for these somewhat more challenging calculations.

496. Limestone, calcium carbonate ($CaCO_3$), is calcined in a lime kiln. For every ton of limestone that is calcined to lime, how much carbon dioxide gas is released to the atmosphere?

$$CaCO_3 \xrightarrow{\triangle} CaO \text{ (lime)} + CO_2$$

Molecular weights: $CaCO_3$ = 100.09 grams/gram-mole

CaO = 56.08 grams/gram-mole $\left.\right\}$ 100.09 grams
CO_2 = 44.01 grams/gram-mole

1 ton = 907,185 grams (or the same number of packets of artificial sweeteners)

For $CaCO_3$: 907,185 grams/100.09 grams/gram-mole = 9063.7 moles

Since 1 mole of $CaCO_3$ yields 1 mole of CO_2, 9063.7 moles of $CaCO_3$ yields 9063.7 moles of CO_2.

9063.7 moles $\times$ 44.01 grams/gram-mole = 398,893 grams of CO_2=879.4 pounds = 0.44 ton of CO_2 for every ton of limestone calcined to lime, or calcite. One can avoid these calculations by studying the balanced chemical reaction. Since mass is conserved, apply molecular weights assuming stoichiometric and a quantitative reaction in the lime kiln. Actually, limestone contains silica and some silicates that, from an industrial hygiene perspective, assume toxicological significance beyond the eye, skin, and respiratory tissue irritant properties of lime and limestone dust.

Answer: 0.44 ton CO_2. Of interest, it was once suggested milk-of-lime $[Ca(OH)_2]$ be used to scrub CO_2 from coal- and fuel oil-fired powerhouses in an effort to help reduce global warming. Milk-of-lime is made by bubbling water through powdered lime to make a slurry: $CaO + H_2O \rightarrow Ca(OH)_2$. CO_2 released to air from calcining limestone equals CO_2 scavenged from fuel combustion gases. There are no free lunches.

497. Industrial hygienists are often confronted with several requests to promptly resolve issues ASAP. Handled on a first-in, first-out basis can lead to equating minimal risks to major risks. Suggest a reasonable method for triage— that is, worst first approach.

Air contaminant occupational exposure, dose	Respiratory irritants with generally **reversible** adverse health effects	Inhaled and skin-absorbed toxicants with **target organ effects or systemically toxic**	Inhaled and skin-absorbed toxicants with **irreversible** adverse health effects (carcinogens, teratogens, reproductive health hazards, pulmonary fibrosis agents, respiratory allergens)
Nondetectable to <10% of TLV, PEL, STEL	Minimum risks	Minimum risks for most healthy, without skin contact workers	**Significant risks for small percentage of unknown workers**
10% to action level (50% of TLV, PEL, STEL)	Moderate risks	**Significant risks to many workers**	**Significant risks for a large percentage of unknown workers**
>Action level	**Significant risks**	**Significant risks**	**Extreme risks for all workers**

Those in bold in the above table should receive priority. One might argue, "If I don't know exposures, how can I assign priority? Agreed, but once the exposures are determined, control methods for multiple overexposed workers can now be given priority. And, seasoned industrial hygienists often have a sense of their priorities before inception of a risk assessment project, and as they carefully observe work practices, measure ventilation, determine exposure durations, and so on.

498. An empty, unventilated room in a building at sea level is 10 feet × 30 feet × 40 feet. Initial relative humidity is very unusual at 0% ("bone dry"). Dry bulb is 75°F. One gallon of water evaporates into this room. What is the new relative humidity?

$$10' \times 30' \times 40' = 12{,}000\,\mathrm{ft}^3 = 340\,\mathrm{m}^3$$

$$1\text{-gallon } H_2O = 8.33\,\mathrm{pounds} = 3778\,\mathrm{mL} = 3778\,\mathrm{g}$$

$$3778\,\mathrm{g}/340\,\mathrm{m}^3 = 11.11\,\mathrm{g/m}^3 = 11{,}110\,\mathrm{mg/m}^3$$

$$\mathrm{ppm}_v\ H_2O \text{ vapor} = (11{,}110\,\mathrm{mg/m}^3 \times 24.45)/18 = 15{,}091\,\mathrm{ppm}_v = 1.509\%$$

$$1\% = 10{,}000\,\mathrm{ppm}_v$$

At 75°F, vapor pressure of water = 22.3 mm Hg.
(22.3 mm Hg/760 mm Hg) × 10^6 = 29,342 ppm = 2.934%, saturation concentration—maximum amount of water vapor this room atmosphere can hold at 75°F and 760 mm Hg (as long as walls and windows are greater than dry bulb temperature when condensation would occur—the dew point).

$$(1.509\%/2.934\%) \times 100 = 51.4\% \text{ relative humidity}$$

499. A study was performed to determine an acceptable air dilution ventilation rate for a low inhalation toxicity solvent with a relatively low vapor pressure and evaporation rate. Several workers separated by several feet dispensed this solvent. Although local exhaust ventilation for each worker would be ideal, costs were deemed prohibitive. After completing 13 studies, it was found that solvent vapor concentrations in ppm_v and the air exchange rate in air changes per hour were linearly correlated as shown in equation:

$$y = 5174.9\ e^{-0.4263x},$$

where
$x = \mathrm{ppm}_v$ solvent vapor, and
$y = $ air changes per hour.

The TLV for the solvent is 500 ppm_v. Traditionally, 10% of the exposure limit is used as the worst acceptable concentration to allow for vagaries in the air movement (uniformity, pedestrian traffic and activities, portable fans, etc.), or, in this case, $50\,\text{ppm}_v$. Recent reports in the medical and toxicological literature revealed increase in toxicity from inhalation of this solvent vapor over ingestion. The TLV was outdated, and it was based on ingestion of the solvent by rats and mice, years ago. So, $25\,\text{ppm}_v$ was selected as a maximum 8-hour exposure limit. How many air changes per hour are required?

$$\frac{25\,\text{ppm}_v}{5174.9} = e^{-0.4263\,x}$$

$$0.00483\,\text{ppm}_v = e^{-0.4263x}$$

$$\ln 0.00483\,\text{ppm}_v = -0.4263\,x$$

$$-5.333\,\text{ppm}_v = -0.4263\,x$$

$$x = \frac{5.333}{0.4263} = 12.5$$

Answer: 12.5 air changes per hour. The energy costs for such a system in a very cold or hot climate will be high if the work area is large. The collective work space should be reduced as functionally possible to allow for this while not compromising the workers' health. The preventative maintenance program should include regular air sampling, measuring air flow at work stations, and servicing fans, filters, belts, and so on.

500. In the previous problem, let us see what the solvent vapor exposures are if the air exchange rate drops to 10.3 air changes per hour. Assume work practices have not changed, and solvent consumption and application methods remain the same.

    The solvent vapor concentration is directly proportional to the reduced air flow.

$$\frac{12.5\,\text{ach}}{10.3\,\text{ach}} = 1.2136$$

$$25\,\text{ppm}_v \times 1.2136 = 30.34\,\text{ppm}_v$$

Answer: $30.3\,\text{ppm}_v$. My first hunch is fan belts are slipping and/or filters are dirty. Filters, of course, can be visually observed and changed as required. Measuring the static pressure drop across filters is more accurate. The air supply fans and the exhaust fans' revolutions per minute are checked with a calibrated tachometer.

> Nothing is more devastating to an opinion than a good number.
> **Frank Wabeke**, *my father*

501. Again, let us consider Problem 499 if solvent consumption use is reduced by 37% through improved work practices. What are the solvent vapor exposures now?

This is a no-brainer: $25\,ppm_v \times (1 - 0.37) = 15.75\,ppm_v$.

Or, $25\,ppm_v \times 0.63 = 15.75\,ppm_v$.

Exposures were reduced to less than $16\,ppm_v$. This makes a strong case that fan speeds can now be reduced by 37%. Savings are remarkable: tempered air needs are down, and solvent consumption is reduced—both "green" accomplishments. Air sampling in the breathing zones of workers is the best measure of exposures. Once done carefully, testing the ventilation and studying any changes in the work practices can be applied. Study of solvent purchase records over time is a crude way of estimating increases or decreases in exposures once a baseline of worker exposures is established by careful air sampling. In general, the author does not rely on this indirect method of characterizing exposures, however it can be helpful in modeling past exposures.

502. The modeling assumptions for a hypothetical diacetyl vapor-exposed mixing room worker are: a large commercial bakery orders four 1-gallon bottles of an equal mixture of diacetyl, vanilla, and cinnamon flavors—all liquids to be mixed and shipped. These are mixed in a small room with no ventilation with dimensions of $12' \times 12' \times 18'$. Room's contents are 15% of room's volume. The bakery will add appropriate amounts of mixed concentrate to each batch of dough to infuse baked goods with flavors and aromas. Mixing room worker adds equal volumes of the three flavors by pouring through a funnel into the 1-gallon shipping containers. What is the average diacetyl vapor concentration in the mixing room?

$$gallon = 3785\,mL$$

$$1/3\ gallon = 1262\,mL = 1.262\,L$$

$$Sea\text{-}level\ atmospheric\ pressure = 760\,mm\ Hg$$

$$Floor,\ walls,\ ceiling\ temperatures = 68°F$$

$$Containers'\ temperature = 68°F$$

$$68°F = 293.15\,kelvin$$

$$Diacetyl\ vapor\ pressure\ at\ 68°F = 52.2\,mm\ Hg$$

$$Diacetyl\ molecular\ weight = 86.09\,grams/mole$$

$$Mixing\ room\ volume = 12' \times 12' \times 18' = 2592\,ft^3$$

$$15\%\ mixing\ room\ volume = 2592\,ft^3 \times 0.15 = 389\,ft^3$$

Net volume of mixing room $= 2592\,ft^3 - 389\,ft^3 = 2203\,ft^3$

Net volume of mixing room $= 62.38\,m^3$

Diacetyl vapor saturation concentration at 68°F $= (52.2\,mm\ Hg/760\,mm\ Hg) \times 10^6 = 68{,}684\,ppm_v = 6.8684\,vol.\ \%$

>LEL and <UEL   built-in match

converting to mg/m³: $(68{,}684 \times 86.09\,g/g\text{-mole})/24.04\,L/g\text{-mole} = 245{,}965\,mg/m^3 = 246\,mg/L.$

Pouring 1.262 L of diacetyl by splash filling into the 1-gallon container releases $1.262\,L \times 246\,mg/L = 310.45\,mg$ diacetyl vapor into the mixer's breathing zone air.

Average concentration of far-field diacetyl vapor throughout the mixing room is

$$[(310.45\,mg/62.38\,m^3) \times 24.04/86.06] = 1.39\,ppm_v$$

There is a small diacetyl vapor cloud concentration substantially higher than this enveloping the worker's near-field breathing zone particularly if she is short with her breathing zone closer to the work bench than would be for a taller worker. As the vanilla and cinnamon liquids are added to the container, more diacetyl vapors are displaced. The three remaining bottles are filled increasing the diacetyl vapor concentration in the mixing room and, especially, in the mixer's breathing zone.

If the ambient and liquid temperatures increase by 10°C (to 86°F), the far-field average diacetyl vapor concentrations will approximately double ($2.78\,ppm_v$) as well as near-field diacetyl vapor exposures.

Other variables increasing diacetyl vapor exposures include a lower ceiling height, more container filling, heated mixing tanks, workers' heights, work practices, spill response, larger containers, larger liquid diacetyl surface areas. Moreover, vanilla and cinnamon vapors must be monitored as well along with the ethanol, the carrier solvent.

503. The safety engineer at a plant in Kansas calls his corporate industrial hygienist in California saying, "My velometer is broken, and the MSDS for *TM-768* states 'Use with adequate ventilation'. What's 'adequate'? Can you help me—now?" Or:

Material Safety Data Sheets, consumer products, and industrial products that have hazardous components typically have the following phrase in the MSDS and/or on the package label: "Use with adequate ventilation." Some products may have the phrase, "Ensure ventilation is adequate to comply with the TLV® or PEL." Both are worthless and meaningless because "adequate" is not described in narrative text, by diagrams that apply to product, and by reference to standard-of-care ventilation engineering books, treatises, or regional consulting industrial hygienists and good ventilation engineers. Moreover, few customers will have no clues what TLVs® or PELs are. Sure, they can call the supplier to ask, but few will undertake this extra "cost of compliance." And, therefore, they remain at risk of inhalation or fire harm.

Sadly, suppliers of products with volatile flammable and toxic ingredients will give detailed instructions on applying their product, but often little or no instructions on how to achieve "adequate" ventilation is offered.

You, as the industrial hygienist for an adhesive manufacturer, determines 400 fpm (4.5 mph) is needed as the minimum cross-draft ventilation to dilute solvent vapors from your company's adhesive product to ensure reasonably safe exposure levels during application and curing. You want to include a simple safety practice for this with your MSDS to give guidance to your customers. Describe a simple procedure to help achieve this.

These are often tough calls. After determining solvent in product has a relatively low inhalation toxicity, general dilution ventilation is acceptable because product will be used in a large area for only 1 hour, and less than a pint will be used. A water-based cleaner will not work. You might suggest:

Using four sheets of any toilet tissue, wrap one sheet around a long pencil so that three sheets remain vertically suspended from the pencil. Hold the pencil at arm's length perpendicular to direction of air flow. Movement of the tissue and angle of repose of three tissues are used to determine the direction and the approximate air flow velocity. Portable pedestal fans can be used for dilution ventilation.

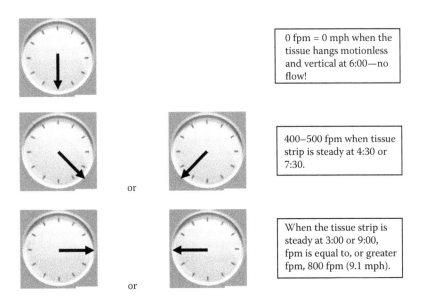

0 fpm = 0 mph when the tissue hangs motionless and vertical at 6:00—no flow!

400–500 fpm when tissue strip is steady at 4:30 or 7:30.

When the tissue strip is steady at 3:00 or 9:00, fpm is equal to, or greater fpm, 800 fpm (9.1 mph).

$$\frac{450\,\text{feet}}{\text{minute}} \times \frac{\text{mile}}{5280\,\text{feet}} \times \frac{60\,\text{minute}}{\text{hour}} = 5.1\,\text{mph}$$

The direction of air flow, obviously, is the deflection direction of the tissue paper. This is helpful to adhesive and paint applicators to position themselves and others upwind of the effusing solvent vapors. Professional painters should have a high-quality velometer in their toolbox and use it regularly during and after application.

Original research for this semi-quantitative air velocity measurement method used 12 brands of toilet paper. The thickest, double-ply paper gave the above results. Therefore, using thinner sheets gives air velocities at least at high as those above.

Instruct your safety engineer to place a pedestal fan on either side of this worker blowing fresh air toward and past the worker as she/he applies the product. Make measurements at the worker's head so that a minimum pre-selected air velocity is used based on the solvent vapor dilution calculations of the industrial hygienist. This simple instruction can be shared with this safety engineer who is miles away. Problem solved expeditiously, quickly.

504. Thermoanemometers used to measure air velocity (and, to some extent, direction) are calibrated to reference conditions of 70°F and 29.98 inches of mercury. Air density at these conditions is 0.075 lb/ft³. That is, they measure mass velocity of air. Air velocity = 460 fpm at an air density of 0.062 lb/ft³. What is the actual velocity?

$$V_a = V_i \times \left[\frac{d_s}{d_a}\right],$$

where
d_s = air density at standard calibration conditions of 0.075 lb/ft³
d_a = actual air density at local temperature and barometric pressure
V_a = actual air velocity
V_i = indicated velocity on thermoanemometer

$$V_a = 460\,\text{fpm} \times \left[\frac{0.075\,\text{lb/ft}^3}{0.062\,\text{lb/ft}^3}\right] = 556\,\text{fpm}$$

Thermoanemometer correction factor = 556 fpm/460 fpm = 1.209.

505. Air at a sooty combustion process was sampled for 15.3 min at 1.43 L/m. The soot collected on the filter was extracted with a mixture of methylene chloride and *n*-hexane. This mixed solvent extract contained 17.8 mcg benzene. The charcoal tube located after the filter contained 119.9 mcg benzene. What was the average air concentration of benzene vapor in ppm$_v$? Assume NTP.

$$15.3 \text{ minutes} \times 1.43 \text{ L/m} = 21.88 \text{ liters}$$

$$119.9 \text{ mcg} + 17.8 \text{ mcg} = 137.7 \text{ micrograms benzene total}$$

$$137.7 \text{ mcg}/21.88 \text{ L} = 6.29 \text{ mcg benzene per liter of air}$$

$$\text{ppm}_v = \frac{(6.29 \text{ mcg/L}) \times 24.45 \text{ L/gram-mole}}{78.11 \text{ g/gram-mole}} = 1.97 \text{ ppm}_v$$

We should always be mindful that airborne particulates can adsorb and absorb some gases and vapors. Failure to consider this can significantly underestimate workers' exposures. Moreover, if the particles are respirable, once deposited in the deepest recesses of the lungs, they can act as a reservoir for vapor release into blood perfusing the lung, or be phagocytized into the lymphohematopoietic system—a target organ for benzene. Benzene is a genotoxic carcinogen for humans with a specific DNA marker. There is no level of benzene in inhaled air that can be claimed "safe."

506. Once two sets of air samples are taken in a worker's breathing zone several days apart below the action level, we can be assured at a 95% confidence level that all exposures of that worker will be below the exposure limit (TLV, PEL, REL, WEEL, etc.) as long as conditions do not change. The advantage of this powerful statistic is that if we carefully note conditions during the two sampling periods, we do not have to repeat sampling unless conditions change. What are the major conditions we must consider to determine if resampling is required to assess potentially higher (or lower) exposures?

- Increased production rates
- Different work practices, tools, equipment
- Overtime, "moonlighting"
- Decreased ventilation
- Spills; eruptions
- Impaired engineering controls; process upsets
- Increase in concentrations of chemicals (decrease in solvent or suspendant
- Different chemicals and materials, and amount
- Increases in chemical process temperature, pressure, reaction rates, addition sequences
- More or fewer production workers; maintenance flaws
- Cofactors (e.g., heat strain, skin absorption, newly acquired health disorder)
- New supervisor
- Change in management culture
- Shift from preventative maintenance to "repair or replace when broken"
- New job responsibilities
- Day, afternoon, or midnight ("skeleton crew") shift

507. An understanding of the settling velocities of dust particles in air is necessary to select proper air pollution collection devices and to help dispel myths such as, "Oh, in time, the particles will fall to the floor. The night janitor will sweep them away." The following table demonstrates the settling velocities of various particles in still air. For respirable particles (those less than five microns), it is seen how normal air currents keep these tiny, invisible particles suspended indefinitely, or until they agglomerate and coalesce into larger particles, substantially hydrate by becoming condensation nuclei for water vapor, or are physically removed from the air by an air pollution control device (highly efficient filter, electrostatic precipitator, etc.).

Particle Diameter (microns)	Settling Velocity (cm/second)	Settling Velocity (ft/hour)
0.01	0.000007	0.0008
0.1	0.00008	0.009
1	0.004	0.5
5	0.08	9
10	0.3	35
20	1.2	142
50	7.5	886
100	25	2953

508. What factors determine the rate at which an organic solvent evaporates into open air?

- The *temperature of the solvent* that, in turn, determines the *solvent's vapor pressure*. As the liquid solvent's temperature increases, the evaporation rate increases. For approximately every 10°C (18°F) increase in liquid solvent temperature, the vapor pressure doubles.
- As the *temperature of air* above the liquid solvent increases, evaporation will increase. Warm air can hold more solvent vapor than cooler air.
- As *air flow* and turbulence increases over liquid solvent's surface, evaporation increases.
- As *surface area of solvent* increases, evaporation rate increases. If sprayed as a mist or fog, surface area increases greatly, and, hence, evaporation is greatly accelerated.
- Is the solvent in a *mixture of other solvents*? Other solvents (and solutes) will dilute the solvent of interest at the liquid and air interface thereby reducing the number of solvent molecules available for evaporation into air.
- *Raoult's law*, which states that the presence of other solvents can hinder or accelerate evaporation.
- *High atmospheric pressure* suppresses the vapor pressure, and, conversely, a solvent in a *low atmospheric pressure* environment evaporates more rapidly.

- *Henry's law*, which states that the amount of material in solution and in vapor or gaseous phase is dependent on the atmospheric pressure above the liquid.
- Does the solvent form an *azeotrope* with water or other organic solvents?

Regardless of the above variables, every organic solvent will evaporate into the air over time. Entropy prevails, and the second law of thermodynamics holds true. Mass is conserved, but, with evaporation, organic solvent distributes into different phases. Even solids, such as iodine crystals or naphthalene flakes, have vapor pressures and sublime directly from the solid phase into their vapor phase. Did you know that frozen water evaporates? I learned while a Navy recruit in Great Lakes, Illinois my freshly washed clothing dried outside during January to March—albeit it took more than a day.

509. A fossil fuel power plant stack gas sample contained 49.5 ppm$_v$ SO_2 corrected for temperature. The stack gas was 27.2% water vapor. What is SO_2 concentration (dry basis) corrected for the water vapor? Corrections were made by the chemist to account for SO_2 dissolving in water vapor forming sulfurous acid (H_2SO_3).

$$\text{dry gas concentration} = (\text{wet basis concentration, } w)/(1 - w),$$

where
w = decimal fraction of stack gas by volume—in this example, 1 − 0.272 = 0.728.

$$49.5\,\text{ppm}_v/(1 - 0.272) = 70\,\text{ppm}_v\ SO_2$$

510. An air sample is taken somewhere west of Denver in the Rocky Mountains where the elevation above the sea level is 2124 meters. What is a concentration of 319 mg/m³ at this altitude corrected to sea level?
      The formula for this calculation is

$$0.9877^a = C_a/C,$$

where
C_a = concentration at altitude a, in mass per unit volume
C = concentration at sea level, in mass per unit volume
a = altitude in 100 meter increments
$0.9877^{21.24} = (319\,\text{mg/m}^3)/C$
$C = (319\,\text{mg/m}^3)/0.9877^{21.24} = 414.9\,\text{mg/m}^3$

511. Poor Peruvian natives extract gold from river sediments by first crushing riverbed soil and rocks into fine dust. After screening, about 2 cubic feet of wet sediment are shoveled into steel pails about 2 feet in diameter. 200 grams liquid mercury is poured into the pails. Barefoot workers wearing cutoff

pants step into the pail, and over about the next 10 minutes continuously stomp and mix wet sediment while the mercury amalgamates with gold and other precious metals such as silver. The supernatant liquid and soil fines are decanted back to the flowing river. Chunks of mercury amalgam are retrieved from the pails. The amalgam is heated outside in open air by a propane torch evaporating mercury into the surrounding atmosphere. Assuming no wind, what is the average mercury vapor concentration in an atmosphere 30 feet high with a 500 feet radius from only one extraction pail?

$$\text{Volume of air cylinder} = \pi\, r^2 h = \pi\, (500')^2 \times 30'$$
$$= 23{,}562{,}000\,\text{ft}^3 = 667{,}252\,\text{m}^3$$

200 grams Hg = 200,000 mg Hg assuming all added mercury is boiled, evaporated

$$200{,}000\,\text{mg Hg}/667{,}252\,\text{m}^3 = 0.3\,\text{mg Hg/m}^3$$

Persons within this space are at significant risk of mercury poisoning with chronic exposures. This practice is a major environmental hazard and must be stopped. Moreover, elemental mercury is directly absorbed through intact healthy skin and more so through damaged skin. Those using the propane torches and workers in the pails are at greatest risk because skin of their feet is expected to be damaged from numerous micro-abrasions from sharp sand and stone granules.

The addition of mercury and the amalgamation process must be automated in a tightly sealed enclosure with excellent mechanical exhaust ventilation. Mercury amalgam must then be heated in an enclosed furnace with an efficient mercury vapor recovery system. The pure condensed mercury can then be recycled to the amalgamation process. A strict mass balance of $\text{mercury}_{in} = \text{mercury}_{out}$ must be developed to ensure minimal losses to the environment. Other elements of the industrial hygiene program must include education and training of all, personal protective equipment, air sampling and analysis, biological monitoring, medical surveillance, warning posters, regular checks of integrity of process enclosures and ventilation equipment efficiency, and recordkeeping. Environmental testing of the community atmosphere, hydrosphere, biosphere, and geosphere must be done to ensure preservation of health and containment of mercury. Mercury is expensive, and the recovered mercury will help defray engineering and operating costs.

512. Ajax gold smelter produces 413 gold ingots at greater than 99.99% purity weekly. Such precious metal must be closely accounted to ensure no losses. The smelter receives gold alloys, but does not balance $\text{gold}_{in} = \text{gold}_{out}$. Mass is conserved, but if smelter receives alloys with equivalent amount of gold equal to 413.01 ingots, where does the missing gold go?

An engineering study reveals loss is due to gold volatilizing as fume. Standard gold bars (ingots) weigh 12.4 kilograms.

413.01 gold$_{in}$ × 12.4 kg/ingot × 0.9999 = 5120.811868 kg received based upon calculations, density, and qualitative and quantitative analysis of the alloys and materials presented to the smelter

413.00 gold$_{out}$ × 12.4 kg/ingot × 0.9999 = 5120.68788 kg produced

$$5120.811868 - 5120.68788 = 0.123988 \text{ kg}$$

gold at \$45,191/kg × 0.123988 kg/week × 52 weeks/year
= \$291,363 per year.

This yearly amount results in a return on investment of less than 5 months for a high-efficiency filtration ventilation system to capture and process gold fume. Gold does not have a TLV; however, silver, interestingly, does at 0.1 mg/m^3 to prevent argyria, a cosmetic disorder wherein skin takes on an irreversible gray color from chronic deposition of microscopic silver particles in the subdermal tissues.

513. Assume gold will one day have a TLV of 0.1 mg/m^3. A smelter is 85 meters long, 8.5 meters high, and 40 meters wide. Assume 20% smelter contents, an average gold fume and dust concentration at 10% TLV (0.01 mg Au/m^3), 12.8 air changes per hour, 9 hour operation per day, and 5 days operation per week. At the current price of \$45,191 per kilogram, what is the yearly dollar loss of gold to the atmosphere?

$$(85 \times 8.5 \times 40 \text{ m}) \times 0.80 = 23{,}120 \text{ m}^3$$

$$23{,}120 \text{ m}^3 \times 0.01 \text{ mg Au/m}^3 = 231.2 \text{ mg Au in smelter atmosphere}$$

231.2 mg Au/air change × 12.8 ach × 9 hours × 5 days/week × 52 weeks/year = 6,924,902 mg = 6.924 kg

6.924 kg Au × \$45,191/kg = \$312,902.48/year—not
exactly chump change

Good process enclosures, excellent local exhaust ventilation, regular preventive maintenance, and, perhaps, lower smelting temperatures will save most of this staggering financial loss. Remember the second law of thermodynamics and entropy. Once that gold is spewed into the atmosphere, it will never be recovered. Sooner or later, as we keep refining gold in smelters, it will all be out in the ether save what is stored in Fort Knox and we wear as jewelry and in dental fillings. It has been estimated that all of the gold ever smelted and refined since antiquity throughout the world would occupy a cube 60 feet on each side.

514. A 10-inch circular free-standing exhaust duct has average duct velocity of 2200 fpm. What is capture velocity 10 inches from the un-flanged duct inlet assuming there are no disruptive cross drafts?

In this case, the capture velocity is 10% of the duct velocity, or 220 feet/minute. This is a helpful "rule of thumb" to keep in mind during facility inspections.

515. In the preceding problem, what is the volumetric flow rate of air inside the duct?

$$\text{Area of circle} = \pi \, r^2 = \pi \, (5")^2 = 78.54 \text{ in}^2$$

$$78.54 \text{ in}^2/144 \text{ in}^2/\text{ft}^2 = 0.5454 \text{ ft}^2$$

$$Q = AV = 0.5454 \text{ ft}^2 \times 2200 \text{ ft}^3/\text{minute} = 1200 \text{ cfm}$$

516. The vapor pressure of an aliphatic hydrocarbon is 43 mm Hg. What is the vapor saturation concentration inside a closed vessel such as a tank, bin, pipeline, or silo situated at sea level?

$$(43 \text{ mm Hg}/760 \text{ mm Hg}) \times 10^6 = 56{,}579 \text{ ppm}_v = 5.6579\%.$$

Such a concentration can occur in an open pit, say 6 feet deep, a few feet above the liquid surface if there is no air movement. These locations must be treated like confined spaces before entry is allowed: worker and supervisor education and training, air sampling and analysis, ventilation, buddy system, self-extraction gear, personal protective equipment, means of communication, and ample lighting are necessary.

517. A worker with a total body surface area of 1.8 m² fell into an open tank and was completely immersed in an organic chemical that is absorbed through intact skin at an average of 0.011 micrograms/square centimeter/second. He is extracted and decontaminated with a total exposure time of 23 minutes. What was his dermally absorbed dose? If this man weighed 81 kilograms, and LD_{50} in rats is 37 mg/kg, what is his systemic dose?

1.8 m² × 10,000 cm²/m² × 0.011 mcg/cm² × 60 seconds/minute × 23 minutes = 273,240 micrograms = 273.24 mg.

273.24 mg/81 kg = 3.37 mg/kg. This is slightly less than 10% of the rat LD_{50}. However, if we do not have the lowest observed adverse effect level and assume toxicity to man to be the same as the rat, this man should be admitted to the hospital and observed after consultation with a toxicologist and pharmacologist.

To this must be added that which was absorbed through his gastrointestinal tract and his respiratory tract to obtain his approximate total dose. If this chemical has an antidote, a medical toxicologist or pharmacist can adjust prescriptive treatment accordingly.

518. A 181-pound, 28-year-old man, and 5'-11" tall consumed three shots of 80-proof whiskey in rapid succession on an empty stomach. What is his most likely blood alcohol concentration (BAC) 45 minutes later?

First, determine his body water volume, the so-called volume of distribution, using the equations of Watson (a) and Hume (b). These equations apply to men—not to boys, women, or girls. Other Watson and Hume algebraic equations apply to them.

   a. V (liters of H_2O) = 2.447 + (0.336 × weight, kg) + (0.107 × height, cm) – (0.095 × age, years)

$$181\,lb = 82.1\,kg$$

$$5' = 11" = 71" = 180.34\,cm$$

$$V = 2.447 + (0.336 \times 82.1\,kg) + (0.107 \times 180.34\,cm) - (0.095 \times 29\ \text{years}) = 52.084\,liters = 52.084\,kilograms$$

$$(52.084/82.2\,kg) \times 100 = 63.44\%\ H_2O$$

   b. V (liters of H_2O) = −14.013 + (0.297 × weight, kg) + (0.195 × height, cm)

Plugging and chugging the values into this equation results in 45.537% body water content. This value appears too low. Besides, the Watson calculation method accounts for age, whereas the Hume method does not. This man is relatively young and does not have a high body fat content. Therefore, with all due respect to Professor Hume, I rely on Watson's method with a 63.44% body water content.

$$\text{three "shots"} = 4.5\ \text{ounces}$$

$$80\ proof = 40\%\ \text{by volume ethyl alcohol}$$

A standard "drink" of alcohol is considered to be 1.5 ounces of 80 proof alcohol—less than 1 ounce of pure alcohol. 80 proof alcohol = 40% ethanol. So, three "shots" = 4.5 ounces of 40% ethanol. 4.5 ounces × 0.4 = 1.8 ounces ethanol.

On an empty stomach, one can assume 100% absorption within 45 minutes.

$$\text{Density of pure, 100\% ethanol} = 0.789\,g/mL$$

$$1.8\ \text{ounces} = 53.23\,mL$$

$$(0.789\,g/mL) \times 53.23\,g = 42.0\,\text{grams of EtOH}$$

BAC are expressed in grams EtOH/100 mL blood.

$$42.0\,\text{grams}/63.44\,\text{L} = 0.662\,\text{g/L} = 0.062\,\text{g}/100\,\text{mL}.$$

This does not exceed the 0.08 g/100 mL BAC limit throughout the United States, although this man is clearly under the influence and unfit to operate machinery.

As a comment, binge drinkers often consume large amounts of alcohol quickly to get to a rapid state of inebriation. In this example, with consumption at this rate, a very high level of intoxication is achieved in less than 2 hours. Bartender must suspend serving and then call a taxi or turn patron's car keys over to a designated driver.

One might wonder why this type of problem is in a book with this title. Industrial hygienists are called on to determine dose of toxicants by all routes of exposure.

519. A full 275,000 gallon aviation fuel storage tank develops a leak at the base. It is a choked flow leak measured at 9.7 gallons per second. A breather tube to prevent tank explosive rupture during filling or implosion during emptying is located on the roof of tank. The internal diameter of the tube is 2 inches. What is the make-up air flow velocity through this tube into the tank as the oil drains?

$$9.7\,\text{gallons/second} \times 1\,\text{ft}^3/7.4805\,\text{gallon} \times 60\,\text{seconds/minute}$$
$$= 77.802\,\text{ft}^3/\text{minute}$$

$$A = \pi\,(1")^2/144\,\text{in}^2/\text{ft}^2 = 0.0218\,\text{ft}^2$$

$$V = Q/A = 77.802\,\text{ft}^3/\text{minute}/0.0218\,\text{ft}^2 = 3569\,\text{feet/minute}.$$

This air velocity is rapid enough to prevent condensation of oil vapors during cold weather tank filling. The tube should be placed at a 30° angle from perpendicular to prevent intrusion of rain, snow, or sleet.

520. People typically underestimate the time before one collapses in inert atmospheres where the normal concentration of oxygen is 21% by volume (precisely 20.9476% by volume) has been diluted with gases other than oxygen. Such gases typically are nitrogen, argon, carbon dioxide, helium, and steam. Yes, steam because this is a gas that does not contain oxygen. Inhalation of water vapor does not support life. A dangerous and common notion is that if one can hold their breath for 30 seconds or more, their survival in an inert atmosphere should be as long.

The normal inspiratory period in an adult is approximately 2 seconds, and the expiratory period is also 2 seconds. A complete cycle of one respiration is, therefore, 4 seconds or, on average 15 breaths per minute. During anxiety, the total cycle can be reduced to 1 second. So, 15 breaths per minute now become 60 per minute. Doctors call rapid breathing tachypnea or hyperventilation.

The total lung capacity of an average adult is 5.6 liters. At a breathing cycle of one inspiration and expiration every 4 seconds, about 0.5 L (500 milliliters) pass in and out of the alveoli and the tracheobronchial tree. Of this 500 mL, 150 mL is in the tracheobronchial tree where oxygen exchange does not occur. Therefore, only 350 mL of air is exchanged with each inspiration–expiration cycle. A normal rate of breathing in an adult is about 30 L/minute, or 500 mL/inhalation + exhalation.

After oxygen exchange in the alveoli, the oxygen gas concentration falls from 21% by volume to 12% by volume. This mixes with the air in the tracheobronchial tree so that the net exhaled concentration is nearly 16%. Although 16% oxygen is less than the normal 21%, it is sufficiently high in oxygen to support mouth-to-mouth breathing during CPR. The balance of the exhaled volume is carbon dioxide, CO_2.

Assuming a worker anxious about his condition and experiencing tachypnea is in a 100% nitrogen atmosphere. How long before he collapses and is unable to effect self-rescue?

Assume the initial gas mixture concentration in his lungs is 16% O_2 by volume. His first breath adds 500 mL of inert N_2 to 5.0 liters of air mixture containing 16% O_2.

1st inhalation: (5.0×0.16) L/5 L $+ 0.5$ L $= 14.5\%$ O_2

2nd inhalation: (5.0×0.145) L/5.5 L $= 13.2\%$ O_2

3rd inhalation: (5.0×0.132) L/5.5 L $= 12.0\%$ O_2

4th inhalation: (5.0×0.12) L/5.5 L $= 10.9\%$ O_2

5th inhalation: (5.0×0.109) L/5.5 L $= 9.9\%$ O_2

Continuing the above iterations, after 11 seconds, the oxygen concentration falls to 5.5%, and breathing stops. He would have collapsed after his fourth inhalation (4 seconds with tachpnea, 16 seconds with normal breathing). Death follows quickly without prompt emergency intervention.

521. What is the partial pressure of solvent X at 620 ppm$_v$ and at atmospheric pressure of 710 mm Hg and 25°C?

$$X_{pp} = (620 \times 10^{-6}) \times 710\,\text{mm Hg} = 0.4402\,\text{mm Hg}$$

522. The accuracy of an air sample that relies on collecting air contaminant(s) within a device such as a filter or an adsorbent/absorbent tube depends on the collection efficiency and subsequent laboratory analysis. These determinants can be called E, where

$$\%E = (C_{measured}/C_{actual}) \times 100$$

If the actual concentration in an air sample is 14.76 mg/m³, collection efficiency of the air-sampling device is 100%, and the measured concentration is 13.15 mg/m³, what is E as a percent?

$$(13.15\,mg/m^3/14.76\,mg/m^3) \times 100 = 89.09\%$$

523. Referring to the previous problem, if the laboratory results are 157.8 micrograms of X in a 10.34 liter air sample, what is the adjusted true result?

$$157.8\,mcg\ X/10.34\,L = 15.26\,mcg\ X/L = 15.26\,mg\ X/m^3$$

$$\text{Correction factor} = 1/0.8909 = 1.122$$

$$15.26\,mg\ X/m^3 \times 1.122 = 17.12\,mg\ X/m^3$$

524. Industrial hygienists encounter situations where the amount of a hazardous agent decreases uniformly over time. Typically, these data can be plotted linearly on a sheet of semi-logarithmic graph paper. That, is one axis has uniform spacing, and the other axis is a logarithmic scale—usually the vertical, or ordinate, axis. Some examples of these situations are the linear first-order decay of a radioisotope over time, the decrease in air contaminant concentration in a room with excellent air movement and mixing, and the detoxification of a material over time. Recall that removal of a toxicant from the body depends on excretion (urine, feces, exhaled air, bile, hair) and the metabolic conversion of the toxicant to another, ideally less toxic material. Write an equation, with some constant k, that describes toxicant concentration then and now.

These data can be described mathematically with a first-order decay equation:

$$E = kP^a$$

Integrating: $\ln E = \ln k + a\,(\ln P)$.

One normally solves for a: $a = (\ln E - \ln k)/\ln P$

substituting, let $k = 3$, $P = 4$, and $a = 2$

$$E = 3 \times 4^2 = 48$$

verifying

$$(\ln 48 - \ln 3)/\ln 4 = (3.871 - 1.099)/1.386 = 2$$

Once data are known, one can use this equation to very helpfully predict what the amount of the toxicant will be, or once was, *in vivo*.

525. What is the percent concentration and partial pressure of oxygen in air at sea level and an altitude with a barometric pressure of 682 mm Hg?

Air is 20.9476% oxygen at any altitude. It is a common misunderstanding that the percentage of oxygen in air decreases with altitude. So,

at sea level: $760 \text{ mm Hg} \times 0.209476 = 159.202 \text{ mm Hg}$

at 682 mm Hg: $682 \text{ mm Hg} \times 0.209476 = 142.863 \text{ mm Hg}$

That is, the oxygen concentration has decreased to $(142.863/159.202) \times 100 = 89.737\%$, or $20.9476\% \times 0.89737 = 18.80\%$.

Our respiration rate and heart rate increase substantially due to reduced oxygen tension in our blood. For the unacclimated, the slightest amount of work induces great fatigue.

The O_2 partial pressure is now $159.202 \text{ mm Hg} \times 0.89737 = 142.863 \text{ mm Hg}$.

This lower amount of oxygen in the ambient air can be a significant risk factor for those with asthma, emphysema, and chronic obstructive lung disease and those with angina and ischemia from coronary artery disease.

526. The total pressure in a low-pressure gas system is 1388 mm Hg. If the pressure in the system of nitrogen gas is 982 mm Hg, and the partial pressure of CO_2 is 109 mm Hg, what is the O_2 gas pressure?

$$P_{total} - P_{nitrogen} - P_{CO2} = P_{O2}$$

$1388 \text{ mm Hg} - 982 \text{ mm Hg} - 109 \text{ mm Hg} = 297 \text{ mm Hg } O_2 \text{ pressure}$

527. State why helium is used in SCUBA systems.

Atmospheric breathing air is approximately 21% by volume O_2 and 79% by volume nitrogen and argon. At the increased pressures of deep water diving, nitrogen in the inhaled air is more soluble in body fat than is helium gas. With an uncontrolled rapid ascent, the nitrogen "boils" out quickly from adipose tissue increasing the risks of nitrogen bubbles in blood perfusing lungs. This, in turn, raises risks of a cerebral and/or a cardiac embolism that can be fatal. Moreover, rapid release of nitrogen from fat can causes the "bends," a painful physiological process. Additionally, since brain tissue is rich in fats and lipids, nitrogen gas accumulates there and can result in nitrogen narcosis with resultant errors in judgment. Helium behaves more like oxygen, is nontoxic, and does not present the hazards and the risks of inhaling high-pressure nitrogen gas. Moreover, while nitrogen is relatively inert, helium gas is physiologically totally inert other than its effects on laryngeal vocal cords.

528. The manufacturer of a direct-reading instrument for hydrogen cyanide gas (HCN) claims a false-positive reading of 10 ppm_v HCN if any of the following gases are present at the indicated concentrations: 10 ppm_v chlorine (Cl_2), 20 ppm_v hydrogen chloride (HCl), and 2 ppm_v hydrogen sulfide

(H$_2$S). Explain how the industrial hygienist can rightfully claim a reading of 10 ppm$_v$ HCN knowing these interfering gases.

HCN has a ceiling TLV of 10 ppm$_v$, so (1) applying the precautionary principle, immediate intervention is necessary because of rapid acute inhalation toxicity of this gas irrespective of the possibility of the other gases being present, (2) the probability of HCN being present with the other gases is remote particularly if the industrial hygienist is knowledgeable of processes nearby the "HCN" reading, (3) HCl and Cl$_2$ gases at the reported interfering concentrations are very irritating to the upper respiratory tract, and their physiological effects would be known to the experienced industrial hygienist, and (4) hydrogen sulfide gas at 2 ppm$_v$ is very malodorous to most people (rotten eggs up to olfactory anesthesia). It would be the better part of valor to assume the reading is indeed HCN instead of debating possible positive interferences of these three gases. Action should be imminent. The gravity of failure to initiate immediate evacuation, sending emergency team with SCBAs to seal any leaks and to staunch the source, apply dilution ventilation, and to locate antidote kits for symptomatic persons should be apparent.

529. The short-term exposure limit (STEL TLV®) for 1,2-dichloroethane is 15 ppm$_v$ (60 mg/m^3) for 15 minutes. Assume the laboratory analytical detection limit for DCE is 2 micrograms. If air is sampled at 82 mL per minute for 15 minutes, will 50% of the STEL be detected?

$$50\% \text{ STEL} = 60\,\text{mg/m}^3 \times 0.5 = 30\,\text{mg/m}^3 = 30\,\text{mcg/L}$$

$$2\,\text{mcg/30 mcg/L} = 0.067\,\text{L, the minimum air volume required}$$

$$0.082\,\text{L/minute} \times 15\,\text{minutes} = 1.23\,\text{L}$$

$$1.23\,\text{L/0.067 L} = 18.36 \text{ times the minimum air volume required}$$

Yes, 50% of the STEL will be detected. However, it would be best to rely on the results of a direct-reading instrument for STEL exposure determinations because the long delay in obtaining laboratory test results is inappropriate for air toxicants with adverse short-term health effects.

530. What is the heating requirement for 64,500 cubic feet of air per minute from 10°F to 76°F at 90% efficiency?

$$100/90 = 1.111 \text{ is the heating inefficiency factor}$$

$$\text{Btu/hour} = 1.111 \times 64,500\,\text{ft}^3/\text{hour} \times 1.08 \times (76°\text{F} - 10°\text{F})$$
$$= 5,107,889 \text{ Btu/hour}$$

531. An animal toxicity testing chamber admits a toxicant gas at a rate of 75 mL every minute. The total air flow through the chamber is 625 L/minute.

Assuming good mixing of test toxicant gas with clean air flow, what is the resultant concentration to which animals inhale?

$$75\,\text{mL} = 0.075\,\text{L}$$

$$(0.075\,\text{L}/625\,\text{L}) \times 10^6 = 120\,\text{ppm}_v$$

532. If dimensions of a toxicity testing chamber are 1.5 feet × 3 feet × 6 feet, and C is the gas concentration in the chamber after some time t, what is the concentration after 2 minutes and 10 minutes? Test gas flow rate and the mixing air flow rates are 75 mL/minute and 625 L per minute, respectively. Refer to Problem 531.

$$1.5' \times 3' \times 6' = 27\,\text{ft}^3 = 764.6\,\text{L}$$

$$\text{Use the formula } C = \frac{a}{V}\,(1 - e^{-(V/X)t}),$$

where
  a = flow rate of toxicant
  V = flow rate through test chamber
  X = volume of chamber

$$C = \frac{0.075\,\text{L/minute}}{625\,\text{L/minute}} \times (1 - e^{-(625\,\text{Lpm}/764.6\,\text{L})\times 2}) = 0.00012\,(1 - e^{-1.635})$$

$$= 0.00012\,(1 - 0.195) = 0.0000966 = 96.6\,\text{ppm}_v \text{ after 2 minutes}$$

$$C = \frac{0.075\,\text{L/minute}}{625\,\text{L/minute}} \times (1 - e^{-(625\,\text{Lpm}/764.6\,\text{L})\times 10}) = 0.00012\,(1 - e^{-8.17})$$

$$= 0.00012\,(1 - 0.000283) = 0.00012 \times 0.999717$$

$$= 0.000112 = 112\,\text{ppm}_v \text{ after 10 minutes.}$$

Recall from the previous problem the maximum concentration is 120 ppm$_v$. This is close to the desired concentration after 10 minutes. In other words, if this was an acute toxicity test, animals should not be exposed until at least 20 minutes have elapsed by which time the maximum concentration of 120 ppm$_v$ is reached. The exposure concentration should be periodically verified with a calibrated air-sampling instrument. This increase in the chamber concentration is asymptotic, and it never actually reaches the desired concentration—in this case, 120 ppm$_v$—but, for practical purposes it does after an elapsed time. Moreover, this assumes that the dilution mixing air is free of the gas or vapor toxicant.

533. What is the capture velocity 9 inches on the centerline in front of a 12"-diameter un-flanged exhaust duct moving 8360 cubic feet of air per minute? What is duct velocity?

Duct area, $A = \pi\, (6")^2/144 \text{ in}^2/\text{ft}^2 = 0.7854 \text{ ft}^2$

$$\text{Velocity}, V = \frac{Q}{10x^2 + A} = \frac{8360\,\text{cfm}}{10\,(0.75')^2 + 0.7854 \text{ ft}^2}$$

$$= 1304 \text{ fpm capture velocity}$$

Duct velocity $= Q/A = 8360\,\text{cfm}/0.7854 \text{ ft}^2 = 10{,}644 \text{ fpm}$

This duct velocity is extraordinarily high. It can be reduced substantially without greatly affecting the capture velocity 9 inches away. Check the American Conference of Governmental Industrial Hygienist's book *Industrial Ventilation* to select the appropriate duct velocity for the type of air contaminant being moved. Controlling cross-drafts at the 9-inch point, a capture velocity of 100–300 fpm would suffice for most generated air contaminants with low momentum. Dust and other particles ejected with high momentum require enclosure and/or a higher capture velocity. Adding a wide flange to duct face improves capture and saves energy by reducing fan speed and horsepower. A flanged hood requires about 75% of the air volume of an un-flanged hood.

534. A subterranean platinum ore mine has a branch shaft 8 feet wide, 7 feet high, and with a semi-circular vault above. A 20" Ø exhaust duct is suspended above the miners in the vault and moves air through the duct at 4200 fpm. What is the average velocity of the make-up air passing through the branch shaft where miners work?

Duct area $= \pi\, (10")^2/144 \text{ in}^2/\text{ft}^2 = 2.182 \text{ ft}^2$

$2.182 \text{ ft}^2 \times 4200 \text{ fpm} = 9164 \text{ cfm}$

Area of shaft cross-section $= (8' \times 7') + (\pi \times 3.5^2)/2 = 56 \text{ ft}^2 + 19.24 \text{ ft}^2 = 75.24 \text{ ft}^2$

Make-up air velocity across mineshaft cross section $= 9164\,\text{cfm}/75.24 \text{ ft}^2 = 121.8 \text{ fpm}$

This is a reasonable air velocity because it is low enough to not stir up settled dust, and it exceeds 50 fpm, an air velocity most would claim stagnant, still air. It is also sufficiently fast to promote evaporative cooling from sweat on the skin during hot working conditions, and this air velocity is not so high to chill skin in cold working conditions. With humid conditions at the mine's face, the exhaust fan should have sufficient reserve capacity to increase air flow by up to approximately 20% (150 fpm).

535. Consider an elevator in a tall building that has interior dimensions of a cube 7 feet on a side. The elevator does not have any mechanical ventilation. Instead, it relies on people coming and going to provide displacement air ventilation. Also assume that, on average, 230 people enter and leave

elevator every hour. What is the air exchange rate of the elevator cabin in air changes per hour?

$$(7\,\text{feet})^3 = 343\,\text{ft}^3$$

Assume each person is "standard reference man" weighing 70 kg with an average density of 1.01 kg/L. Therefore, 70 kg/(1.01 kg/L) = 69.307 L.

Entering air displacement: 230/hour × 69.307 L = 15,941 L/hour
$$= 563\,\text{ft}^3/\text{hour}$$

Exiting air displacement: 230/hour × 69.307 L = 15,941 L/hour
$$= 563\,\text{ft}^3/\text{hour}$$

Total air displacement/hour = 563 ft³/hour + 563 ft³/hour = 1126 ft³/hour

1126 ft³/hour/343 ft³ = 3.28 air changes per hour

This is low. Check the latest ASHRAE standards for commercial elevators. This must be supplemented with mechanical air supply and/or exhaust because people have body odors and wear perfumes and cologne some might find offensive. Any person with chronic obstructive lung disease (chronic bronchitis, emphysema) and asthma can be affected and at risk. Moreover, a building custodian or a package delivery person transferring toxic and/or odiferous materials might wrongly use the passenger elevator instead of the freight elevator. Finally, passengers enter and exit in a random pattern so that the air exchange rates change nonuniformly as the elevator population changes.

536. Consider the same elevator in preceding problem at 3:00 a.m. when a passenger departs the top floor. He inadvertently drops a bottle that breaks and spills 50 mL of a highly volatile organic chemical that evaporates at 0.035 grams per second into the elevator's atmosphere. Nobody enters the elevator as it descends to the ground floor over a 40 second period. The molecular weight of the chemical is 78 grams/gram-mole. What vapor concentration can an ascending passenger expect when he enters this elevator? Will he be surprised? Assume no ventilation.

0.035 grams/second × 40 seconds = 1.4 grams = 1400 milligrams

343 ft³ = 9712.7 liters = 9.712 m³

$$\text{ppm}_v = \frac{(\text{mg/m}^3) \times 24.45\,\text{L/g-mole}}{\text{molecular weight}} = \frac{(1400/9.712) \times 24.45}{78\,\text{g/g-mole}}$$

$$= 45.2\,\text{ppm}_v$$

Determine if the chemical has a ceiling concentration and compare the result with this. A more likely reality check is that the passenger might report a strange odor to the security guard who, knowing nothing about the solvent, contacts the local emergency response team. The elevator passenger will certainly be surprised if he is not anosmic, and the organic chemical is malodorous below 45.2 ppm$_v$ in air.

537. Of the large variety of air pollution control devices commercially available, there are five broad categories for airborne particles such as dust, fumes, mist, chips, and fibers. List them. Identify which are virtually worthless for respirable particles such as smoke and welding fume.

    Fabric collectors, electrostatic precipitators, wet scrubbers, centrifugal and inertial separators, and gravity chambers. Centrifugal separators such as cyclones or gravity chambers even connected in series are typically ineffective for respirable particles—essentially worthless.

538. An air contaminant collection device is attached to the inlet of a Mariotte bottle. This bottle contained 68.92 liters of water at the start of air sampling and 21.67 liters at completion. The duration of sampling was 56 minutes, 40 seconds. What air volume was sampled, and what was the average air-sampling rate?

$$68.92\,L - 21.67\,L = 47.25\,L$$

$$47.25\,L/56.67\,minutes = 0.8338\,liters\,per\,minute = 833.8\,mL/minute$$

539. The velocity pressure in an exhaust duct is 0.86 inches of water at 70°F. This ventilation system is in Portland, Oregon where the barometric pressure is 1.0 atmosphere. What is the velocity (in feet per minute) in the duct at this point?

$$V = 4005 \times (0.86^{0.5}) = 4005 \times 0.8345 = 3714\,fpm$$

540. The concentration of cyclohexane vapor in the air of a laboratory after a spill is 466 ppm$_v$. The laboratory exhaust hoods are not operating, but the general ventilation system designed at 3.5 air changer per hour is. What is the vapor concentration after 90 minutes? Assume excellent mixing of supply air with the exhausted air.

    With excellent air mixing, and no recirculation of the exhaust air into the makeup air inlet, the first-order concentration decay equation applies:

$$C_t = C_o\,e^{-xt},$$

where
    C_t = the concentration in the room at time t (in this case, 90 minutes),
    C_o = original concentration in the room at time $t = 0$ (466 ppm$_v$),
    e = the base of natural logarithms,

x = air changes per hour (3.5), and
t = elapsed time (90 minutes).

The negative sign for exponent indicates there is reduction in vapor concentration over the elapsed time and as time progresses.

$$\text{Elapsed time of 90 minutes} = 1.5 \text{ hours}$$

$$3.5 \text{ ach} \times 1.5 \text{ hours} = 5.25 \text{ air changes}$$

$$C_t = (466 \text{ ppm}_v)e^{-5.25} = 466 \text{ ppm}_v \times 0.00525 = 2.45 \text{ ppm}_v$$

2.45 ppm$_v$ is substantially below the TLV of 100 ppm$_v$. However, for some this might still be odiferous. Operating the ventilation system for 3 h before allowing laboratory personnel to reoccupy the laboratory will reduce the solvent vapor concentration to

$$C_t = (466 \text{ ppm}_v)e^{-10.5} = 466 \text{ ppm}_v \times 0.0000275 = 0.013 \text{ ppm}_v$$

541. An 85 cubic feet of air per minute ceiling exhaust fan operates in a bathroom that does not have a fixed, forced ventilation system from a central heating system. Make up enters through a crack at the bottom of the door that is 7/16" × 28". What is the average air velocity through this opening?

$$7/16" = 0.4375 \text{ in}^2$$

$$85 \text{ cfm}/[(0.4375 \text{ in}^2 \times 28")/144 \text{ in}^2/\text{ft}^2] = 999 \text{ fpm}$$

542. A Dewar flask tips onto the floor of an empty chemical storage room releasing 2 liters of liquid nitrogen. The room has no ventilation and is 10' × 14' × 22'. What is the concentration of nitrogen gas after the liquid nitrogen evaporates eventually producing an homogenous atmosphere? To simplify the calculation, assume that there is no significant back pressure of the gas upon the room and on the liquid nitrogen.

$$10' \times 14' \times 22' = 3080 \text{ ft}^3$$

$$3080 \text{ ft}^3 = 87.216 \text{ m}^3$$

First, determine the mass of nitrogen in the room before the spill. Molecular weight of nitrogen is 28.01 grams/gram-mole. Percentage of nitrogen in normal air is 78.084% by volume = 780,840 ppm$_v$.

$$\text{mg/m}^3 = \frac{\text{ppm}_v \times \text{molecular weight}}{24.45} = \frac{780,840 \text{ ppm}_v \times 28.01}{24.45}$$

$$= 894,533 \text{ mg/m}^3$$

$894,533\,\text{mg/m}^3 \times 87.216\,\text{m}^3 = 78{,}017{,}590\,\text{mg}$ of nitrogen in room air before spill.

The density of liquid nitrogen at its boiling point and 1 atmosphere pressure is $50.47\,\text{lb/ft}^3$.

$$50.47\,\text{lb} = 22{,}892{,}807\,\text{mg}$$

$$1\,\text{ft}^3 = 28.317\,\text{L}$$

Density of liquid nitrogen $= 22{,}892{,}807\,\text{mg}/28.317\,\text{L} = 808{,}447\,\text{mg/L}$.

Two liters spilled and evaporated, so: $2\,\text{L} \times 808{,}447\,\text{mg/L} = 1{,}616{,}894$ mg in air.

mg N_2 before spill + mg N_2 from spill = total mg N_2 in room air after spill $= 78{,}017{,}590\,\text{mg} + 1{,}616{,}894\,\text{mg} = 79{,}634{,}484\,\text{mg}$ total after spill and total evaporation

$$79{,}634{,}484\,\text{mg}/87.216\,\text{m}^3 = 913{,}072\,\text{mg/m}^3$$

$$\text{ppm}_v = \frac{(\text{mg/m}^3) \times 24.45}{\text{molecular weight}} = \frac{(913{,}072\,\text{mg/m}^3) \times 24.45}{28.01} = 797{,}023\,\text{ppm}_v$$

The concentration of nitrogen gas increased from 78.084% to 79.7023%. If a larger amount of liquid nitrogen evaporated in a smaller room, the risk of physical asphyxiation increases proportionately. It is good industrial hygiene practice to install fixed, regularly calibrated oxygen gas sensors and alarms at strategic fixed locations where liquefied gases and gases stored in pressurized systems are used and stored.

543. Large building fires—conflagrations—present severe asphyxiation hazards to fire fighters, emergency medical personnel, and building occupants. Briefly describe the four major asphyxiant gases and a physical asphyxiation factor.

The fire itself consumes prodigious amounts of surrounding oxygen gas that can cause physical asphyxiation. The major combustion gas—carbon dioxide—is a physical asphyxiant. CO_2, moreover, is a strong respiratory stimulant promoting rapid breathing. Medically termed tachypnea, this effect hastens collapsing from asphyxiation. Two chemical asphyxiant gases are generated in fires. The major one is carbon monoxide. Less prevalent, but rapidly chemically asphyxiating, hydrogen cyanide is another. Burning wool produces ample amounts of HCN, a potent chemical asphyxiant. Victims of such fires also inhale copious amounts of soot. These particles physically clog small airways in the deep tissue airways further hastening asphyxiation.

544. Wet bulb and dry bulb thermometer psychrometric readings demonstrate a relative humidity reading equivalent to nine grains of moisture per pound of air in an office building. The design objective is 56 grains/pound of

air when outdoor temperature is 0°F, and we want 50% relative humidity at 70°F. Assume one air change every hour. The building volume is 100,000 ft³. How much water vapor must be added to ambient air to achieve this workplace comfort objective?

$$56 \text{ grains} - 9 \text{ grains} = 47 \text{ grains of } H_2O \text{ required}$$

$$\frac{100,000\,ft^3 \times (1\,air\,change/hour) \times 47\,grains}{13.5\,ft^3/lb\,of\,air \times 7000\,grains/ft^3} = 50\,lb\,H_2O\,per\,hour$$

The author recommends humidifying the air by evaporative techniques—never by misting methods because, especially with hard water, minerals in water deposit on surfaces creating spots. Moreover, the dissolved and suspended particles in water including bacteria and viruses are inhaled.

545. Throughout this book are examples of how industrial hygienists and toxicologists can make technical data meaningful to lay persons such as workers, jurors, and our courts. The author has used the following with a tad of levity to help explain what one part per million is:

A man enters his favorite bar with ½ cup of dry vermouth asking the bartender to use it to make a very dry martini—in fact, a one part per million martini. "Think you can you do it?" asks the patron. The bartender says, "Sure, but, I'll have to make some calculations first. Would you like a gin or vodka martini?" "Vodka is fine," the customer replies. "Well," says the bartender, after calculations, "It will be some time before you get your drink because we need to construct a railroad spur to bring in tank cars of vodka. I need 31,250 gallons! Two jumbo cars and one huge mixing tank should do it. And, I'm sorry, but I have to limit you to one drink. Olive or onion? Stirred it will be. If you want it shaken, go to California for your drink during an earthquake."

How about a one part per trillion martini? One milliliter equals approximately 20 drops. How many parts per trillion is only a single drop of vermouth in a 30,000-gallon railroad tank car of gin or vodka?

$$1\,mL/20 = 0.05\,mL/drop$$

$$30,000 \text{ gallons} = 113,562,354\,mL$$

$$113,562,354\,mL \times 20 \text{ drops/mL} = 2,271,247,080 \text{ drops}$$

$$1 \times 10^{12}/2,271,247,080 = 440 \text{ parts per trillion—a very dry martini}$$

The previous problem "gets third grade" and speaks many volumes. The ability of today's analytical chemists and their increasingly sensitive instruments to detect parts per billion and parts per trillion for many analytes in matrices of all sorts has outstripped the toxicologists' abilities to make sense of it all. How about 1 ppb and 1 ppt?

The bartender in previous problem could save some real estate and his customer a huge amount of money if the customer agreed to a less dry martini. Standard railroad liquid tank cars contain 30,000 gallons. So, instead of calculated 31,250 gallons of vodka, by using 30,000 gallons, the customer's martini becomes less dry by a factor or $30,000/31,250 = 0.96$, or 4.0% "wetter." It is fair to assume the customer would not notice this after his first drink or so. "Hey, drinks are on Joe!"

This rail tank car is 59.75 feet long coupler end to coupler end. So, a 1 ppb martini would require 1000 of these tank cars. And this 1 ppt martini would require 1,000,000 tank cars. Excluding the locomotives, how long would this 1 part per trillion martini train be?

$59.75' \times 10^6 = 59,750,000$ feet $\times$ 1 mile/5280 feet $= 11,316$ miles, a contiguous land mass on Earth only met by the distance from the tip of southern Africa to the most northeasterly tip of Siberia.

546. An empty building has interior dimensions of $37.5' \times 28.0' \times 18.0'$. At NTP, how many pounds of air are in this room?

$(37.5' \times 28.0' \times 18.0') \times 0.075$ pounds air/ft^3 = 1417.5 lb of air. One can wonder if small building movers include this mass in their calculations for weight bearings of their equipment. How about aircraft designers for the engines' and the fuselage lift capabilities?

547. The interior of an empty steel building was blasted with dry ice to remove rust and scale. 550 pounds of dry ice was used. The interior dimensions of this building are $36' \times 36' \times 12'$. The abrasive blaster wore an abrasive blasting suit with a full face air line respirator operated in continuous flow mode. The dry ice evaporated into the air from the rubble accumulated on the floor. As the dry ice evaporated, an increase in atmospheric pressure was relieved through a tiny pipe in the ceiling. Carbon dioxide, being denser than air, at first hovered near the floor but eventually mixed with the air after abrasive blaster left for the day. The room does not have ventilation. (1) Calculate the relative density of CO_2 to air. (2) What is the average concentration of CO_2 in the atmosphere? Assume 500 ppm$_v$ CO_2 at the beginning of blasting.

$$36' \times 36' \times 12' = 15,552 \, ft^3 = 440.4 \, m^3$$

$$550 \, pounds = 249,475,804 \, mg$$

The relative density of CO_2 to air can be calculated by comparing their molecular weights: The "apparent" molecular weight of dry air is 28.941 (see Problem 58), and CO_2 has a molecular weight of 44.01 grams/gram-mole. That is, CO_2 gas is (44.01/28.941) 1.52 times denser than air.

First, calculate the mass of CO_2 in the building air before the abrasive blasting:

$$\frac{mg}{m^3} = \frac{ppm_v \times molecular\,weight}{24.45\,grams/gram\text{-}mole} = \frac{500 \times 44.01}{24.45} = 900.0\,mg/m^3$$

$900 \, \text{mg/m}^3 \times 440.4 \, \text{m}^3 = 396{,}360 \, \text{mg of } CO_2$ in the building's atmosphere. Total CO_2 gas in this building's atmosphere is $249{,}475{,}804 \, \text{mg} + 396{,}360 \, \text{mg} = 250{,}121{,}640 \, \text{mg}$.

$$\text{ppm}_v = \frac{(250{,}121{,}640 \, \text{mg/}440.4 \, \text{m}^3) \times 24.45}{44.01} = 315{,}523 \, \text{ppm}_v \, CO_2$$

$$= 31.55\%$$

This CO_2 gas concentration is immediately dangerous to life and health. All doors to the building must be locked and posted with bold warning signs. Entry must not be permitted until the air samples demonstrate the inside CO_2 gas concentration is identical to the ambient CO_2 gas concentration. Workers shoveling the rubble into wheelbarrows must be informed of the CO_2 gas inhalation risks to life. Portable pedestal fans should be strategically placed to dilute and mix CO_2 with air as it is safely exhausted from the building. Because one can reasonably predict the CO_2 gas concentrations using calculation methods above before the work begins, special industrial hygiene methods can be invoked to prevent the accumulation of such a dangerous atmosphere as work proceeds: buddy system and rescue personnel, a communication system because airborne dust will also be high decreasing visibility for the standby personnel, and special ventilation.

548. Consider previous Problem 547. If the atmosphere inside this room is now 31.55% carbon dioxide gas, what is oxygen gas concentration by volume?
   Normal air is 20.9476% oxygen by volume.

$$100\% \text{ air} - 31.55\% \, CO_2 = 68.45\% \text{ air remaining}$$

$$68.45\% \text{ air} \times 0.209476 = 14.3386\% \, O_2$$

549. Let us turn Problem 461 into a fire forensics issue. The dairy farmer claims that the carpenter contractor he hired to repair cattle stalls was negligent by having excess gasoline in an open container for his electricity generator that when spilled caused a fire destroying his barn's interior. He claims feed hay, straw bedding, and wood stalls burned interior including expensive milking machines, hoses, valves, pumps. The carpenter denies negligence and hires a defense lawyer. Is the farmer's claim plausible? His cow, Elsie, by the way, escaped from the barn unharmed. She developed a posttraumatic stress disorder and an anxiety neurosis never producing milk again. 20 chickens fled never to return. The farmer's claim also alleges loss of income from disrupted milk, cheese, and egg production.
   Sure, because only a small amount of burning gasoline on dry hay and straw will cause a rapid conflagration. Let us further say that the farmer maintains that there was an explosion sufficiently strong to damage the barn's interior but weak enough not to cause collapse of the barn walls and the roof. Is this plausible?

Let us use calculations from Problem 461 to see:

$$\frac{999\,\text{ppm}}{2.5\,\text{gallons}} = \frac{14{,}000\,\text{ppm}}{x\,\text{gallons}}$$

Solving for x, we see that 35.0 liquid gallons of gasoline would have to evaporate and mix evenly throughout the barn's interior to achieve the LEL for the gasoline and air mixture. It appears highly unlikely to most reasonable men that one cow could kick over a drum containing that much gasoline. Therefore, the defense lawyer files a motion for summary judgment by the Court. Let us verify the 35.0 gallons of gasoline with calculations as in Problem 461:

$$35.0\,\text{gallons} = 132{,}489.4\,\text{mL}$$

$$132{,}489.4\,\text{mL} \times 0.8\,\text{g/mL} = 105{,}991.52\,\text{g} = 105{,}991{,}520\,\text{mg}$$

$$\text{ppm} = \frac{(105{,}991{,}520\,\text{mg} / 1{,}715.7\,\text{m}^3) \times 24.45}{108\,\text{g/g-mole}} = 13{,}985.7\,\text{ppm} \cong 1.4\%$$

Again, 1.4 volume percent is the average vapor concentration throughout the barn. Pockets of explosive gasoline vapor between LEL and UEL are foreseeable near a gasoline spill. Therefore, prudent judges would see this as a "battle of the experts" and turn the decision making over to a jury with the Court's oversight.

550. Determine dilution ventilation requirements for an indoor work area where organic solvent-containing adhesive is used at the rate of 3 gallons over an 8-hour workday. The density of the solvent (X) is 0.87 g/mL. The adhesive contains 40% by volume X. 100% of the solvent evaporates essentially uniformly over the 8-hour work period.

The plant manager stipulated that workers must not be exposed over 80% of the TLV TWAE of 100 ppm$_v$. While this plant manager's attitude is commendable, he is not an expert in the application of TLVs, to wit: "ACGIH® states that TLVs® and BEIs® are guidelines to be used by professionals in the practice of industrial hygiene." The industrial hygienist brought this to the plant manager's attention recommending that the control objective should be as low as possible and never more than 80% of the action level, or 40 ppm$_v$. He cited the statistical and foundational basis for this recommendation. The TLV, in this case, is based on the prevention of headache and eye irritation.

This dilution ventilation air requirement is

$$Q = K(q_e/C_a)$$

where
Q = the dilution air flow rate

K = a dimensionless mixing factor accounting for less than perfect mixing of air contaminant in the room and the contaminant inhalation toxicity. Usually, K varies between 3 and 10 with 1 being perfect mixing. 7–10 values are used if mixing conditions are poor. Normally, before assigning higher K factors, attempt to improve mixing by adjusting damper throw directions and volume and rearranging supply outlets and exhaust fan locations and whatever else promotes turbulence and good mixing of work place air with fresh, tempered air.

q_c = volumetric flow rate of pure contaminant vapor or gas, c

C_a = acceptable air contaminant concentration, volume or mole fraction (ppm$_v$ × 10^{-6})

Solving

$$C_a = [0.40\,(100\,\text{ppm}_v)] \times 10^{-6} = 40 \times 10^{-6} \text{ (volume fraction)}$$

The mass flow rate of X is

$$m_X = \left[\frac{3\,\text{gallons}}{8\,\text{hours}}\right] \times \left[0.4\frac{\text{gallons}\,X}{\text{gallons adhesive}}\right] \times \left[\frac{(0.87)(8.34\,\text{lb})}{1\,\text{gallon}\,X}\right] = 1.09\,\text{lb/hour}$$

$$\left(\frac{1.09\,\text{lb}}{\text{hour}}\right) \times \left(\frac{454\,\text{g}}{\text{lb}}\right) \times \left(\frac{1\,\text{hour}}{60\,\text{minutes}}\right) = 8.24\,\text{grams}\,X/\text{minute}$$

$$\frac{8.24\,\text{grams/minute}}{92\,\text{grams/gram-mole}} = 0.0896\,\text{gram-mole/minute}$$

The resultant solvent X volumetric flow rate is calculated directly from the ideal gas law:

$$\frac{(0.0896\,\text{g-mole/minute}) \times [0.08206\,\text{atm-L/g-mole-K})] \times (293\,\text{K})}{1\,\text{atmosphere}}$$

$$= 2.15\,\text{L/minute}$$

Therefore, the minimum required dilution ventilation flow rate with the K mixing factor of 5 is

$$Q = \frac{(5)(2.15\,\text{L/minute})}{40 \times 10^{-6}} = 268{,}750\,\text{liters/minute}$$

$$\left(\frac{268{,}750\,\text{liters}}{\text{minute}}\right) \times \left(\frac{1\,\text{ft}^3}{28.312\,\text{L}}\right) = 9492\,\text{cfm}$$

551. A 3'-6" wide × 28" laboratory exhaust bench hood has the following face velocity measurements in fpm: 125, 135, 110, 140, 150, 140, 145, 155, and 130. There are two slots in the plenum each $1\text{-}\frac{1}{2}$ inches × 2.5 feet. The plenum is 6 inches deep, 2.5 feet high, and 3'-6" wide. Air is exhausted through a 10-inch diameter duct. What volume of air does this hood move? What are the average plenum velocity, average slot velocity, and average duct velocity for this system?

$$\text{Average hood face velocity} = (125 + 135 + 110 + 140 + 150 + 140 + 145 + 155 + 130)/9 = 136.7 \text{ fpm}$$

$$\text{Hood face area} = (42" \times 28")/144 \text{ in}^2/\text{ft}^2 = 8.167 \text{ ft}^2$$

$$\text{Hood exhaust volume} = 136.7 \text{ fpm} \times 8.167 \text{ ft}^2 = 1116.4 \text{ ft}^3/\text{minute}$$

$$\text{Total slot area} = 2(1.5" \times 30")/144 \text{ in}^2/\text{ft}^2 = 0.625 \text{ ft}^2$$

$$\text{Average slot velocity} = 1116.4 \text{ cfm}/0.625 \text{ ft}^2 = 1786.2 \text{ fpm}$$

$$\text{Average plenum velocity} = 1116.4 \text{ cfm}/(6" \times 42")/144 \text{ in}^2/\text{ft}^2 = 637.9 \text{ fpm}$$

Because the average plenum velocity is less than ½ of the slot velocity, there will be uniform air flow through both slots, and if the slots are properly spaced, there should be fairly uniform air flow across the hood face. Slight variations in the hood face velocities could be attributable to cross drafts and/or improper positioning of the velometer. A picture frame air foil around the hood face offers uniform air flow.

$$\text{Duct velocity} = 1116.4 \text{ cfm}/(\pi r^2)/144 \text{ in}^2/\text{ft}^2$$
$$= 1116.4 \text{ cfm}/(\pi \times 5"^2)/144 \text{ in}^2/\text{ft}^2 = 2046.9 \text{ fpm}$$

This duct velocity is acceptable for gases and vapors. It should be increased to >3000–4000 fpm for most medium density dusts and airborne particles.

552. An empty room 12' × 20' × 40' with 20% relative humidity is connected with a single closed door to an adjacent empty room 18' × 20' × 50' with 65% relative humidity. Both rooms are tightly sealed and without ventilation. What is the relative humidity when the door is opened and after the air in both rooms is completely mixed? Assume the air temperature in both rooms is identical at 70°F and outside walls are warmer than 70°F.

Calculate mass of water vapor in each room. Add these masses. Then calculate mass of air in each room. Add the masses. The solution requires use of humidity tables or a psychrometric chart. Assume 70°F at sea level.

$$12' \times 20' \times 40' = 9600 \text{ ft}^3$$

"Standard air" is 70°F at 760 mm Hg and 0% relative humidity. One cubic foot of standard, dry air weighs 0.075 pound mass (lb_m).

$$9600 \ ft^3 \times 0.075 \ lb/ft^3 = 720 \ pounds \ of \ air$$

$$18' \times 20' \times 50' = 18{,}000 \ ft^3$$

$$18{,}000 \ ft^3 \times 0.075 \ lb/ft^3 = 1350 \ pounds \ of \ air$$

20% relative humidity air at 70°F contains 0.0032 lb water/lb dry air

65% relative humidity air at 70°F contains 0.0103 lb water/lb dry air

$$720 \ pounds \ air \times 0.0032 \ lb/lb = 2.304 \ pounds \ of \ water \ vapor$$
$$in \ a \ small \ room$$

$$1350 \ pounds \ air \times 0.0103 \ lb/lb = 13.905 \ pounds \ of \ water \ in \ a \ large \ room$$

$$Total \ pounds \ water \ vapor \ in \ both \ rooms = 2.304 \ lb + 13.905 \ lb$$
$$= 16.209 \ pounds$$

$$Total \ pounds \ air \ in \ both \ rooms = 720 \ lb + 1350 \ lb = 2070 \ pounds \ of \ air$$

$$16.209 \ lb \ H_2O/2070 \ lb \ air = 0.00783 \ lb \ H_2O/lb \ air$$

Using a psychrometric chart, this is equivalent to 50% relative humidity at 70°F. The larger room at 65% relative humidity was at risk of mold growth amplification. Now, after mixing, both rooms are virtually free of this risk.

553. Three biologist spelunkers plan to study a bat cave in southern Texas. It has been claimed 20,000,000 bats live in this cave based on previous studies. Biologists' inspections will be at midnight after the bats leave to feed on aerial insects. Prior studies reported 25 ppm_v ammonia gas in cave atmosphere during bat occupancy. The ammonia gas emanates from guano bats' fecal droppings onto the cave floor. This enormous colony of bats departs *en masse* in only a few seconds. As they depart, fresh air sweeps in to dilute ammonia gas. The cave dimensions are 300-feet long, 200'-wide, and nine-feet high. Average bat volume is 250 mL. What ammonia gas concentration can the spelunkers expect? The cave temperature is 70°F. The atmospheric pressure is 760 mm Hg this evening.

$$Cave \ volume = 300' \times 200' \times 9' = 540{,}000 \ ft^3 = 15{,}291 \ m^3$$

$$Total \ bat \ volume = 20{,}000{,}000 \ bats \times 0.25 \ L/bat = 5{,}000{,}000 \ L = 5000 \ m^3$$

The net cave atmosphere volume during bat occupancy is 15,291 m^3 – 5000 m^3 = 10,291 m^3.

$$\frac{mg}{m^3} = \frac{ppm_v \times molecular \ weight}{24.45 \ L/g\text{-}mol} = \frac{25 \ ppm_v \times 17.03}{24.45} = 17.41 \ mg/m^3$$

17.41 mg/m³ × 10,291 m³ = 179,166 mg NH_3 is in the cave atmosphere during bat occupancy. It's no wonder they leave!

So, almost instantaneously as bats fly out, this 179,166 mg of ammonia expands into the now empty cave volume, and the NH_3 concentration is reduced to:

$$ppm_v\ NH_3 = (179,166\ mg/15,291\ m^3) \times 24.45/17.03 = 16.8\ ppm_v\ NH_3$$

Moreover, this 16.8 ppm_v ammonia gas is further diluted by the in-rushing volume of 5000 m³ of fresh air.

Without going into the tedious dilution ventilation calculations, the now-empty cave volume of 15,291 m³ is further diluted with 5000 m³ of fresh air to about:

$$15,291\ m^3/(15,291\ m^3 + 5000\ m^3) = 0.7536.$$

$$16.8\ ppm_v\ NH_3 \times 0.7536 = 12.7 \pm 3\ ppm_v\ NH_3$$

Most workers find this ammonia gas concentration irritating. The author suggests powered air-purifying respirators with fresh cartridges for NH_3 and organic amines or full-face air-purifying respirators with the same cartridge type. These respirators provide eye protection from ammonia gas and any materials falling off the ceiling. Disposable full body suits, hard hats, fall protection, rubber boots and gloves, and scrupulous skin and tool sanitation hygiene after departing the cave are critical. The biologists must be educated in hazards of pulmonary and systemic fungal disease by inhaling moist cave air (e.g., histoplasmosis) and carefully trained in the robust protective measures.

554. What is the density of dry air at 312°F? Use the Rankine temperature scale. At 70°F, the density of dry air is 0.075 lb/ft³.

$$\text{Density of hotter air} = 0.075\ lb_m\ /ft^3 \times \frac{(459.7 + 70)°F}{(459.7 + 312)°F} = 0.05148\ lb/ft^3$$

This simple problem is one of demonstrating that as a fixed mass of gas is heated in elastic systems (the surrounding air is considered elastic), it expands as a function of increase in absolute temperature, and accordingly, its density decreases. That is, there is the same mass in larger volume.

555. Ambient air was sampled at 83.7 L/minute for 137.5 hours through a large EPA-approved high-efficiency filter. The exposed filter weight was 780.2 mg, and the weight of the blank presampling filter was 702.3 mg. What was the concentration in micrograms per cubic meter?

$$137.5\ \text{hours} = (137.5\ \text{hours} \times 60\ \text{minutes/hour}) + 30\ \text{minutes}$$
$$= 8280\ \text{minutes}$$

Air volume sampled = 83.7 Lpm × 8280 minutes = 693,036 liters
$$= 693.036 \text{ m}^3$$

Difference in filter weight = 780.2 mg − 702.3 mg = 77.9 mg

$$77.9 \text{ mg}/693.036 \text{ m}^3 = 0.1124 \text{ mg/m}^3$$

Total ambient airborne particulates = 112.4 micrograms per cubic meter.

556. Air was sampled for total calcium in a community downwind from a lime kiln. The air temperature and atmospheric pressure were steady at 35°F and 705 mm Hg, respectively. The air-sampling duration was 27 hours and 12 minutes, and the uncorrected sampling rate was 28.3 Lpm. What volume of air was sampled?

Correct for absolute temperature: 35°F = 274.817 Kelvin

27 hours, 12 minutes = 1632 minutes

28.3 liters/minute × 1632 minutes = 46,185.6 liters
$$= 46.185.6 \text{ m}^3 \text{ (uncorrected)}$$

46.185 m³ × (274.817 kelvin/273.16 kelvin) × (760 mm Hg/730 mm Hg)
$$= 49.063 \text{ m}^3$$

557. In Problem 556, 32.7 mg total particulate was collected by the filter. 59.8% of this was analyzed as calcium. What was the total airborne particulate concentration? Assuming the calcium was due to lime, what was the lime dust concentration in the community air?

$$32.7 \text{ mg}/49.063 \text{ m}^3 = 0.666 \text{ mg TSP/m}^3$$

Lime (CaO) is obtained by calcining (roasting) calcium carbonate in a lime kiln by the reaction:

$$CaCO_3 + \text{intense heat} + \text{time} \rightarrow CaO + CO_2$$

Transfer of crushed limestone to the kiln was not observed to be a dusty process. However, copious amounts of fugitive lime dust emanated from product transfer points and out of the top of the lime kiln. Therefore, it was presumed virtually all calcium dust collected on the filter was from calcining limestone. Moreover, the wind direction throughout sampling period was virtually constant from the lime kiln to community air-sampling point.

$$0.666 \text{ mg TSP/m}^3 × 0.598 = 0.398 \text{ mg Ca/m}^3$$

Molecular weight calcium = 40.08 grams/gram-mole

Molecular weight calcium oxide (lime) = 56.08 grams/gram-mole

Therefore, the amount of lime dust in the air sample was 0.398 Mg $Ca/m^3 \times (56.08/40.08) = 0.557$ mg CaO/m^3

TSP of 0.666 mg/m³ − 0.557 mgCaO/m³ = 0.109 mg/m³ background airborne dust

Limestone contains various amounts of silica and silicates. The kiln's calcining temperature, while very hot, is insufficient to convert a-quartz into cristobalite or tridymite or amorphous silica (glass). All lime dust and limestone dust exposures should include measurements of respirable silica dust.

558. Properly disposing even modest amounts of hazardous waste is pricey and highly technical. Industrial hygienists are often consulted regarding HW disposal. One, employed by a large chemical company, was asked about disposal of a solution of potassium permanganate and a separate tank of sulfuric acid. He remembered a waste storage tank of ferrous sulfate on site. Recalling his knowledge and training in inorganic chemistry and, specifically, reduction–oxidation reactions, he writes the chemical equation. What is it?

$$2\ KMnO_4 + 10\ FeSO_4 + 8\ H_2SO_4 \rightarrow 5\ Fe_2(SO_4)_3 + K_2SO_4$$
$$+ 2\ MnSO_4 + 8\ H_2O$$

This is a win-win. The relatively innocuous ferrous sulfate is oxidized to the also innocuous ferric sulfate. Manganese sulfate is also much less harmful than the potassium permanganate. Importantly, both hazardous wastes ($KMnO_4$, H_2SO_4), are destroyed *in situ*—no off-site transportation and no treatment, storage, and disposal facility required. The reaction products should be considered for sale as a plant fertilizer additive. Fe, K, and Mn are essential plant nutrients, and sulfates promote soil acidification.

Each mole of $KMnO_4$ requires 5 mol of ferrous sulfate, H_2O, and 4 mol of sulfuric acid. From this, and his training in chemical engineering, he stipulates the necessary equipment, piping, mixing motor, feed rates, temperature controls, and industrial hygiene precautions.

559. An air sample filter was chemically digested and diluted to 100 mL with distilled water. A 10 mL aliquot was analyzed at 12.4 mcg Co/mL. The air sample volume was 1390 L. The distilled water was analytically free from cobalt. What was the airborne cobalt concentration?

Total amount of cobalt = (100 mL/10 mL) × 12.4 mcg Co = 124 mcg Co

$$mcg/M^3 = \frac{(C \times V) - B}{V_{corr} \times F},$$

where
    V_{corr} = air volume sampled corrected for atmospheric temperature and pressure
    C = concentration of analyte in the aliquot (mcg/mL)

V = volume of aliquot (mL)
B = total amount of analyte in the blank (mcg)
F = fraction of total sample in the aliquot used for analysis (dimensionless).

Since the blank was Co-free, and the aliquot correction was made as above, the above equation is not now necessary:

$$\text{mcg Co/m}^3 = 124 \text{ mcg}/1.390 \text{ m}^3 = 89.2 \text{ mcg Co/m}^3$$

560. Look at Problem 161 where the approximate density of methyl chloroform vapor was calculated at 4.55 (dimensionless, and air = 1.00). Can the density of a pure vapor be calculated more easily than the stated method?
    Yes, calculate the ratio of their respective molecular weights. Molecular weight of methyl chloroform = 133.42 g/gram-mole (precisely), and the "apparent" molecular weight of air = 28.966. Since the chemical with a higher molecular weight is more dense (the numerator):

$$133.42/28.966 = 4.606$$

Again, to reemphasize, this demonstrates that a pure vapor cannot settle out of air because it is mixing with air and becoming less dense over time as the following problem demonstrates.

561. What is the density of a mixture of 500 ppm$_v$ methyl chloroform vapor in air?

$$500 \text{ ppm}_v = 0.005\%_v$$

$$0.005 \times 4.606 = 0.023$$

$$0.995 \times 1.000 = 0.995$$

$$0.023 + 0.995 = 1.0108$$

That is, the density of this air gases and methyl chloroform vapor mixture is only 1.08% times greater than air alone. This slight difference is insufficient to exceed thermal air current effects, ventilation motion, and pedestrian traffic air movement. This mixture will not stratify on the floor.

562. Converse to the previous problem, what is the density of 500 ppm$_v$ helium in air?

The density of helium gas is 0.138 (dimensionless, where air = 1.000).

$$500 \text{ ppm}_v = 0.005\% \text{ (i.e., } 1,000,000 \text{ ppm}_v = 100\%, 10,000 \text{ ppm}_v$$
$$= 1\%, \text{ and so on)}$$

$$0.005\% \times 0.138 = 0.00069$$

$$0.995 \times 1.000 = 0.995$$

$$0.00069 + 0.995 = 0.99569 \ (air = 1.000)$$

A balloon filled with this gas mixture will never ascend, whereas the balloon filled with 100% helium rapidly ascends by overcoming gravity to a point and balloon's mass.

563. Consider even higher concentrations of vapor or gas mixtures in air than provided in previous problems. What is specific gravity of 100% toluene vapor at its LEL?

The LEL for toluene vapor is variously reported, but the most conservative value of 1.27% by volume at NTP air is used for fire risk management purposes.

$$1.27\% \ toluene \times 3.17 \ specific \ gravity = 4.026$$

$$100\% \ air - 1.27\% \ toluene = 98.73\% \ air \ in \ this \ LEL \ mixture$$

$$98.73\% \ air \times 1.00 \ specific \ gravity = 98.73$$

Specific gravity of this mixture $= (4.026 + 98.73)/100 = 1.02756$.

That is, the density of this mixture is only about 2.8% heavier than air alone. Also, one must be mindful in shutdown conditions and no ventilation and activities in the area, explosive vapors can accumulate in low spots where sources of ignition may be present. This explains why it is important to maintain a modicum of air motion in areas with flammable-explosive solvents during shutdown and during periods of low activity and low natural air movement.

564. Capture of airborne dust as dry material falls from one point onto a moving belt conveyor is typically required to protect health of nearby workers and to prevent accumulation of fugitive (and maybe explosive and/or expensive) dust on surfaces. Dry materials cascade 14 inches onto an 18-inch wide belt conveyor moving at 300 fpm. What volume of exhaust ventilation air is needed to capture the airborne dust?

Committee on Industrial Hygiene of the American Iron and Steel Institute reports an empirical equation to derive exhaust ventilation requirements for such material transfer processes:

$$Q = S \times W \times \sqrt{\frac{H}{3}},$$

where
  Q = required air volume, cfm
  S = 350 cfm for belt speeds below 250 feet per minute (fpm)
  S = 550 cfm for belt speeds from 250 to 500 fpm
  S = 750 cfm for belt speeds exceeding 500 fpm

W = conveyor belt width, feet
H = height of material fall, feet

$$Q,\text{cfm} = 550\,\text{cfm} \times 1.5\,\text{feet} \times \sqrt{\frac{1.17\,\text{feet}}{3}} = 73.7\,\text{cfm}$$

This exhaust volume appears very low for an open hood. Most likely this is the exhaust rate for conveyor belt with an enclosing hood and location of the exhaust duct take-off near the discharge point. Because this is an empirical equation, the author recommends modest experimentation with the type of material, distance of drop, and so on. Exhaust fan should have flexibility to handle various flow volumes. The use of blast gates has been applied by some to balance flow at multiple take-off points in a long conveyor hood. Regardless, provide cleanout doors at appropriate locations. Robust ventilation engineering is needed for this type of processing for combustible dusts. Be mindful for accumulations of explosive dust in conveyor systems. Control ignition sources among which could be reckless torch cutting of sheet metal without a hot work permit. The simple drilling of a metal screw into the exhaust hood might be a sufficient ignition source. Hammer tapping an exhaust duct with accumulated explosive dust could initiate an explosion.

From an energy conservation standpoint, a shorter drop of material and narrower conveyor belt helps. The industrial hygienist should work with plant engineers to see if wet methods of slightly moist dust will hinder dust transfer to the air. These engineering controls, if successful, will reduce the size (or even need) for a dust collection system such as a cyclone, wet scrubber, baghouse, and so forth. The author, with certain dusts, has found a gentle water spray mist at dust generation points typically highly successful for conveyor belt transfers. Addition of a wetting agent (e.g., a surfactant) to the wet mister can promote reduction in airborne dust concentrations.

565. Exhaust ventilation air is passing through a 12-inch internal diameter duct at 2350 feet per minute. There is contraction in the duct to an 8-inch internal diameter duct. What is the new duct velocity in this narrower section?

$$\text{Area of circle} = \pi r^2 = \pi\,(6"^2/144\ \text{in}^2/\text{ft}^2) = 0.785\ \text{ft}^2$$

$$\pi\,(4"^2/144\ \text{in}^2/\text{ft}^2) = 0.349\ \text{ft}^2$$

$$Q = AV \text{ for 12-inch duct} = 0.785\ \text{ft}^2 \times 2350\ \text{fpm} = 1844.75\ \text{cfm}$$

$$V = Q/A \text{ for 8-inch duct} = 1844.75\ \text{cfm}/0.349\ \text{ft}^2 = 5285.8\ \text{fpm}$$

This is an exceptionally high duct velocity. This velocity is rarely needed for even highly dense dusts. This duct velocity could be a significant generator of noise. One should consider a lower exhaust volume to achieve the appropriate transport velocity in the duct for materials of concern. This will also be a significant savings in energy costs. See the next problem (566).

566. Assume in the previous problem that the sheet metal worker had to reduce the 12" diameter duct to an 8" diameter duct because of space limitations. Let us see if an 8 inch square duct—instead of an 8 inch diameter duct—would also work.

$$8" \times 8" = 64 \text{ in}^2 = 0.444 \text{ ft}^2$$

$$V = Q/A = 1844.75 \text{ cfm}/0.444 \text{ ft}^2 = 4192.6 \text{ fpm}$$

This 1093.2 fpm reduction in duct velocity helps and is acceptable for many types of airborne dusts, but it is about 2000 fpm excessive for most gases and vapors in the ppm_v to low percent$_v$ range. The reader is encouraged to calculate different duct transport velocities with different duct cross-sectional areas.

567. An industrial hygienist obtains 11 direct-reading instrument values of toluene vapor in the breathing zone throughout a worker's 8-h workshift recording them as 17, 21, 22, 23, 15, 19, 29, 30, 13, 19, and 18 ppm_v. What is the average or mean concentration of toluene vapor for this worker on this day?

$$(17 + 21 + 22 + 23 + 15 + 19 + 29 + 30 + 13 + 19 + 18) \text{ ppm}_v/11$$
$$= 226/11 = 20.55 \text{ ppm}_v$$

This slightly exceeds the TLV for toluene vapor. Implement rigid controls to lower exposures well below the action level of 10 ppm_v.

568. What is the standard deviation of the preceding vapor concentration samples? The standard deviation provides a statistic of the average variability in a series of data from a population. Hand-held scientific calculators permit one to do this very quickly, but it is helpful to occasionally revert to the "old fashioned" way as follows:
The standard deviation $SD_x = \sqrt{[1/n - 1]\sum_{i=1}^{n}(x_i - \bar{x})^2}$. Set up a table as follows:

x	Average	x – average[a]	$(x - \text{average})^2$
17	20.55	3.55	12.60
21	20.55	0.45	0.20
22	20.55	1.45	2.10
23	20.55	2.25	5.06
15	20.55	5.55	30.80
19	20.55	1.55	2.40
29	20.55	8.45	71.40
30	20.55	9.45	89.30
13	20.55	7.55	57.00
19	20.55	1.55	2.40
18	20.55	2.55	–6.50
			Σ, total = 279.76

[a] Since the difference between the value and the average will be squared, the ± signs may be disregarded. That is, all values assume a positive number.

SD for this data set = $(279.76/11 - 1)^{0.5} = 27.976^{0.5} = 5.29$.

Note the geometric standard deviation is a better descriptor of air contaminants than the normal standard deviation. See Problem 315 for an example to calculate the geometric mean of a population data set. These data also point out the importance of obtaining several spot-reading samples throughout exposure period and during the work shift. An excellent way of characterizing this worker's exposure would be full-shift sampling with a personal air-sampling pump along with spot direct-reading measurements of air contaminants in the breathing zone. The latter help to characterize peak exposures—particularly helpful for chemicals that have STEL® or C® TLV values.

569. The concentration of toxicant in a person's blood at 1:42 p.m. is 17.6 micrograms per deciliter. Without medical intervention, the concentration at 5:19 p.m. is 11.7 mcg/dL. What is the half-life of this toxicant in this person if we assume first-order detoxification kinetics?

$$t_2 - t_1 = 5:19 - 1:42 = 217 \text{ minutes}$$

$$c_1 - c_2 = 17.6 \text{ mcg/dL} - 11.7 \text{ mcg/dL} = 5.9 \text{ mcg/dL decay}$$

$$T_{1/2} = \ln\left[\frac{t_2 - t_1}{\ln c_1 - \ln c_2}\right] = \frac{217 \text{ minutes}}{\ln 17.6 - \ln 11.7}$$

$$= \frac{217 \text{ minutes}}{2.868 - 2.60} = 809.7 \text{ minutes} = 13.5 \text{ hours}$$

570. In the previous problem, assuming first-order decay toxicokinetics, when could one expect this person's blood concentration to be 8.8 mcg/dL?

$$17.6 \text{ mcg/dL}/8.8 = 2.0, \quad \text{or exactly one half-life.}$$

$$1:42 \text{ p.m.} + 13.5 \text{ hours} = 3:17 \text{ a.m.}, \quad \text{the next morning}$$

571. Verify the calculations in the two preceding problems by substituting values into first-order decay equation.

$$k \text{ (the decay constant)} = 0.693/T, \quad \text{where } T = \text{elapsed time} = 217 \text{ minutes}$$

$$k = 0.693/217 = 0.003194/\text{minute}$$

$C = C_0 \times e^{-kT} = 17.6 \text{ micrograms/dL} \times e^{-0.003194 \times 217} = 17.6 \text{ mcg/dL} \times e^{-0.693}$
$= 17.6 \times 0.500 = 8.8 \text{ mcg/dL}$, the biological half-life concentration for this person

However, we should be mindful that while two points determine a straight line, they do not necessarily determine first-order decay. Two more blood samples obtained, say, 5 hours apart from the first would determine if there is linearity. Toxicants can have second-order decay where, for example, there is rapid decay from aqueous or the mineralizing (bones and teeth) compartments, but a slower decay from lipid and adipose tissues.

572. Assume the clinic's or hospital's policy is a patient cannot be discharged until their blood level for this toxicant is less than 6 micrograms per deciliter. How long will this take for this person?

This could be plotted on semi-logarithmic graph paper, but the danger for this is others might apply the graph to everyone. Detoxification and excretion rates vary from one person to another.

From Problem 570, we see that the blood concentration took 13.5 hours to reduce 50% to 8.8 mcg/dL. After another 13.5 hours, the blood level should reduce to 4.4 mcg/dL. So, without one resorting to tedious mathematics, the patient may be released approximately 21 hours after the first blood sample was drawn if a convenient time for the clinic or hospital. Regardless, do not discharge until there is a confirmation by the laboratory the blood level of this toxicant is below 6.0 mcg/dL, and the patient is entirely asymptomatic and agrees to being discharged.

573. 650 liters of a nonreactive gas with a density of 1.833 kg/m³ are mixed with 350 liters of another nonreactive gas with a density of 0.678 kg/m³. Both gases are at equal temperature and pressure. What is the density of this gas mixture?

$$0.650 \text{ m}^3 \times 1.833 \text{ kg/m}^3 = 1.191 \text{ kg}$$

$$0.350 \text{ m}^3 \times 0.678 \text{ kg/m}^3 = 0.237 \text{ kg}$$

$$1.191 \text{ kg} + 0.237 \text{ kg} = 1.428 \text{ kg/m}^3$$

574. Work begins at 8:00 a.m. for a house painter. Between 11:00 a.m. and 2:00 p.m., he is exposed to a steady concentration of 45 ppm_v CO. He does not take a break for lunch and quits working at 4:00 p.m. He is exposed for remainder of his work day to 0.6 ppm_v CO. Calculate his TWAE to CO gas.

$$3 \text{ hours} \times 45 \text{ ppm}_v = 135 \text{ ppm}_v\text{-hours}$$

$$5 \text{ hours} \times 0.6 \text{ ppm}_v = 3 \text{ ppm}_v\text{-hours}$$

$$(135 \text{ ppm}_v\text{-hours} + 3 \text{ ppm}_v\text{-hours})/8 \text{ hours} = 17.3 \text{ ppm}_v$$

This exceeds the 12.5 ppm_v action level based on the ACGIH TLV of 25 ppm_v. Such an exposure peak from 11:00 a.m. and 2:00 p.m. appears very strange and warrants investigation to identify source and mitigate. 0.6 ppm_v is not atypical for background, ambient CO concentration in some urban areas or if nearby combustion sources are present (e.g., gasoline fuel electricity generator, salamander).

575. For the previous problem (574), by how much must the CO concentration during the 3-hour interval be reduced so that he is exposed to the CO action level?

$$5 \text{ hours} \times 0.6 \text{ ppm}_v = 3 \text{ ppm}_v\text{-hours}$$

$$\text{Action level} = (8 \text{ hours} \times 25 \text{ ppm}_v)/2 = 100 \text{ ppm}_v\text{-hours}$$

$$100 \text{ ppm}_v\text{-hours} - 3 \text{ ppm}_v\text{-hours} = 97 \text{ ppm}_v\text{-hours}$$

$$97 \text{ ppm}_v\text{-hours}/3 \text{ hours} = 32.3 \text{ ppm}_v \text{ CO}$$

576. Verify the calculations for the previous problem (575).

$$5 \text{ hours} \times 0.6 \text{ ppm}_v = 3 \text{ ppm}_v\text{-hours}$$

$$3 \text{ hours} \times 32.3 \text{ ppm}_v = 96.9 \text{ ppm}_v\text{-hours}$$

$$3 \text{ ppm}_v\text{-hours} + 96.9 \text{ ppm}_v\text{-hours} = 99.9 \text{ ppm}_v\text{-hours}$$

The target organs for CO poisoning are the brain, cranial nerves, heart, and retina. The investigating industrial hygienist must also consider the painter's simultaneous exposures to any organic solvent vapors emanating from any "oil-based" paints, thinners, stains, lacquers, varnishes, and enamels because these organic vapors also affect the brain and the heart as target organs. The TLV additive mixture rule must be applied here.

577. If, as in a previous problem (574), this painter is also exposed to 37.6 ppm$_v$ for the solvent vapor mixture (8-hour TLV = 50 ppm$_v$) including the 32.3 8-hour TWAE to CO, what is his additive mixture TWAE to both CO and organic solvent vapors?
    $TWAE_1/AL_1 + TWAE_2/AL_2 + \cdots + TWAE_n/AL_n$ must be equal to or less than 1.0 (unitless), equal to or less than 0.5 for action level compliance.
    17.3 ppm$_v$/12.5 ppm$_v$ + 37.6 ppm$_v$/50 ppm$_v$ = 1.38 + 0.75 = 2.13 = 213% of the action level for toxicologically additive mixture of CO gas and the organic solvent vapors.
    Note that although the organic solvent vapor mixture exposure is below its action level, the presence of another identical target organ toxicant (CO) substantially raises the additive exposure of their combined action levels. Technically, target organ for CO are the erythrocytes and their hemoglobin, but high oxygen demands upon the brain, myocardium, and retina are, in a sense, secondary targets. CO, therefore, is a systemic toxicant because blood perfuses all organs and tissues.

578. Dealing with alcohol-intoxicated employees can be a delicate issue. Concerns are for their safety and for those they encounter, for whom they have responsibilities, employer's liabilities, aggressive behavior, violence, and impaired decision making. Companies deal with this in different ways.

Some foolishly send them immediately home. This can breed tremendous liability for the employer if, for example, they are involved in an automobile collision resulting in injuries or deaths. Others will call a relative to pick them up, but this, too, can have liabilities. Still others call for a taxi cab or request a company security officer to drive the inebriated worker home. Problems with these approaches have been no one was home to receive the worker, house keys had been lost, violent behavior upon the driver, arguments, and jumping from the moving car. Again, such methods can raise more safety and legal liability issues. As health professionals, we also have public safety duties. Some employers deflate the tires of an inebriated employee's car, place nail strips in driver's path, or block the driver's vehicle in the parking lot with a massive truck.

The best approach, in the author's view, is to detain them until sober. When this occurs is a question and depends on liver size, BAC, percent water distribution in the body, and the liver's blood perfusion rate. If the employee refuses the BAC test or to remain on company property, detain him long enough to secretly call police so that a patrol car officer arrests him the moment he leaves and enters a public street.

One can calculate a safe release time. Let us say an employee's BAC is 0.21% at 1:00 pm. Under most state's driving laws, this person is almost "triple drunk" (DUI ≥ 0.08%). Company policy is to not release their worker until his BAC is 0.02%, and he remains in custody of the security department. He is not paid for time required to "sober up." If his BAC decreases to 0.17% in one half-hour (1:30 pm), and he stopped drinking at 1:00 pm, when may he go home? Detoxification of ethanol follows zero-order kinetics, but for this calculation, assume the catabolism of ethyl alcohol is a first-order toxicant decay (–), a reasonable approximation. Hard data supersede medical and or a police officer's opinions of sobriety.

$$\text{Integrating, } \frac{dC}{dt} = -kC_0$$

$$\text{at } t = 0.5 \text{ hour, } \ln \frac{C}{C_0} = -kt$$

$$\frac{C}{C_0} = \frac{0.17}{0.21}$$

$$\text{thus, } k = \frac{\ln(0.17/0.21)}{0.5 \text{ hour}} = 0.423 \text{ h}^{-1}$$

$$\text{at other concentrations and times } \ln \frac{C}{0.21} = -0.423t$$

$$\text{half-life, } t_{1/2} = \frac{\ln 2}{k} = \frac{0.693}{0.423} = 1.64 \text{ hours}$$

$$\text{for } C = 0.02\%, t = \frac{\ln(0.02/0.21)}{(-0.423) \text{ hour}^{-1}} = 5.56 \text{ hours}$$

Therefore, this worker may leave for home about 6:30 p.m.—not any sooner. Admonish him! Never again! Inform him his future employment could be toast.

The employer's choice to not release until the man's BAC drops to 0.02% is wise because above 0.4%, unconsciousness, possible coma, and verging near death exists; one is stuporous between 0.3% and 0.4%; obvious impairment exists when between 0.1% and 0.3%, definitive impairment exists at 0.08–0.1%; some impairment exists at 0.05–0.08%; and possible impairment exists below 0.05%. Elimination rates and effects of ethanol vary greatly from person to person, but the average is 0.00028% per minute.

579. Let us see how the solution to Problem 578 agrees with preceding calculations:

$$5.56 \text{ hours} \times 60 \text{ minutes/hour} = 333.6 \text{ minutes}$$

$0.21 \times 0.00028\%/\text{minute} \times 333.6 \text{ minutes} = 0.0196\%$. This agrees with the calculated % BAC and shows his/her ethanol detoxification rate is nearly average. Verify this with a venous BAC analysis before safely releasing this employee. Regardless, conduct the sobriety tests and carefully question the employee about his/her driving abilities. Carefully document your calculations, measurements, and observations.

580. A full circular open surface barrel with internal diameter of 2 feet, 9 inches contains warm ethyl benzene. Three days, four hours, and 17 minutes later the surface level of this organic chemical decreased 4 and 3/8 inches. What was the average evaporation rate in milligrams per square centimeter per minute?

$$\emptyset = 2'\text{-}9'' = 33''$$

$$r = 16.5''$$

volume of evaporated ethyl benzene is a cylinder of $\pi r^2 h$

$$h = 4 \text{ inches} + 3/8 \text{ inches} = 4.375''$$

$$\pi (16.5'')^2 \times 4.375'' = 3741.9 \text{ in}^3 = 61{,}318.75 \text{ mL}$$

$$61{,}318.75 \text{ mL} \times 0.8665 \text{ g/mL} = 53{,}132.7 \text{ g} = 53{,}132{,}700 \text{ mg}$$

$$3 \text{ days} \times 24 \text{ hours/day} \times 60 \text{ minutes/hour} = 4320 \text{ minutes}$$

$$4320 \text{ minutes} + 257 \text{ minutes} = 4577 \text{ minutes}$$

$$53{,}132{,}700 \text{ mg}/4577 \text{ minutes} = 11{,}608.6 \text{ mg/minute}$$

$$\text{Surface area of tank} = \pi \ (16.5")^2 = 855.3 \text{ cm}^2$$

$$11{,}608.6 \text{ mg/minute}/855.3 \text{ cm}^2 = 13.57 \text{ mg/cm}^2/\text{minute}$$

581. In the previous problem (580), assume evaporation is into the air of a building 30 feet long and 20 feet wide. A wall-mounted exhaust fan is at one end. A make-up air louver is at the opposite end. The building was designed for a ventilation rate of 1.5 cfm/ft² of floor surface area. What is the average steady-state concentration of ethyl benzene vapor in this small building?

$$\text{ppm}_v = \frac{\text{ER} \times 24.45 \times 10^6}{Q \times \text{molecular weight}},$$

where
ER = evaporation or generation rate in g/minute
Q = ventilation rate in L/minute

$$Q = 30' \times 20' \times 1.5 \text{ cfm/ft}^2 = 900 \text{ cfm} = 25{,}485 \text{ L/minute}$$

$$\text{Ethyl benzene molecular weight} = 106.17 \text{ grams/mole}$$

$$11{,}608.6 \text{ mg/minute}/1000 \text{ mg/g} = 11.6086 \text{ grams evaporate per minute}$$

$$\text{ppm}_v = \frac{11.6086 \times 24.45 \times 10^6}{25{,}485 \times 106.17} = 104.9 \text{ ppm}_v$$

The TLV for ethylbenzene at the time of this writing is 100 ppm$_v$ as an 8-hour TWAE. The *American Conference of Governmental Industrial Hygienists* in their *2010 Notice of Intended Changes* proposes to reduce the TLV to 20 ppm$_v$ based on new toxicity information. Moreover, ethylbenzene has small, but significant, amounts of benzene in it. Benzene, a genotoxic human carcinogen, has no "safe" limit of exposure. That is, lower is safer. Less frequent exposure is safer.

Exposure would be excessive for those occupying this building for prolonged time. Industrial hygiene control options include: Place a lid on this barrel! Reduce the solvent's temperature. Limit occupancy time in the building. Provide appropriate respirators. In this simple case, increase the ventilation as a last resort only until mechanical local exhaust ventilation can be installed. Benzene's high toxicity does not permit use of dilution ventilation.

582. Compare the arithmetic mean and the geometric mean of the following values: 3, 5, 6, 6, 7, 10, 12. The arithmetic mean and the geometric means are measures of central tendency in a group of values.

$$\text{Arithmetic mean, AM} = \frac{3 + 5 + 6 + 6 + 7 + 10 + 12}{7} = 7$$

$$\text{Geometric mean, GM} = \sqrt[7]{3 \times 5 \times 6 \times 6 \times 7 \times 10 \times 12} = \sqrt[7]{453,600}$$

$$\log \text{GM} = \frac{1}{7} (\log 453,600) = \frac{1}{7} (5.6567) = 0.808$$

GM = 6.43 (Enter 0.808 into calculator. Enter "INV" or "2nd". Enter "log.")

The geometric mean of a set of unequal positive values is always less than the arithmetic mean. The geometric mean is normally a better statistic and measure for central tendency for a set of air-sample results.

There are other measures of central tendency: the harmonic mean, the median, mode, weighted arithmetic mean, and root mean square. Each has advantages and disadvantages depending on the data and the intended purposes of analysis.

583. Industrial hygienists and toxicologists rely on animal model toxicological testing, and practitioners know that these have limitations and are often the only way to assess and manage risk. One significant shortcoming is that the numbers of test animals is often limited, and, therefore, extrapolations from these data to humans are tenuous. The following table demonstrates how animal test population size can affect the reliability of test outcomes.

Probability of Toxic Effects in Man	Number of Animals Used in Experiments[a]	
	95% Probability	99% Probability
100%	1	1
80	2	3
60	4	6
50	5	7
40	6	10
20	14	21
10	29	44
5	59	90
2	149	228
1	299	459
0.1	2995	4603
0.01	29,956	46,050

[a] Number of animals to be included in an experiment in order to find at least one subject with the toxic effect (assuming identical incidence of toxic effect in animals and man). G. Zbinden: *Progress in Toxicology*, New York, Springer, No. 1, 1973 (T. Marthaler, Biostatistics Centre, University of Zurich).

584. Industrial hygienists and toxicologists who serve as expert witnesses are asked how much did cigarette smoking contribute, for example, in causing the plaintiff's primary lung cancer. The following example with supporting laboratory animal testing statistics helps address this:

Mr Robert Smith smoked two packs of cigarettes daily for 30 years. According to available epidemiological evidence, this means that he increased his chances of developing lung cancer (by cigarette smoking alone) by 10×, that is, from baseline prevalence of about 10 per 100,000 for nonsmokers *per annum* to about 100 per 100,000 *per annum*. Multiplying this by Mr Smith's adult life of 35 years, we arrive at a cumulative probability a man with Mr Smith's smoking history would develop lung cancer at any time during this period from smoking alone is 3.5% at most. While this is higher than one's chances of developing lung cancer if one did not smoke, it remains a statistical fact that 96.5% of all men who smoke two packs of cigarettes daily do not die from lung cancer in a 35-year period. We then arrive at the conclusions that he might be one of the unlucky 3.5%, or he had exposures to other pulmonary carcinogens to place him at even higher risk of developing his lung cancer.

Lung Cancer Epidemiology/Biostatistics for Cigarette Smoking in Men:	
Nonsmoker:	10/100,000 *per annum*
One pack/day:	60/100,000 *per annum*
Two packs/day:	100/100,000 *per annum*

$(100/100,000) \times 35$ years $= 0.035 = 3.5\%$

585. A flexible container (e.g., a large balloon) contains 17.6 liters of an ideal gas. The temperature of the gas is 25°C, and the container is attached to a pole in Miami. The gas is slowly heated to 46°C at constant pressure. Common sense suggests to most of us that this gas will expand, but to what volume? Disregard back pressure from the balloon.

This is an application of Charles' Law.

$$V_2 = 17.6\,\text{L} \times \frac{(46 + 273)}{(25 + 273)} = 18.84\,\text{L}$$

586. An open face floor level air supply register to a bedroom is 6 inches × 12 inches. The ventilation design engineer stipulated 320 cfm. What is the velocity of the air supply? Later, someone installs a safety grill on the register with a grill factor of 0.6. What is the new air supply velocity?

$$6'' \times 12''/144\ \text{in}^2/\text{ft}^2 = 0.5\ \text{ft}^2$$

$$V = Q/A = 320\ \text{cfm}/0.5\ \text{ft}^2 = 640\ \text{fpm}$$

The grill reduces the open face area. Therefore, air velocity increases, although the air volume remains constant. The grill adds to the system's total pressure.

$$640\ \text{fpm}/0.6 = 1067\ \text{fpm}$$

587. Standard air supply or air return ducts installed between studs 16 inches on center are 3 inches × 12 inches. Now and then, space limitations require a circular duct to replace a rectangular duct. Calculate equivalent diameter of a circular duct to replace the rectangular duct.

> $Ø = (4ab/π)^{0.5}$, where a and b are the dimensions of the rectangle (or a square)

$$Ø = (4 × 3" × 12"/π)^{0.5} = 6.77 \text{ inches}$$

Let us verify the dimension: Area of circle = $πr^2 = π (6.77"/2)^2 = 35.997$ in^2 = (for the rectangular duct) 3" × 12" = 36 in^2.

This is an odd dimension not available as a stock item. Moreover, it is too wide for 2" × 4" studs (actually, 1.5" × 3.5"). Perhaps two smaller rectangular ducts could divide the air volume to two air supply or return registers. The carpenter and sheet metal worker must collaborate.

588. It is often necessary to calculate horsepower of ventilation system fan motors. HP is an index of the amount of work a machine can perform over a time period. One HP equals 33,000 foot-pounds of work per minute. This, in turn, equals 0.746 kilowatt. Calculate the horsepower of a fan operating at 1750 rpm with a torque of 16.8 foot-pounds. This calculation is also required to determine a fan's electricity requirements.

HP = (torque (ft-lbs) × rpm) 5250 = (16.8 × 1750)/5250 = 5.60 horsepower

> 5.60 HP × 0.746 kilowatt/HP = 4.18 kilowatts of electricity
> are needed (4180 watts).

589. Consider the previous problem (588). Torque is a force that produces, or attempts to produce, rotation—usually measured in foot-pounds, or inch-pounds. A force of 1 pound applied to a handle of a crank, the center of which is 1 foot from the center of the shaft, produces a torque 1 foot-pound on the shaft if the force that is provided is perpendicular to (not along) the crank. What is the torque on a shaft of a ventilation exhaust fan with brake horsepower of 8.4 operating at 1600 rpm?

> Torque, in foot-pounds = (HP × 5250)/rpm = (8.4 × 5250)/1600
> = 27.56 ft-lbs

Such data are needed to ensure selected fan has sufficient mechanical strength to operate without shaft fracture failures. Fan catalogues stipulate these safety data.

590. An industrial combustion device steadily consumes 11.2 gallons of no. 2 fuel oil every hour. Assume this oil is 86.1% carbon, 13.8% hydrogen, 0.1% sulfur, and with negligible free oxygen. The fuel oil weighs 6.8 pounds per gallon. Calculate the combustion air (oxygen) requirements.

The amount of air required for perfect (stoichiometric) combustion of hydrocarbon fuels with carbon, hydrogen, and sulfur (pounds of air per pounds of combustibles) are: carbon (11.5), hydrogen (34.3), and sulfur (4.3).

Consider 1 pound of fuel:

0.861 C × 11.5 × 6.8 = 67.3 pounds of air
0.138 H × 34.3 × 6.8 = 32.2 pounds of air
0.001 S × 4.3 × 6.8 = 0.03 pound of air
67.3 lb + 32.2 lb + 0.03 lb = 99.53 lb of air
99.53 pounds of air × 1.10 (10% excess*) × 11.2 gallons = 1226.2 lbs air/h
1226.2 lbs of air per hour/60 minutes/hour = 20.437 pounds of air/minute
20.437 lbs air/minute/0.075 lbs/ft³ standard air/minute = 272.5 cfm

591. A natural gas water heater has a capacity of 44,000 Btu/hour. Using the standard method of determining combustion air requirements, calculate minimum volume of fresh air required for this water heater.

The minimum required volume is 50 cubic feet per 1000 Btu/hour.

For each minute of operating time:

$$(50 \text{ ft}^3/1000 \text{ Btu/hour}) \times 44{,}000 \text{ Btu/hour} = 50 \text{ ft}^3/\text{hour} \times 44$$
$$= 2200 \text{ cfh} = 36.7 \text{ cfm}$$

The author investigated a fully preventable CO asphyxiation fatality in a residence that resulted from a combination of several factors:

- Death occurred between 12:30 a.m. and about 4:30 a.m. when wind velocity was nonexistent (steady at 0 mph throughout this 4-hour period).
- The vent stack was located several feet below the upper roof line (a building code violation).
- The weather cap was only approved for a fan-assisted exhaust stack (a building code violation). This stack was a gravity vent relying solely on the "chimney effect" and wind motion. Absent wind motion to create a Venturi suction on the stack, this weather cap created a CO death trap.
- The water heater (50 gallons, 44,000 Btu/hour) was located in a tightly sealed wood box with insufficient combustion make-up air and no make-up air louvers (a serious building code violation). At least two openings (e.g., 5" × 10") were needed—one located no more than one foot above floor and the other located no more than one foot from the top of the water heater cabinet.
- The exhaust stack had two 90° elbows instead of two 45° elbows (a building code violation).
- The lateral exhaust stack section was sloped downward instead of at least $\frac{1}{4}$-inch rise over 12" run (a building code violation).

---

* If the process and operating engineer can guarantee perfect mixing of supply air with atomized fuel oil mist, the 10% "safety factor" can be reduced to 5% or less. This will reduce operating costs for our increasingly "green" society. Two percent is probably the lowest amount of excess air to ensure carbon monoxide gas does not form and combustion is nearly stoichiometric.

- The bedroom where the decedent was found did not have cold air returns to the furnace (a building code violation).
- There was a high hot water demand shortly before the decedent retired for the night (laundering clothing and using a large whirlpool tub).
- There was a clear pathway for exhaust gases to ascend from the box where the water heated was located to the decedent's bedroom.
- The decedent slept on the floor on an air mattress with her head located within 3 feet from where exhaust gases entered her bedroom. As the exhaust gases cooled, they became denser and hovered over her body because of the lack of air circulation.

The combination of these multiple factors is analogous to the many factors leading to the foul weather catastrophe portrayed in the movie *The Perfect Storm* wherein absent one or more of these weather factors, the storm would not have occurred.

592. Vapor pressure of a volatile liquid can be determined by slowly bubbling a known volume of a pure gas through the liquid at a known temperature and pressure. In an experiment, 4.21 liters of dry air were bubbled through 7.6621 grams of liquid benzene at 24.4°C and atmospheric pressure (760 mm Hg). Benzene remaining after the experiment weighed 5.5230 grams. Assuming the air became saturated with benzene vapor and temperature of liquid benzene and air remained constant, what is the vapor pressure of benzene in torr?

$$\text{Difference in weight} = 7.6621 \text{ g} - 5.5230 \text{ g} = 2.1391 \text{ g} = 2139.1 \text{ mg}$$

$$4.21 \text{ L} = 0.00421 \text{ m}^3$$

$$2139.1 \text{ mg}/0.00421 \text{ m}^3 = 508,099.8 \text{ mg/m}^3$$

$$ppm_v = \frac{mg}{m^3} \times \frac{22.4 \text{ L/gram-mole}}{\text{molecular weight}} \times \frac{\text{absolute temperature}}{273.16 \text{ K}}$$
$$\times \frac{760 \text{ mm Hg}}{\text{atmospheric pressure, mm Hg}}$$

$$ppm_v = 508,099.8 \text{ mg/m}^3 \times \frac{22.4 \text{ L/gram-mole}}{78.11 \text{ grams/mole}} \times \frac{24.4 + 273.16 \text{ K}}{273.16 \text{ K}}$$
$$\times \frac{760 \text{ mm Hg}}{760 \text{ mm Hg}} = 158,725$$

$$(\text{vapor pressure, mm Hg}/760 \text{ mm Hg}) \times 10^6 = \text{saturation concentration, } ppm_v$$

$$(\text{vp, mm Hg}/760 \text{ mm Hg}) \times 10^6 = 158,725 \text{ } ppm_v$$

$$\text{Vapor pressure} = 120.6 \text{ mm Hg} = 120.6 \text{ torr}$$

This does not agree with calculation by the Antoine equation (91.2464 mm Hg at 24.1°C). Perhaps the bubbling produced small mist particles suggesting a higher vapor pressure because of greater loss of mass into the vapor phase over time. I would repeat, but would not bubble the air through the benzene. Simply allow the air to pass very slowly (e.g., 5 mL/minute) over surface of the benzene to ensure vapor saturation.

593. You find an old 4.5 liter pressure bottle labeled "Stink Damp" at a dump site near a coal mine. The pressure gauge reads 2750 lbs/in². The ambient temperature is 12°C. What is the mass of H_2S in this gas bottle?

$$PV = nRT$$

$$n = PV/RT$$

Molecular weight of H_2S = 34.08 grams/gram-mole

R = universal gas constant = 0.0821 L-atm/mole-K

2750 lbs/in² = 187.126 atmospheres

$$n = [(187.126 \text{ atm})(4.5 \text{ L})]/[(0.0821)(285.16 \text{ K})] = 35.97 \text{ moles}$$

35.97 moles × 34.08 grams H_2S/gram-mole = 1225.86 grams of H_2S

594. Without wearing a SCBA, you foolishly open the valve (problem 593), but it breaks while twisting the gas release handle instantly enveloping you inside an 8-foot diameter hemisphere bubble of H_2S gas. What is the average gas concentration inside this hemisphere? Disregard your body volume. Assume no wind, 760 mm Hg, and 25°C.

$$\text{Volume of hemisphere} = [(4/3) \pi r^3]/2 = (1.333 \times 3.1416 \times (4')^3)/2 = 134.04 \text{ ft}^3$$

$$134.04 \text{ ft}^3 = 3.796 \text{ m}^3$$

$$ppm_v = [(1,225,860 \text{ mg}/3.796 \text{ m}^3) \times 24.45]/34.08$$
$$= 231,683 \text{ ppm}_{v5} H_2S \ (>23\%!)$$

You will collapse instantly and die unless you are safely rescued, administered oxygen, CPR, and antidotes. The antidote kit for H_2S poisoning is available by prescription from Eli Lilly Company. Their kit contains amyl nitrite, sodium nitrite, sodium thiosulfate, and various needles, syringes, and a nasogastric tube. You also might have been seriously, maybe fatally, injured by the ejected gas valve.

595. A worker in Portland (Oregon or Maine) is exposed to 20 ppm_v CO gas and 19.8% by volume of oxygen gas throughout his work shift with both gases at a steady concentration. Is he overexposed?

The TLV for CO is 25 ppm_v. The percent by volume of oxygen in standard air at any elevation is 20.95%. OSHA's lower limit of oxygen exposure is 19.5% by volume. Therefore, at first blush, it appears that he is not over-exposed to CO or underexposed to O_2.

However, CO competes with O_2 for binding sites on hemoglobin, the red protein in erythrocytes that transfers oxygen from blood perfusing the lungs to tissue sites throughout the body. This is a fully reversible biochemical reaction:

$$HgB + O_2 \rightleftharpoons HgBO_2$$

However, CO also bonds with hemoglobin. This binding strength is 210–280 times greater than oxygen. That is, a small amount of CO binds with hemoglobin thereby hindering oxygen transport from the inhaled air to the tissues. Carbon monoxide is a chemical asphyxiant that can fatally poison even though the inhaled air is 20.95% by volume. The biochemical reaction equation is

$$HgB + CO \rightleftharpoons COHgB$$

CO has additive toxicological effects with a diminished oxygen concentration in the inspired air. The toxicological additive mixture rule appears to apply:

Both Portlands on opposite coasts of the United States are at sea level (i.e., they are ports), so we may assume an average barometric pressure of 760 mm Hg.

$$(20 \text{ ppm}_v/25 \text{ ppm}_v) + (19.8\%/20.95\%) = 0.8 + 0.945 = 1.745$$

This worker is exposed at 174.5% for the combined gases. Prompt intervention is required. Persons with previous myocardial infarctions, cardiac ischemia, risk of stroke, hyperthyroidism, diabetes, obesity, and several other medical risk factors are at greatest risk. Physiological heat strain is another risk factor that is additive to CO. Other chemical asphyxiants include H_2S, HCN, soluble cyanide salts, aniline, soluble sulfide salts, and certain aniline derivatives.

596. Air at 20.9476% O_2 by volume is what weight percent O_2 at 760 mm Hg and 70°F?

Normal, dry air weighs 0.075 lb/ft³ at 760 mm Hg and 70°F. The four major gases in air are nitrogen, oxygen, argon, and carbon dioxide. Standard air contains 350 ppm$_v$ CO_2 and smaller amounts of other trace gases. These and CO_2 are omitted in the following table because of their minor weight percentage contributions:

$$(0.017384/0.074564) \times 100 = 23.31\%$$

Standard air contains 20.9476% oxygen by volume and 23.31% oxygen by weight. This makes sense because the molecular weight of oxygen exceeds the apparent molecular weight of air.

597. An automated manufacturing process requires the gas-phase oxidation of certain materials in a small, securely closed process room. Stoichiometrically equivalent amounts of ozone and chlorine are mixed. Four separate ozone and chlorine gas alarms are installed side by side at strategic locations outside this room to protect workers nearby. A significant breach occurred in the door gaskets to the room that subsequently released large amounts of both gases. The alarms failed to provide audio-visual signaling. Several workers developed severe chemical pneumonitis. Why did these alarms fail?

Perhaps alarms failed to detect gas-phase reaction product of ozone with chlorine. Ozone, being a stronger oxidant than the weaker oxidant chlorine, could arguably oxidize chlorine gas to chlorine dioxide. Moreover, there are 33% more ozone molecules than chlorine molecules which tend to shift reaction to chlorine dioxide:

$$4O_3 + 3Cl_2 \rightarrow 6ClO_2$$

The gas-phase reaction product, chlorine dioxide, in this hypothetical example has a TWAE TLV of 0.1 ppm_v and a STEL of 0.3 ppm_v to prevent lower respiratory tract irritation and bronchitis. Assuming these fixed locations of gas monitors were recently calibrated with certified span gases, and they did not lose power, there might be other reasons why the alarms failed, but it appears that they may not be responsive to chlorine dioxide. Regardless, this example points out importance of calibrating gas alarms under foreseeable conditions to help rule out false-positive, false-negative, and zero response gas readings. In other words, with the example, one could conclude both chlorine and ozone gases did not reach their alarm signal triggers while excessive amounts of chlorine dioxide migrated into breathing zones of workers. Consult with your gas alarm manufacturers and suppliers so that they test reasonable interaction scenarios with gases of your concern. Request their calibration data curves for various gas mixture scenarios.

Gas	Density (lb/ft³) at 70°F and 760 mm Hg	Volume %/100	Weight (lb/ft³)
Oxygen	0.08279	0.209476	0.017384
Nitrogen	0.072	0.78084	0.05622
Argon	0.103	0.00934	0.00096
		Total weight =	0.074564

Another possible example could be a reaction of the oxidizing gas chlorine dioxide with the reducing gas hydrogen sulfide:

$$Cl_2 + H_2S \rightarrow 2HCl + S_o$$

and $Hg_o + Cl_2 \rightarrow HgCl_2 + Hg_2Cl_2$. And phosgene gas with water vapor (high relative humidity):

$$COCl_2 + H_2O \rightarrow 2HCl + CO_2$$

The danger of the last example is if the phosgene gas alarm was calibrated under high humidity conditions, low humidity conditions might under-alarm for high levels of phosgene gas. Alarms must be set for phosgene gas, not for less acutely toxic hydrogen chloride gas. Better, have alarms for both with different audible tones.

598. On suspicion, a local public health agency analyzed water supplied to local homes and businesses for methylene chloride, a once-common, highly volatile chlorinated organic solvent (bp = 40°C = 104°F). Results were 120 parts CH_2Cl_2 per billion by weight/volume. 40 gallons of this water were heated to 200°F and maintained at this temperature for 2 hours in an empty, unventilated enclosed room with the inside dimensions of 20' × 14' × 7.5' (equivalent to nearly a 1-$\frac{1}{2}$ car garage with flat roof). What was the average methylene chloride vapor concentration after mixing with the air? Assume NTP. Disregard back pressure from water's vapor pressure at this tempera-ture and the small back pressure from methylene chloride vapor.

Assume, reasonably because of its low boiling point and high vapor pres-sure, all methylene chloride evaporated.

$$40 \text{ gallons} = 151.416 \text{ L}$$

$$120 \text{ ppb}_w = 0.120 \text{ ppm}_w$$

$$0.120 \text{ ppm}_w/L = 0.120 \text{ mg/L}$$

$0.120 \text{ mg/L} \times 151.416 \text{ L} = 18.17 \text{ mg } CH_2Cl_2$ evaporated into the air.

20' × 14' × 7.5' = 2100 ft³ = 59.47 m³. (Disregard volume of water, water container, and the heating apparatus, etc.)

Molecular weight of methylene chloride = 84.94 grams/mole

$$\text{ppm}_v = \frac{(18.17 \text{ mg}/59.47 \text{ m}^3) \times 24.25 \text{ L/gram-mole}}{84.94 \text{ grams/mole}}$$

$$= 0.0879 \text{ ppm}_v = 87.9 \text{ ppb}_v$$

Although concentration of methylene chloride vapor in air is very low, the relative humidity would have been very high.

599. Consider the previous problem (598) and assume an initial humidity of 0% before heating the water container began and a final volume of 39.6 gallons after the 2 hours of heating elapsed. Again, assuming no ventilation or con-densation, what is the new relative humidity? Assume atmospheric pressure of 760 mm Hg.

$$40 \text{ gallons} - 39.6 \text{ gallons} = 0.4 \text{ gallons of water evaporated}$$

$$0.4 \text{ gallon} = 1.514 \text{ L} = 1514 \text{ mL}$$

$$1514 \text{ mL} \times 1.00 \text{ g/mL} = 1514 \text{ g}$$

$$1514 \text{ g} = 23{,}364.6 \text{ grains}$$

$2100 \text{ ft}^3 \times 0.075 \text{ lb/ft}^3 = 157.5$ pounds of dry air in building before heating began. (0.075 lb/ft^3 is the mass of dry air per cubic foot of air at NTP.)

$$23{,}364.6 \text{ grains}/157.5 \text{ lbs} = 148.3 \text{ grains water/lb of dry air}$$

From a high-quality psychrometric chart (available from the *InterNet* or from the *American Society of Heating, Refrigeration, and Air Conditioning Engineers*), this amount of water vapor at a dry bulb temperature of 75°F exceeds 100% relative humidity, so there will be condensation of liquid water upon cooler surfaces and/or precipitation from the air.

At a dry bulb temperature of 85°F, the relative humidity is 80%, and at a dry bulb temperature of 100°F in this room, this amount of water vapor produces a relative humidity of 52%. All of this demonstrates that at a constant atmospheric pressure, warm air holds more moisture than cooler air—not surprising once one crunches the numbers and understands the mass and energy transfer concepts.

600. A worker spills 0.25 L chloroform on his clothing. He undresses immediately and promptly and tightly wraps all wet clothing in double-layer aluminum foil. Taking the solvent-saturated clothing home, he decides not to launder his soiled clothing. Instead, he places clothing in his clothes dryer to tumble until dryness occurring in exactly 9 minutes because of chloroform's high vapor pressure and volatility. His clothing is just tumbled dry without heating. His clothes dryer has a 165 cfm exhaust fan. What is average concentration of chloroform vapor in the exhaust?

$$9 \text{ minutes} \times 165 \text{ cfm} = 1485 \text{ ft}^3 = 42.05 \text{ m}^3$$

$$\text{Molecular weight CHCl}_3 = 119.38 \text{ grams/mole}$$

$$\text{Density CHCl}_3 = 1.48 \text{ g/mL}$$

$$0.25 \text{ L} = 250 \text{ mL}$$

$$250 \text{ mL} \times 1.48 \text{ g/mL} = 370 \text{ g} = 370{,}000 \text{ mg}$$

$$\text{ppm}_v = \frac{(370{,}000 \text{ mg}/42.05 \text{ m}^3) \times 24.45 \text{ L/gram-mole}}{119.38 \text{ grams/mole}} = 1802 \text{ ppm}_v$$

601. For the previous problem (600), sketch a diagram of the most likely appearance of the chloroform vapor concentration in the exhaust duct as a time function from the start of the dryer until clothing dryness.

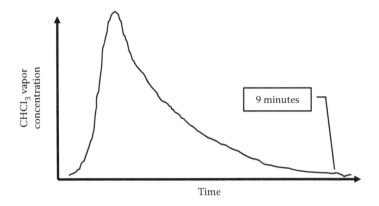

602. An air flow of 3750 cfm is heated from 23°F to 70°F. How much sensible heat is added to the air? Said another way, how much heat must be added to the air?

Sensible heat, $h_s = 1.08 \times 3750$ cfm $\times (70°F - 23°F) = 190,350$ Btu/hour

This information, among other factors (e.g., heating efficiency, heat transfer), can be used to select the best furnace.

As air, or any object, is heated, its temperature, of course, increases in proportion to the heat added. This increase in heat, h_s, is known as sensible heat. Similarly, as heat is removed from air, its temperature decreases, and the heat removed is also called sensible heat. Simply stated, heat that changes the temperature of an object is sensible heat. Air has mass; therefore, air is an object.

The factor 1.08 is the specific heat of air $\times$ density of inlet air at 70°F $\times$ 60 minutes per hour.

603. A home owner agrees with his HVAC contractor to design the heating system for 70°F and 50% relative humidity in a northern climate city. Assume that make-up air is solely by infiltration and the house net volume is 120,600 ft³. Assume 0.6 air change per hour for this home. How much water vapor must be added hourly to maintain this humidity?

The formula for pounds of water per hour is

$$\frac{\text{building net volume} \times \text{air changes/hour} \times \text{grains of moisture required}}{\text{specific volume of water} \times 7000 \text{ grains water/pound}}$$

From a psychrometric chart, 56 grains of moisture (water) is needed per pound of air at 70°F and 50% relative humidity minus the amount of moisture already in the air, or: 56 grains − 9 grains = 47 grains of water.

$$\frac{120,600 \text{ ft}^3 \times 0.6/\text{hour} \times 47 \text{ grains}}{13.5 \text{ ft}^3 \times 7000 \text{ grains/lb}} = 36 \text{ pounds of water per hour}$$

36 pounds of water per hour/8.33 lbs/gallon = 4.3 gallons/hour

The humidification system must be capable of evaporating 4.3 gallons of water hourly. However, if air temperatures exceed 0°F, less humidification, obviously, is required.

604. Unlike the previous problem (603) where ventilation is by infiltration (cracks in the doors and windows, people coming and going, and combustion make-up air for appliances), commercial buildings typically have mechanical air supply systems. Calculate humidification requirements for a small industrial facility that has 7800 cfm outside air introduced under the above conditions with a building air exchange rate of 0.8/hour. How much humidification is required?

$$7800 \text{ cfm} \times 60 \text{ minutes/hour} = 468,000 \text{ cfh}$$

$$\frac{468,000 \text{ cfh} \times 0.8/\text{hour} \times 47}{13.5 \text{ ft}^3/\text{lb} \times 7000 \text{ grains/lb}} = 186.2 \text{ pounds of } H_2O/\text{hour}$$

186.2 lbs/hour/8.33 lbs/gallon = 22.35 gallons must be evaporated per hour into the building's air-handling system (assuming 0% relative humidity for the make-up air).

605. For the preceding problem (604), what is the air exchange schedule on a cfm/ft² of floor area basis if the building's length is 42 feet and the width is 32 feet?

7800 cfm/42' × 32' = 5.8 cubic feet of air per minute per square foot of floor area. For most industrial buildings this is high and suggests that a reduction might be achievable while still maintaining employee comfort if processes are not present requiring mechanical local exhaust ventilation. Michigan, for example, requires a minimum of 1 cfm per square foot of floor area supplied mechanically and/or by natural air flow (wind, thermal gradient effects).

606. A polymer chemist developed a new plastic film and wants to test its permeability for benzene. He fills a sphere of this film with benzene. The sphere's diameter is 11.5 inches. The sphere is in a stainless-steel box with interior dimensions of 16" × 16" × 16". He immediately seals the box lid. He obtains a direct reading of 1250 ppm$_v$ of benzene vapor through a gas-sampling port after 8 hours. Assuming air in the box was free of benzene at the start of the experiment, what was the perfusion rate of liquid benzene into vapor phase through membrane in mcg/cm²/minute? Assume NTP conditions.

S, surface area of sphere $= 4\pi r^2 = \pi d^2 = 12.57 \, r^2$
$S = \pi \, (11.5 \text{ in})^2 = 415.48 \text{ in}^2 = 2680.5 \text{ cm}^2$
V, volume of sphere $= (4/3) \, \pi \, r^3 = (1/6) \, \pi \, d^3 = 4.189 \, r^3$
$V = (1/6) \, \pi \, d^3 = (1/6) \, \pi \, (11.5 \text{ inches})^3 = 796.3 \text{ in}^3$
Volume of box $= (16")^3 = 4096 \text{ in}^3$
Net volume of box with sphere $= 4096 \text{ in}^3 - 796.3 \text{ in}^3 = 3299.7 \text{ in}^3$

$$\frac{mg}{m^3} = \frac{ppm_v \times molecular\ weight}{24.45\ L/gram\text{-}mole} = \frac{1250\ ppm_v \times 78.11\ g/mole}{24.45}$$

$$= 3993.4\ mg/m^3$$

$$3299.7\ in^3 = 0.05407\ m^3$$

$3993.4\ mg/m^3 \times 0.05407\ m^3 = 215.92$ mg of benzene was in the space between the sphere and the box

$$8\ hours \times 60\ minutes/hour = 480\ minutes$$

$$215.92\ mg = 215,920\ micrograms$$

$$215,920\ mcg/480\ minutes = 449.8\ mcg/minute$$

$$449.8\ mcg/minute/2680.5\ cm^2 = 0.1678\ mcg/minute/cm^2$$

Permeation would continue over a greater time; however, he was investigating this polymer as a possible fabric for making personal protective clothing for benzene-exposed workers, and it is unlikely that one would wear a suit from this material longer than a normal 8-hour work shift. Forces and stretching on sphere's surface might affect permeation rate. Repeating the experiment with a box made of this polymeric film inside the larger box should be considered. Another test procedure follows in the next problem.

607. The chemist in the previous problem (606) attaches a box with a window 2 cm × 6 cm to a ventilation tube. This time he tests permeation of toluene by filling the box with toluene. The polymeric membrane covers the window. Toluene-free air flows through the tube and over the box containing liquid toluene at 0.5 Lpm for 8 hours at NTP. The exit end of the ventilation tube is attached to a large activated charcoal tube that was analyzed for toluene at the end of the experiment. 0.468 mg of toluene was detected and 0.0 mg found in a backup charcoal tube. The 6 cm dimension of the window aligned with air flow axis. Calculate toluene vapor that permeated through the film in mcg/cm²/minute.

$$0.468\ mg\ \varnothing\text{-}CH_3 = 468\ mcg\ toluene$$

$$8\ hours = 480\ minutes$$

$$468\ mcg\ toluene/480\ minutes = 0.975\ mcg/minute$$

$$2\ cm \times 6\ cm = 12\ cm^2$$

$$0.975\ mcg/minute/12\ cm^2 = 0.0815\ mcg/cm^2/minute$$

The permeation rate for toluene is roughly one-half that for benzene. The higher molecular weight of toluene might account, in part, for this. Also, this is a dynamic vapor generation system whereas the benzene permeation was done in a static system that might account for some variability. Undoubtedly, there are many other physicochemical factors (e.g., gravitational effects of the mass of the toluene on membrane, evaporative cooling as the liquid toluene passes into the vapor phase, uniformity of polymer film thickness, humidity). Another set of such experiments with xylene isomers and ethyl benzene appears warranted.

608. An empty abrasive blasting room $12' \times 12' \times 8'$ has 3.77 mg respirable silica dust suspended per cubic meter. The room must be exhausted until dust level is less than 10% of the TLV. The make-up air and the exhaust systems are configured so that $K = 1$. The exhaust system is 1350 cfm. The air supply system is 1270 cfm. That is, this abrasive blasting room is under negative pressure with respect to the adjacent work areas. Any leakage in gaskets and seals is inward. How long must this exhaust system operate? The TLV for respirable silica dust is 0.025 mg/m^3 = 25 mcg/m^3.

$$10\% \text{ TLV} = 2.5 \text{ mcg/m}^3$$

$$\text{Room volume} = 12' \times 12' \times 8' = 1152 \text{ ft}^3$$

$$3.77 \text{ mg/m}^3 = 3770 \text{ mcg/m}^3$$

Apply the "10% Rule." That is, for every 2.3 volumes of well-mixed clean air that passes through the contaminated dusty air, the dust concentration dilutes by 10% (precisely 10.0259%). This assumes perfect mixing of clean air with dusty air.

1152 ft^3 × 2.3 (= 2650 ft^3 of total dilution air) reduces the respirable silica dust concentration to 377 mcg α-SiO$_2$/m^3.

Another 2650 ft^3 reduces the dust concentration to 37.7 mcg/m^3.

Another 2650 ft^3 reduces the dust concentration to 3.77 mcg/m^3.

Another 2650 ft^3 reduces silica dust concentration to 0.377 mcg/m^3. This is below design objective. However, since respirable silica is a likely human carcinogen—besides being an aggressive agent of progressive pulmonary silicosis, it would be prudent to operate ventilation system until the room air is equivalent in particulate silica quartz concentration to the ambient air.

$$(2650 \text{ ft}^3) \times 4 = 10{,}600 \text{ ft}^3$$

$$10{,}600 \text{ ft}^3/1350 \text{ cfm} = 7.85 \text{ minutes} = 7 \text{ minutes, } 51 \text{ seconds}$$

Abrasive blasting rooms should be no larger than needed for the largest part to be abraded so that the ventilation purge times are kept brief. An electrical interlock on the door such that workers cannot enter until the exhaust and supply fans stop operating at specified time is suggested. Otherwise, workers might be tempted to enter the blasting room without an air-supplied abrasive blasting helmet.

609. Confirm results in the previous problem (608) with a first-order decay equation.

$$C = C_0 \times e^{-[Q/V]t} = 3770 \text{ mcg/m}^3 \times e^{-[1350 \text{ cfm}/1152\text{ft}^3]7.85\text{ minutes}} = 0.38 \text{ mcg/m}^3$$

If, for example, K was estimated to be 3, the ventilation operating time becomes 7.85 minutes $\times$ 3 = 23.55 minutes.

610. What is the partial pressure of oxygen in dry air at sea level and at 16.5% O_2 by volume?

Sea level barometric pressure = 760 mm Hg

Normal air contains 20.9476% by volume at any altitude or partial pressure.

760 mm Hg $\times$ 0.209476 = 159.20176 mm Hg for oxygen at sea level

760 mm Hg $\times$ 0.165 = 125.4 mm Hg

This partial pressure of oxygen should not have notable effect on a healthy person at rest—that is, no cardiac ischemia, hyperthyroidism, and so on, although OSHA regards any atmosphere below 19.5% by volume as oxygen deficient to account for a wide range of health status in workers. However, when oxygen level partial pressure falls to 90–120 mm Hg, there is increased respiration and a slight diminution of coordination in healthy people at rest. Loss of ability to think clearly occurs at an oxygen partial pressure of 76–90 mm Hg, and loss of consciousness and death result at 45–76 mm Hg. Of course, if one is simultaneously exposed to a chemical asphyxiant gas (e.g., CO, H_2S, HCN), adverse health effects will be more rapidly profound.

611. A closed tank contains 58.3 vol. % propane. The balance is normal air. What is the partial pressure of oxygen in this tank located in San Pedro Terminal near Long Beach, California?

$100\%_v - 58.3\%_v = 41.7\%_v$ air

760 mm Hg $\times$ 0.417 = 316.92 mm Hg for air

316.92 mm Hg $\times$ 0.209476 = 66.387 mm Hg for oxygen

To reduce the oxygen concentration to a fatal level, a gas or vapor level of about 50% by volume must be attained. This is only possible with a gas or a highly volatile liquid. In such concentrations, direct effects of most gases are in themselves fatal. In this problem, propane (C_3H_8) acts promptly as a narcotic gas causing rapid collapse followed by fatal anoxia subsequent to hypoxia.

612. Problems 610 and 611 are based on dry air. What if the partial pressure of water vapor in air for Problem 610 is 12.1 mm Hg?

Since the total pressure does not change, the composition of the atmosphere at sea level becomes:

Oxygen: $0.209476 \times (760 \text{ mm Hg} - 12.1 \text{ mm Hg}) = 156.667 \text{ mm Hg}$

Nitrogen, argon, other gases: $0.790524 \text{ mm Hg} \times (760 \text{ mm Hg} - 12.1 \text{ mm Hg}) = 591.233 \text{ mm Hg}$

$$\text{Total pressure} = 12.1 \text{ mm Hg} + 156.667 \text{ mm Hg} + 591.233 \text{ mm Hg}$$
$$= 760 \text{ mm Hg}$$

613. What is oxygen partial pressure at sea level and 100% relative humidity at 45°C (113°F)?
    Vapor pressure of water at these conditions is 71.88 mm Hg. Vapor pressures of water and volatile organic chemicals are available from handbooks of chemistry and physics.

$0.209476 \times (760 \text{ mm Hg} - 71.88 \text{ mm Hg}) = 144.145 \text{ mm Hg}$ versus 159.202 mm Hg for dry air, or $(144.145/159.202) \times 100 = 90.54\%$.

This is almost a 10% decrease in oxygen because of the high water vapor content. Dry bulb temperature, high humidity, and diminished oxygen concentration would be taxing to healthiest workers particularly under high work load conditions. Add CO gas, and the physiological and toxicogical challenge increases.

614. A process attendant works in the ASA department (acetylsalicylic acid, *Aspirin®*) of a large pharmaceutical company. The air is heavy with ASA dust with an average concentration at the TLV of 5 mg/m^3. He inhales 10 m^3 of air every work shift and does not wear a respirator. His cardiologist prescribed one baby *Aspirin* tablet daily to help prevent heart attack and stroke from blood clots. Is he toxicologically overexposed to ASA?
    No. We can reasonably assume that all ASA dust he inhales is either absorbed by respiratory tract and/or his gastrointestinal tract. This dose would be 5 mg/m^3 × 10 m^3 = 50 mg of acetylsalicylic acid. The TLV for ASA was selected to help prevent eye and respiratory tract irritation. One regular ASA tablet is 325 mg, and a baby ASA tablet is 81 mg. However, dusty workplace air requires attention to prevent release of ASA dust to the air by better process enclosures and isolation and good mechanical local exhaust ventilation. A wet scrubber or dry filters would work. His cardiologist, armed with this knowledge, might consider reducing his oral dose or instructing him to swallow the tablet late in his work shift so that he has benefit of a pharmacologically sustained dose during evening hours as well as at work. But, of course, because cardiologist errs, we must substantially reduce this worker's dust exposures. A hematologist should be consulted if the worker experience abnormal bleeding because ASA is a blood thinner.

615. Industrial hygiene reports are often significantly improved by inclusions of graphs, pie charts, histograms, diagrams, photographs, and other graphic representations. Graphs with x–y coordinates can be particularly illustrative. For example, graphs depicting the increased dilution ventilation with decreased air contaminant levels; decreased group blood lead levels with air concentrations of lead dust and fume; and decrease or increase of toxicant in body fluid over time are typical examples. Graphs for x and y coordinates for curve fitting are of four general types:

   a. Polynomial function (rarely used in industrial hygiene, toxicology, ventilation)
   b. Straight line: $y = mx + b$
   c. Exponential curve: $y = ab^x$, or $\log y = \log a + x (\log b)$ ⎫ commonly
   d. Power function: $y = ax^b$, or $\log y = \log a + b (\log x)$ ⎬ used
      ⎭

   Before computers, plotting of paired data (often using squares) was tedious. Now with programs such as *MicroSoft Excel*® and others, curve fitting is a breeze. We only need to type paired data; punch a few keystrokes; and computer selects best fit, supplies equation, regression factor, coefficients of correlation, and a plot of data so it can be cut and pasted into one's report. Exponential curves and power function curves can be likened to cooling rate of a cup of coffee and growth rate of a tree where the temperature and growth, respectively, can be plotted as a time function. See Problem 544 and the following Problem 616.

616. A worker collapsed after inhaling an organic chemical vapor. After admission to the hospital, venous blood samples were taken for this chemical with the following values (time of blood specimen withdrawal was converted to military time): 1:30 p.m. (1350) = 35.3 mcg/dL; 3:00 p.m. (1500) = 25.2 mcg/dL; 4:30 p.m. (1630) = 10.1 mcg/dL; 6:30 p.m. (1830) = 2.5 mcg/dL; and 7:00 p.m. (2100) = 1.5 mcg/dL. Plot these data (concentration *versus* time) using the best fit graph.

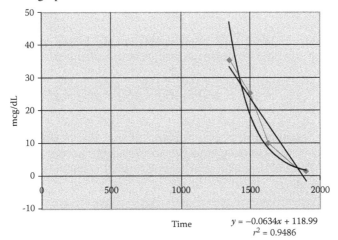

$y = -0.0634x + 118.99$
$r^2 = 0.9486$

Note display chart provides linear fit and an asymptotic fit. The asymptotic display is most accurate because the concentration of the toxicant in blood cannot be less than zero. The linear regression of $r^2 = 0.9486$ demonstrates the good correlation between blood level as a function of time. More important than this graph is the regression equation given in the chart. These data demonstrate the toxicant is being removed from his body through metabolic and/or excretory pathways. The close, but nonlinear, asymptotic regression suggests partial zero- and first-order decay metabolic transformation kinetics coupled with excretory kinetics of removal. For example, ethanol follows zero-order excretion kinetics (essentially, this is first-order excretion kinetics). That is, EtOH can be measured in exhaled breath (excretion kinetics) while EtOH is metabolized by the liver to CO_2 and H_2O. Combinations of excretory pathways are not uncommon (exhaled breath, hair and nails, urine, bile, breast milk, menstrual blood, dead epidermal cells and flakes, feces).

617. Using the equation for line of best fit in the previous problem (616), what was the man's most likely blood toxicant concentration at 5:12 p.m.?

$$5:12 \text{ p.m.} = 1712 \text{ military time} = x$$

$$Y = -(0.0634)\, x + 118.99 \text{ (the classic linear equation: } y = mx + b)$$

$$y = -0.0634\,(1712) + 118.99 = -108.5408 + 118.99 = 10.45 \text{ mcg/dL}$$

Such calculations are valuable in forensic investigations because one can easily calculate the concentration of toxicant at the time of an adverse event (e.g., a fall from elevated work space, murder, missed judgment, assault, automobile crash, time of last exposure, etc.).

618. A 12.5 liter partially evacuated gas-sampling bottle is used to sample workplace air for a gas contaminant that is absorbed and chemically reacted with a reagent with a distinctive color. The color intensity is measured with a spectrophotometer and compared to laboratory-prepared standards. The atmospheric pressure at time of sampling was 712 mm Hg. The residual partial pressure in the gas bottle was 690 mm Hg. In this example, 25 mL of reagent was used. How can the volume of this sample be calculated?

$$V_s = (V - A) \times \frac{P_2 - P_1}{P_2}$$

where V_s = sample volume, V = volume of bottle, A = volume of absorbent added to the gas-sampling bottle, P_1 = residual partial pressure in the bottle, and P_2 = the atmospheric pressure at the location and time of gas or vapor sampling.

$$V_s = (12.5 \text{ L} - 0.025 \text{ L}) \times \frac{712 \text{ mm Hg} - 690 \text{ mL}}{712 \text{ mm Hg}} = 0.385 \text{ L}$$

Evacuated gas-sampling bottles are rarely used now for industrial hygiene survey purposes because they are cumbersome and often limited to brief grab samples. They find wide applications in gas and vapor stack sampling when concentrations may be substantially higher than in employee work zones.

619. In the previous problem (618), 12.4 mcg of phenol were detected. What was the phenol vapor concentration in ppm_v? Assume NTP.

$$12.4 \text{ mcg phenol}/0.385 \text{ L} = 32.21 \text{ mg phenol/m}^3$$

Molecular weight of phenol = 94.11 grams/gram-mole

$$ppm_v = \frac{(mg/m^3) \times 24.45}{molecular\ weight} = \frac{32.21 \times 24.45}{94.11} = 8.37 ppm_v$$

The TLV for phenol vapor is 5 ppm_v as 8-hour time-weighted average. This air sample, however, is a "grab" sample that might not be representative of this worker's exposures over a work shift even if it was collected in his breathing zone. Phenol is absorbed readily through intact skin—sometimes fatally. The industrial hygienist must be especially mindful of this—particularly for this chemical which has been fatal subsequent to brief and limited skin surface area dermal contact.

620. At one time, there was a convenient "rule of thumb" to evaluate the ventilation supplied to rooms where there was open stick electrode welding (not fine wire MIG or TIG welding, or torch cutting). This equation for general dilution ventilation is

$$Dilution\ air, cfm = \frac{100,000 \times (electrode\ diameter,\ in)^2}{minutes\ per\ electrode}$$

The "minutes per electrode" in this equation refers also to the idle time both during welding and changing new electrodes for consumed electrode stubs—but only if welding was done continuously by several simultaneous welders. In such cases, this would help to ensure total welding fume did not exceed 10–15 mg/m³. The author presents this equation only for historical purposes and disagrees with this approach to control welders' exposures to metallic fumes for several reasons:

a.   Some modern electrodes contain more toxic components than years ago.
b.   The equation does not account for welding-generated gases.
c.   General dilution ventilation for large volume stick welding is energy wasteful. Mechanical local exhaust ventilation located at the point of generation of fume and gases is the control method of choice. Otherwise, where feasible, weld inside mechanically exhausted booth enclosures.
d.   This approach does not meet current, hygienic welding fume control practices.

e. Modern industrial hygiene engineering strives to control welders' exposures to far below 10–15 mg/m³ total fume in their breathing zones. Absent highly toxic components in welding fume, prudent industrial hygienists attempt to control total welding fume exposures to less than 1 mg/m³.

621. Galena (lead sulfide, PbS) is the major ore from which lead is obtained. To lower the amount of sulfur from galena, the ore is sintered with air (oxygen) to convert it to litharge (lead oxide, PbO). For each ton of galena charged in sintering furnace, how much sulfur dioxide is produced? Assume stoichiometric oxidation at NTP.

$$2\ PbS + 3O_2 \rightarrow 2\ PbO + 2\ SO_2$$

Molecular weight PbS = 239.26 grams/gram-mole

Molecular weight SO_2 = 64.06 grams/gram-mole

Molecular weight PbO = 223.20 grams/gram-mole

Molecular weight O_2 = 31.9988 grams/gram-mole

1 ton = 907.185 kg = 907,185 grams

907,185 grams/239.26 grams PbS/gram-mole = 3791.628 moles PbS

Feed: 2×239.26 grams = 478.52 grams

3×31.9988 grams = 95.9964 grams

478.52 grams + 95.9964 grams = <u>574.52 grams</u>

Product: 2×223.20 grams = 446.40 grams

2×64.06 grams = 128.12 grams

446.40 grams + 128.12 grams = <u>574.52 grams</u>

This verifies mass balance ($mass_{in} = mass_{out}$).

If there are 3791.628 moles of PbS per ton of galena feed, 3791.628 moles of SO_2 can be produced—the stoichiometric gas yield.

3791.628 moles $SO_2 \times$ 64.06 grams/mole = 242,891.69 grams of SO_2 produced

242,891.69 grams SO_2 = 0.2677 ton SO_2

622. How much air (oxygen) is required to oxidize the litharge in the previous problem (621)?

We learned from Problem 596 that standard, dry air contains 20.9476% O_2 by volume and 23.31% O_2 by weight.

$$\left. \begin{array}{l} \text{For PbS: 2 moles} = 478.52 \text{ grams} \\ \text{For } O_2\text{: 3 moles} = 95.9964 \text{ grams} \end{array} \right\} \begin{array}{l} \text{That is, each mole} \\ \text{of PbS requires } 1\text{-}\frac{1}{2} \\ \text{moles of } O_2 \end{array}$$

So, each gram of PbS requires $478.52/95.9964 = 4.9848$ grams of oxygen.

$$1 \text{ ton of galena} = 907{,}185 \text{ grams}$$

$$907{,}185 \text{ grams} \times 4.9848 \text{ grams of } O_2/\text{gram of galena} = 4{,}522{,}135.8 \text{ grams}$$
$$O_2 \text{ required}$$

$$0.075 \text{ lb air/ft}^3 \times 0.2331 \; O_2 = 0.01748 \text{ lb } O_2/\text{ft}^3$$

$$0.01748 \text{ lb } O_2/\text{ft}^3 = 7.9288 \text{ grams } O_2/\text{ft}^3$$

$$4{,}522{,}135.8 \text{ grams} \times 1 \text{ ft}^3/7.9288 \text{ grams} = 570{,}343 \text{ ft}^3 \text{ of oxygen required}$$

Air requirements are $570{,}343 \text{ ft}^3 \; O_2 \times 100/23.31 = 2{,}446{,}773.9 \text{ ft}^3$ of air/ ton. The size of the air blower fan(s) depends on the oxidation rate, galena feed rate, lead oxide extraction rate, amounts of bottom ash and fly ash generated, the galena's moisture content of ore and air, if air is preheated, and among other factors. To help ensure complete oxidation, this typically will require surplus oxygen (air), say, by 10–20%.

623. From the previous problem (622), apply 15% excess air. Assume a process feed rate of 2.3 tons of galena per hour. How many cfm will this furnace require?

$$2{,}446{,}773.9 \text{ ft}^3 \times 0.15 = 367{,}016.1 \text{ ft}^3 \text{ excess air}$$

$$2{,}446{,}773.9 \text{ ft}^3 + 367{,}016.1 \text{ ft}^3 = 2{,}813{,}790 \text{ ft}^3/\text{ton/hour}$$

$$2{,}813{,}790 \text{ ft}^3/\text{ton/hour} \times 2.3 \text{ tons/hour} = 6{,}471{,}717 \text{ cfh}$$

$$6{,}671{,}717 \text{ cfh}/60 \text{ minutes/hour} = 107{,}862 \text{ cfm}$$

Waste heat from the process can be used to preheat the air ensuring oxidation efficiency.

624. Some organizations (e.g., *American Petroleum Institute*) and a few jurisdictions (e.g., Delaware, New Jersey) endorse the *Substance Hazards Index* (SHI) for volatile and toxic liquids. This can be used to very roughly compare the relative hazards from spills of liquids based on their inhalation toxicity and volatility. The formula is

$$\text{SHI} = \frac{\text{VP} \times 10^6}{\text{BP} \times \text{ERPG} - 3'}$$

where

VP = vapor pressure of the liquid at 20°C (and some solids, e.g., phenol),
mm Hg,

BP = barometric (atmospheric) pressure, mm Hg, and

ERPG is the *Emergency Response Planning Guide* established by a committee of the American Industrial Hygiene Association. The ERPG values are intended to provide estimates of airborne concentration ranges where one reasonably might anticipate observing adverse health effects as a consequence of exposure to the specific substance.

ERPG-3 = maximum airborne concentration (in ppm_v) below which it is believed that nearly all people could be exposed for up to 1 h without experiencing or developing life-threatening health effects. ERPGs have been developed for about 50 chemicals.

Compare SHIs for methanol (ERPG-3 = 5000 ppm_v) and benzyl chloride (ERPG-3 = 25 ppm_v) in Sarasota, Florida.

Vapor pressures of benzyl chloride and methanol at 20°C are, respectively, 0.758 and 97.658 mm Hg.

$$\text{For benzyl chloride: SHI} = \frac{0.758 \text{ mm Hg} \times 10^6}{760 \text{ mm Hg} \times 25 \text{ ppm}_v} = 39.89 \text{ (unitless)}$$

$$\text{For methanol: SHI} = \frac{97.658 \text{ mm Hg} \times 10^6}{760 \text{ mm Hg} \times 5000 \text{ ppm}_v} = 25.70 \text{ (unitless)}$$

In comparing, we see the much less volatile and more toxic benzyl chloride vapor has a greater SHI than the more volatile but less toxic methyl alcohol vapor.

625. 1.2 microliters (μL) of liquid mercury were injected into an empty *Lucite*® box with interior dimensions of 12" × 14" × 14". After evaporation, assume mercury vapor did not deposit on surfaces of the poly-(methylmethacrylate). Assume NTP. What is the mercury vapor concentration?

$$12'' \times 14'' \times 14'' = 2352 \text{ in}^3 = 0.03854 \text{ m}^3$$

$$\text{Mercury density} = 13.534 \text{ g/mL}$$

$$1.2 \text{ μL} = 0.0012 \text{ mL}$$

$$0.0012 \text{ mL} \times 13.534 \text{ g/mL} = 0.0162 \text{ mg}$$

0.0162 mg/0.03854 m^3 = 0.420 mg/m^3, or 16.8 times the TLV for elemental Hg and its inorganic forms (as mercury). This high concentration may be used for calibration of direct-reading mercury vapor detectors. Larger boxes, with smaller injection volumes, and/or serial vapor-phase dilution techniques with clean air can all be used to obtain lower mercury vapor concentrations.

626. How much anhydrous analytical regent grade mercuric sulfate must be dissolved in 10% reagent grade sulfuric acid and diluted to 100 mL with mercury-free water to obtain a stock solution of 100.00 mg/100 mL = 100 mg/ 0.1 L = 1000 ppm$_{m/v}$?

$$\text{Molecular weight } HgSO_4 = 296.65 \text{ grams/gram-mole}$$

$$\text{Molecular weight } Hg = 200.59 \text{ grams/gram-mole}$$

$$(200.59/296.65) \times 100 = 67.6184\% \text{ Hg}$$

$$100/67.6184 = 1.478887$$

$$147.8887 \text{ mg } HgSO_4 \times 0.676184 = 100.0 \text{ mg Hg}$$

Weigh 147.8887 mg of mercuric sulfate. Carefully transfer to a 100 mL volumetric flask containing about 30 mL 10% sulfuric acid. Thoroughly rinse the weighing pan with 10% sulfuric acid collecting rinsings through a glass funnel into the volumetric flask. Dilute volumetric flask to the "to contain" mark with 10% sulfuric acid. Insert stop tightly. Label and store in a dark cabinet. This stock solution contains 1.000 mg Hg/mL (1.000 ppm$_w$). Ensure dry $HgSO_4$ before weighing. This problem is a refresher because industrial hygienists can also be industrial hygiene analytical chemists and prepare standard stock solutions for instrument calibration.

627. An industrial hygienist wants to demonstrate how minutely small the TWA TLV® for beryllium is to worker's health and safety training class she is teaching (0.00005 mg/m³). To help make this concept most meaningful, she chooses a building with the footprint of a football field excluding the end zones (165' × 300') with a ceiling 18' high—a standard interior height for most single-story industrial buildings. She claims the building is tightly sealed, is empty, and has no ventilation. She poses a calculation based on Problem 488 wherein we calculated mass of a single grain of table salt at 0.0585 mg. Assuming these are grains of beryllium dust, how many— crushed to an ultrafine powder—would have to be dispersed in air of this building to equal the TLV concentration for Be?

$$165' \times 300' \times 18' = 891,000 \text{ ft}^3 = 22,230.31 \text{ m}^3$$

$$22,230.31 \text{ m}^3 \times 0.00005 \text{ mg/m}^3 = 1.1115 \text{ mg Be dispersed in this}$$
$$\text{huge building}$$

One grain of table salt (NaCl) weighs 0.0585 mg.

1.1115 mg/0.0585 mg/grain = exactly—and only—19 grains of table salt (Be).

Respective densities of beryllium and sodium chloride are 1.85 and 2.165 g/cc^3; therefore, the size of the beryllium granules are only 2.165/1.85 = 1.17 times larger than the table salt crystals.

628. A cook at an automobile assembly plant files a grievance with his union claiming the kitchen atmosphere is heavy with "fat and calorie vapors." He believes this is the sole cause of his 14.6 pounds weight gain in 14-½ weeks. He asserts that his diet has not changed over this interval, and his exercise patterns are the same as before employment. Before launching an expensive kitchen ventilation study and an air-sampling regimen, see what can be done mathematically to verify or reject his claim. In fairness to him, do not pre-judge and assume claim is preposterous.

Assuming his scale is accurate, and he correctly reads the dial, one can aver that weight gain must be weight$_{in}$. The body does not "make" mass. Weight$_{in}$ can only be food, beverages, nonnutritional ingested material, and, as claimed, retained air contaminants. The cook denies pica.

Make very liberal assumptions: 9-hour workdays, 6 days/week, inhales 30 m^3 of air per work day (8–12 m^3 is typical for most people), breathing zone air contains 1000 mg of pure fat mist and vapor per cubic meter, 100% absorbtion of fat by his respiratory tract into his body; no catabolism of lipids, and 100% storage in his fat and adipose tissue depots over 14-½ weeks.

$$6 \text{ days/week} \times 14.5 \text{ weeks} = 87 \text{ work days}$$

$$9/8 = 1.125 = \text{the adjustment for work days exceeding 8 hours/day}$$

$$87 \text{ work days} \times 1.125 = 97.875 \text{ equivalent 8-hour work days}$$

$$30 \text{ m}^3/\text{work day} \times 97.875 \text{ work days} = 2936.25 \text{ m}^3$$

2,936.25 m^3 × 1000 mg/m^3 = 2,936,250 mg = 2936.25 g = 6.473 lb < 14.6 pounds—only 44.3% of his claimed weight gain. Under these very extreme assumptions, his claim has no merit and is baseless. Bury the grievance.

Believing this cook, he should be examined by a physician and, perhaps, receive nutrition, diet, and exercise counseling if underlying metabolic, endocrine, renal (water retention), cardiovascular, and gastrointestinal disorders are fully ruled out. Investigate electrolytes and salt intake. Regardless, he must not sample soup or desserts on the job. The assistant cook, "Skinny," must do the taste testing.

What was his most likely weight gain if more realistic values are selected?

$$5 \text{ days/week} \times 14.5 \text{ weeks} = 72.5 \text{ work days}$$

$$10 \text{ m}^3/\text{work day} \times 72.5 \text{ work days} = 725 \text{ m}^3$$

$$10 \text{ mg/m}^3 \text{ (estimated)} \times 70\% \text{ absorbed (est.)} \times 50\% \text{ retained (estimated)} = 3.5 \text{ mg/m}^3$$

725 m^3 × 3.5 mg/m^3 = 2537.5 mg = 2.538 g = 0.0055 pound increase in weight

629. Antimony (Sb) is obtained by roasting, oxidizing its primary ore, stibnite (Sb_2S_3), in furnaces with hot air:

$$Sb_2S_3 + 4O_2 \rightarrow 2SbO + 3SO_2$$

   The antimony oxide is reduced with NaCl and scrap iron to obtain antimony or by reducing it with carbon. Sb can also be obtained by reacting stibnite with a strong acid or a strong base. A chemical engineer and air pollution control engineer who work for the same mining company have excess 30% w/v sulfuric acid in storage tanks awaiting hazardous waste disposal. They are considering using acidification to extract antimony sulfate from stibnite. A key concern, however, is the process generates the more toxic hydrogen sulfide instead of the less toxic sulfur dioxide. Regardless, the H_2S can be burned to SO_2. How much H_2S will be produced from each ton of stibnite so treated with excess sulfuric acid?

   Assume stoichiometric conversion. Reaction times, operating temperatures, and feed rates can be determined later.

$$Sb_2S_3 + 3H_2SO_4 \; Sb_2(SO_4)_3 + 3H_2S$$

$$\text{Molecular weight } Sb_2S_3 = 339.68 \text{ grams/mole}$$

$$\text{Molecular weight } H_2S = 34.08 \text{ grams/mole}$$

$$1 \text{ ton} = 907{,}184.74 \text{ grams}$$

$$907{,}184.74 \text{ g}/339.68 \text{ g/gram-mole} = 2670.7 \text{ moles stibnite/ton of 100\% ore.}$$

   Therefore, since 3 moles of hydrogen sulfide gas are released for each mole of stibnite:

$$2670.7 \text{ moles} \times 3 = 8012.1 \text{ moles of } H_2S$$

$$8012 \text{ mole} \times 34.08 \text{ grams/mole} = 273{,}048.96 \text{ grams } H_2S$$

   $273{,}048.96 \text{ grams} = 0.301$ ton H_2S produced for each ton of stibnite treated with excess sulfuric acid. From this, the engineers can calculate the air requirements to burn (oxidize) the H_2S to SO_2.

630. From the previous problem (629), calculate the amount of oxygen (from air) that is needed to oxidize the hydrogen sulfide gas to sulfur dioxide gas.

$$H_2S + 1.5O_2 \rightarrow SO_2 + H_2O$$

$$0.301 \text{ ton } H_2S = 273{,}048.96 \text{ grams}$$

$$273{,}048.96 \text{ grams}/34.08 \text{ grams/mole} = 8012 \text{ moles of } H_2S$$

Therefore, $8012 \times 1.5 = 12{,}018$ moles of oxygen are required.

$12{,}018$ moles $\times 31.9988$ grams/mole $= 384{,}561.578$ grams of oxygen are required.

Knowing furnace H_2S gas generation rate, one can calculate the rate at which oxygen (air) must be fed to flare. An excess of air is required to ensure complete oxidation of hydrogen sulfide to sulfur dioxide. Stack sampling—at least initially—is needed to establish combustion operating parameters: stibnite ore feed rate, sulfuric acid concentrations, mixing rates and efficiency, solution temperatures, stack gas flow rates (temperature, pressure, water vapor content, etc.).

631. The OSHA *Inorganic Lead Standard* (29 *CFR* 1910.1025) stipulates calculations for overtime exposures. The maximum dose is a PEL of 400 mcg/m^3-hours. That is, 50 mcg Pb/m^3 for 8 hours, 40 mcg Pb/m^3 for 10 hours, and 33 mcg Pb/m^3 for 12 hours are all equivalent doses. Problems 60 and 101 use Brief and Scala's model to calculate maximum exposure limit for unusual work schedules. Their model, however, is not incorporated into air contaminant standards by OSHA in spite of its elegance. If Brief and Scala's model is applied to the OSHA lead standard, the PEL for a 12-hour work day is 25 mcg/m^3, not 33 mcg/m^3. A few additional OSHA standards besides lead require adjustments for overtime work. The OSHA *Air Contaminants* (29 *CFR* 1910.1000) does not specifically state adjustments must be made for overtime work exposures. This has confused some. Explain how the language in 29 *CFR* 1910.1000 can be applied to address overtime exposures and unusual work schedules.

29 *CFR* 1910.1000 gives an example of how to calculate 8-hour time-weighted average exposures. Consider a chemical with a *PEL* of 100 ppm_v. A worker was exposed to a gas at 200 ppm_v for 4 hours and 0 ppm_v for 4 hours. His 8 hour time-weighted average exposure is, obviously, 100 ppm_v. On a 10-hour day, he was exposed to 150 ppm_v for 5 hours and 50 ppm_v for 5 hours. What was his 8-hour time-weighted average exposure?

$$150 \text{ ppm}_v \times 5 \text{ hours} = 750 \text{ ppm}_v\text{-hours}$$

$$50 \text{ ppm}_v \times 5 \text{ hours} = 250 \text{ ppm}_v\text{-hours}$$

(750 ppm_v-hours $+ 250 \text{ ppm}_v$-hours)/8 hours $= 125 \text{ ppm}_v$, or 25% overexposed. Always divide by 8 hours—not by the 10 hours in this case.

The OSHA compliance officer and others—using this approach—can rightfully justify invoking calculations and OSHA regulations for overtime work schedules.

632. Crude oil is considered "sour" when it contains more than one grain of H_2S in 100 ft^3 of air bubbled through a specified amount of oil. If below one grain $H_2S/100$ ft^3, crude is considered "sweet." How many ppm_v is one grain of $H_2S/100$ ft^3 at NTP?

The equation to convert any amount of gas as mass/volume into ppm_v (at 0°C and 760 mm Hg atmospheric pressure) is

$$ppm_v = \frac{mg}{m^3} \times \frac{22.4 \text{ L/gram-mole}}{\text{molecular weight}} \times \frac{\text{absolute temperature}}{273.15 \text{ kelvin}}$$
$$\times \frac{760 \text{ mm Hg}}{\text{barometric pressure, mm Hg}}$$

$$1 \text{ grain} = 64.799 \text{ mg}$$

$$100 \text{ ft}^3 \text{ of gas} = 2.8317 \text{ m}^3$$

$$64.799 \text{ mg}/2.8317 \text{ m}^3 = 22.88 \text{ mg/m}^3$$

Molecular weight of $H_2S = 34.08$ grams/gram-mole

At 75°F and 760 mm Hg, the above equation is

$$ppm_v = \frac{(mg/m^3) \times 24.45 \text{ L/gram-mole}}{\text{molecular weight}} = \frac{22.88 \text{ mg/m}^3 \times 24.45}{34.08} = 16.4$$

Therefore, crude oil gas at more than 16.4 ppm_v H_2S at 75°F and 760 mm Hg is deemed "sour" crude oil. Also, when the total sulfur level in oil exceeds $0.5\%_{w/v}$, the oil is considered "crude." The hydrogen sulfide in crude oil can be reacted with organic amines with subsequent recovery of the commercially valuable sulfur.

633. "Sour gas" is natural gas or other gas containing significant amounts of hydrogen sulfide (H_2S). Natural gas is typically considered "sour" if there are more than 5.7 milligrams of H_2S per cubic meter of natural gas. What is this in ppm_v?

$$ppm_v = \frac{(5.7 \text{ mg/m}^3) \times 24.45 \text{ L/gram-mole}}{34.08 \text{ grams/gram-mole}} = 4.089 \text{ ppm}_v \, H_2S$$

Practically, "sour gas," by definition, contains over four parts hydrogen sulfide gas by volume per million parts of gas. One can correct this calculation to natural gas instead of air.

634. A 14-inch diameter duct has an average duct velocity of 4500 fpm. What is the velocity of the discharged air 30 duct diameters away?
    A "rule of thumb" for discharge ducts and pressure jet openings is that the velocity is 10% of the duct velocity 30 duct diameters away.

$$14" \times 30 \, Øs = 420" = 35'$$

One could expect an air velocity of 450 fpm 35' from the duct discharge if there are no cross drafts—that is, air is discharged into still air. It is often wrongly assumed the concentration of air contaminants in the duct is diluted to ten percent 30 duct diameters away. The concentration is only slightly less than duct concentration. It just moves more slowly as it dissipates energy into the surrounding air.

635. An exhaust duct's air velocity is 2760 fpm. What is the duct velocity pressure?

$$VP_{duct} = (2760/4005)^2 = 0.475" \text{ water gauge}$$

636. A 4.5' diameter circular exhaust hood is 2 feet above a hot source 2 feet in diameter. Ambient air temperature is 72°F, and the temperature of the source is 190°F. What is the induced air flow due to the rising current of warm air?

This is considered a low circular canopy hood. As long as the distance between the hot source and the plane of the hood is less than 3 feet, the diameter of the warm air column is approximately the equivalent diameter of the source. The diameter of the canopy hood must be at least one foot larger than the diameter of the source. The following equation is used for such circular canopy hoods:

$$Q_t = 4.7(D_f)^{2.33}(\Delta t)^{0.42},$$

where
   Q_t = food air flow, cfm
   D_f = hood diameter, feet
   Δt = the difference in temperature between the hot source and ambient air, °F

$$Q_t = 4.7 \ (4.5')^{2.33} \ (190°F - 72°F)^{0.42} = 1160 \text{ cfm}$$

This air flow rate into the low circular canopy hood assumes no disruptive cross drafts.

A similar equation can be used for low rectangular hoods with same stipulations as for low circular canopy hoods:

$$Q_t/L = 6.2 \ W^{1.33} \ \Delta t^{0.42},$$

where
   L = length of the rectangular hood, feet
   W = width of the rectangular hood, feet

Both types of hoods are effective to control humidity from water vapor from steamy processes. Canopy hoods should not be used for high toxicity air contaminants.

637. Side wall and ceiling air supply grilles should be located approximately how many feet above the floor?
   a. Eight feet
   b. 10 feet
   c. 12 feet

  d. 18 feet
  e. Four feet above a shortest standing person's head

  Answer: b. All should have a convenient manual louver adjustment method (pull chains or pull rods so that supply air can be directed away from occupants during the winter in cold weather climates).

638. In a local exhaust hood with slots, maximum plenum velocity should be no more than:
  a. 10% slot velocity
  b. 25% slot velocity
  c. 50% slot velocity
  d. 80% slot velocity
  e. 120% slot velocity

  Answer: c. To ensure uniform air flow through all slots.

639. As a "rule of thumb" to ensure no reentry of stack exhaust air contaminants, the stack height should be at least how high?
  a. 20% greater than building height
  b. 50% greater than building height
  c. 80% greater than building height
  d. Equal to building height
  e. 250% building height

  Answer: b. Architects and building owners often argue against high stacks they consider unsightly. The "effective stack height" can be achieved by increasing discharge velocity through a narrower cone stack head and/or inducing additional ambient air flow into the stack with a roof-mounted fan injecting ambient air into base of the stack. Be mindful, however, excessively high discharge velocities can be noisy, resulting in community complaints particularly at night.

640. Benzene is evaporated into room air at the rate of 0.1 pint during every 8-hour work shift. What is recommended dilution air volumetric rate to control exposures to benzene vapor at no more than 10% of the NIOSH REL of 0.1 ppm_v?

  Benzene is far too toxic to rely on dilution ventilation to protect the health of exposed workers. Mechanical local exhaust ventilation at point of benzene vapor release must be used if benzene cannot be eliminated and substitution with a safer solvent and/or different work practices cannot be achieved. As a genotoxic carcinogen, there is no "safe" level of exposure to this chemical with its wild card behavior. All one can say is lower is safer.

641. Industrial hygienists respond to complaints of "stuffy" air. What explains "stuffy"? D. Jeff Burton, editor, *Hemeon's Plant and Process Ventilation, 3rd Edition* (Lewis Publishers, 1999) describes this nebulous term to assist the practitioner:

  > "*Stuffy*" may, in practice, be interpreted to mean any of the following conditions, or combinations of them:

1. The air temperature may be above the level of comfort, with low air velocities (e.g., $T > 76°F$; $V < 25$ fpm).
2. Radiant heat, perhaps of low intensity, impinges on parts of the body, especially the face, together with a low degree of air motion. (Author's comment: Increasing air volume and velocity will not blow radiant heat away. This can, however, provide some degree of comfort when the relative humidity is low.)
3. An uncomfortably elevated relative humidity may prevail (e.g., >60–70% RH).
4. A concentration of some airborne substance (e.g., body odor, tobacco smoke, chemical gas, or vapor) causes unpleasant or irritating odors to pervade the atmosphere).

Some people, lectures, sermons, and conversations are "stuffy," but these do not reside in the province of industrial hygiene.

642. 0.7 pint of an organic solvent mixture entirely evaporates within 8 hours into the air of a small work room with dimensions of 12' × 20' × 9'. The solvent contains benzene at 750 ppm$_v$. The work room is empty and has no fresh air exchange ventilation. What is the benzene vapor concentration in room air after 8 hours have elapsed?

$$750 \text{ ppm}_v = 0.075\%_v$$

$$0.075\%_v \times 0.7 \text{ pint} = 0.0525 \text{ pint} = 24.842 \text{ mL}$$

$$\text{Benzene density} = 0.8765 \text{ g/mL}$$

$$24.842 \text{ mL} \times 0.8765 \text{ g/mL} = 21.774 \text{ g} = 21,774 \text{ mg}$$

$$12' \times 20' \times 9' = 2160 \text{ ft}^3 = 61.164 \text{ m}^3$$

$$\text{ppm}_v = \frac{(\text{mg/m}^3) \times 24.45}{\text{molecular weight}}$$

$$= \frac{(21,174 \text{ mg}/61.124 \text{ m}^3) \times 24.45 \text{ L/gram-mole}}{78.11 \text{ grams/gram-mole}} = 108.4 \text{ ppm}_v$$

643. Of concern besides the inhalation toxicity of benzene vapor in the preceding problem (642), are fire and explosion hazards by other remaining organic solvent vapors. If the balance of the mixture is toluene, what fire risk management issues exist?

Practically, for fire risk assessment, we may assume the entire 0.7 pint is toluene.

$$\text{Toluene LEL} = 3.3\% \text{ by volume in air (33,000 ppm}_v)$$

$$\text{Upper explosive limit (UEL)} = 19\% \text{ by volume in air}$$

$$\text{Flash point} = 45°F$$

$$\text{Density} = 0.86 \text{ g/mL at } 68°F$$

$$\text{Molecular weight of toluene} = 92.14 \text{ grams/gram-mole}$$

$$0.7 \text{ pint} = 331.224 \text{ mL}$$

$$331.224 \text{ mL} \times 0.86 \text{ g/mL} = 284.853 \text{ g} = 284,853 \text{ mg}$$

$$\text{ppm}_v = \frac{(\text{mg/m}^3) \times 24.45}{\text{molecular weight}}$$

$$= \frac{(284,853 \text{ mg}/61.124 \text{ m}^3) \times 24.45 \text{ L/gram-mole}}{92.14 \text{ grams/gram-mole}} = 1236.6 \text{ ppm}_v$$

Fortunately, the maximum toluene (plus benzene) vapor concentration is below the LEL of 33,000 ppm$_v$. However, if evaporation resulted from a spill, the vapor immediately above the liquid would exceed the UEL and then LEL within a slight distance. A source of ignition at that point will result in vapor flash fire with a most probable ignition of the remaining liquid.

644. One pint of warm toluene containing 0.085% benzene by volume evaporates every 4 hours into workplace air. Although dilution ventilation is entirely unacceptable to control air toxics such as benzene, what ventilation is required to limit workers' exposures to 10% of the NIOSH REL for benzene? Assume an incomplete air mixing factor of 5 [scale of 1 (perfect) to 10 (abysmal)].

$$10\% \text{ REL} = 0.1 \text{ ppm}_v \times 0.1 = 0.01 \text{ ppm}_v$$

$$\text{mg/m}^3 = \frac{\text{ppm}_v \times \text{molecular weight}}{24.45 \text{ L/gram-mole}}$$

$$= \frac{0.01 \text{ ppm}_v \times 78.11 \text{ g/gram-mole}}{24.45 \text{ L/gram-mole}} = 0.03195 \text{ mg/m}^3$$

$$1 \text{ pint} = 473.176 \text{ mL}$$

$$473.176 \text{ mL} \times 0.085 = 40.22 \text{ mL of benzene}$$

$$\text{Benzene density} = 0.8765 \text{ g/mL}$$

$$40.22 \text{ mL} \times 0.8765 \text{ g/mL} = 35.2528 \text{ g} = 35,252.83 \text{ mg benzene evaporate}$$
during the 4 hours (240 minutes)

$$35,252.83 \text{ mg}/240 \text{ minutes} = 146.89 \text{ mg/minute}$$

$$V_{required} = 5 \left[\frac{146.89 \text{ mg/minute}}{0.03195 \text{ mg/m}^3} \right] \times 35.315 \text{ ft}^3/\text{m}^3 = 811,803 \text{ cfm!}$$

The purpose of this problem is to demonstrate the absurdity of attempting to use general dilution air to control highly toxic air contaminants. The energy costs are prohibitive, and the wind storm created by the large volume of dilution air would be disruptive. Moreover, controlling workers' exposures to toluene vapor remains.

645. A room is 8' × 10' × 16'. A 30 cfm exhaust fan is installed in a wall adjacent to the building's 16' axis. A make-up air louver is installed in the opposite wall. Benzene evaporation from a warm source is uniform at 0.01 pint/hour. How long will it take for the benzene vapor concentration to reach 0.05, 0.1, and 10 ppm_v? Assume 0.001 ppm_v in make-up air. Further assume the room is empty but for the solvent, and air mixing is excellent—such as it is. Before we proceed, ventilation by dilution is never acceptable because of benzene's high inhalation toxicity both acutely and chronically. But we will move on to demonstrate the calculation process and the learning points.

$$\text{Room floor} = 10' \times 16' = 160 \text{ ft}^2$$

$$30 \text{ cfm}/160 \text{ ft}^2 = 0.1875 \text{ ft}^3/\text{ft}^2 \text{ of floor area}$$

$$V = \text{volume of room} = 8' \times 10' \times 16' = 1280 \text{ ft}^3$$

$$Q = \text{ventilation rate} = 30 \text{ cfm}$$

$$30 \text{ cfm} \times 60 \text{ minutes/hour} = 1800 \text{ ft}^3/\text{hour}$$

$$\text{Air changes/hour (ach)} = (1800 \text{ ft}^3/\text{hour})/1280 \text{ ft}^3 = 1.406$$

Benzene's specific gravity is 0.8786 g/mL, and its molecular weight is 78.11 grams per gram-mole. NIOSH's REL for benzene is 0.1 ppm_v as an 8-hour time-weighted average.

$$\text{Effective ventilation rate (evr)} = \frac{(403)(10^6)(0.8786)(0.01/60)}{(78.11)(0.1 \text{ ppm}_v)} = 7555 \text{ cfm}$$

This evr controls only NIOSH's REL. Mechanical local exhaust ventilation must be used to control below the action level of 0.05 ppm_v.

The room's cross-sectional area is 8' × 10' = 80 ft². 7555 cfm/80 ft² = 94 fpm—a comfortable air velocity for most people.

$$\Delta t = -\frac{V}{\text{evr}} \left[\ln \left(\frac{G - \text{evr} \times C_2}{G - \text{evr} \times C_1} \right) \right] = -\frac{1280 \text{ ft}^3}{7555 \text{ cfm}} \left[\ln \left(\frac{0.03 - 7555 \times 0.1 \text{ ppm}_v}{0.03 - 7555 \times 0.001 \text{ ppm}_v} \right) \right]$$

$$= 0.78 \text{ minutes} = 49 \text{ seconds}$$

$$\Delta t = -\frac{V}{\text{evr}} \left[\ln \left(\frac{G - \text{evr} \times C_2}{G - \text{evr} \times C_1} \right) \right] = -\frac{1280 \text{ ft}^3}{7555 \text{ cfm}} \left[\ln \left(\frac{0.03 - 7555 \times 1 \text{ ppm}_v}{0.03 - 7555 \times 0.001 \text{ ppm}_v} \right) \right]$$

$$= 1.67 \text{ minutes}$$

0.1 ppm$_v$ benzene vapor after only 49 seconds, and 10 ppm$_v$ after 1.67 minutes. These calculations help to demonstrate the foolishness of attempting to control exposures to high toxicity air contaminants with dilution ventilation.

646. The acute 4-hour inhalation LC$_{50}$ for diacetyl in rats is 2.25 mg/L. Diacetyl's vapor pressure is 52.2 mm Hg, and its molecular weight is 86.09 grams per gram-mole. What is this in ppm$_v$? Compare this to saturation concentration at NTP to determine the vapor hazard ratio. For a discussion of vapor hazard ratios, refer to Problem 416.

$$
\begin{array}{c}
O \qquad\quad CH_3 \\
\backslash\backslash \quad\ / \\
C-C \\
/ \quad\ \backslash\backslash \\
H_3C \qquad O
\end{array}
$$

$$2.25 \text{ mg/L} = 2250 \text{ mg/m}^3$$

$$
\text{ppm}_v = \frac{(2250 \text{ mg/m}^3) \times 24.45 \text{ L/gram-mole}}{86.09 \text{ grams/gram-mole}} = 639 \text{ ppm}_v
$$

A 2001 NIOSH diacetyl inhalation study using rats demonstrated profound injuries in the respiratory tract at 285–371 ppm$_v$ after 4 hours. The lead investigator said these were "... the most dramatic cases of cell death I've ever seen."

Vapor saturation concentration in a sealed container = [52.2 mm Hg/ (760 mm Hg + 52.2 mm Hg)] $\times$ 10^6 = 64,270 ppm$_v$.

VP$_{sat}$/ppm$_v$ = 64,270 ppm$_v$/639 ppm$_v$ = 100.6, a very high vapor hazard ratio. Note that the denominator is typically the TLV or the lowest REL. However, since a limit for the worst acceptable inhalation level has not been established, the author selected LC$_{50}$. The vapor hazard ratio will be substantially greater after an exposure limit is developed because of the high inhalation toxicity.

For reference, the LC$_{50}$ for chlorine in cats and rabbits is 660 ppm$_v$ for 4 hours. Four-hour LC$_{50}$s for ozone and bromine are 10.5 ppm$_v$ (hamsters) and 750 ppm$_v$ (mice), respectively.

Diacetyl is an artificial butter flavoring agent that was used for microwave popcorn and other flavored food products such as ice cream and baked goods. Inhalation of vapors by workers at low concentrations caused bronchiolitis obliterans. These severe lung injuries can be fatal if a total lung transplant is impossible. There was a retail customer who developed bronchiolitis obliterans after eating large amounts of microwaved popcorn daily and deliberately inhaling diacetyl vapors from freshly opened bags.

The slope of the inhalation dose–response curve is unknown. However, when 50% of the test rats died from inhaling only 639 ppm$_v$ after only a 4-hour exposure, whatever permissible exposure level must be promulgated will be very low. With assumption that human response is similar to the pathophysiology in rats, we can expect an inhalation limit far below the LC$_1$ in rats to apply for exposed workers.

The combination of high vapor pressure and high inhalation toxicity requires every workplace exposure be exquisitely controlled by tight enclosures, mechanical local exhaust ventilation, rigorous worker education and training, and full-face airline respirators for spills and emergency response. Routine workplace surveillance and medical surveillance are essential. Openly pouring liquid diacetyl and transfer of dusty, powdery materials containing diacetyl outside of an enclosing hood with mechanical local exhaust ventilation must be barred because a worker wearing a full-face airline respirator pouring diacetyl from one container to another in the open puts workers far away at risk from inhaling fugitive, migrating diacetyl vapors. Transfer of dusty, powdery materials containing diacetyl must be done with utmost caution by ensuring airborne dust is captured by local exhaust hood. Bonding and grounding electrostatic charge accumulation practices must be enforced.

647. Compare vapor hazard ratio of benzene in Denver to Atlantic City. Assume 25°C for both cities.

The vapor hazard ratio is more precisely defined by the equation:

$$(P_{vapor} \times 10^6)/(EL \times BP),$$

where
P_{vapor} = vapor pressure of chemical at 25°C in mm Hg
EL = exposure limit, ppm_v. Apply the lowest recommended limit. In this case use 0.1 ppm_v recommended by NIOSH as 8-hour TWAE limit for benzene. In certain circumstances, the IDLH can be applied (Immediately Dangerous to Life and Health, with "immediate" typically being within 30 minutes).
BP = barometric pressure in mm Hg for the location

Denver's altitude is 5283' equivalent to 12.09 psia = 625.2 mm Hg.

Atlantic City's altitude = 7 feet equivalent to 14.69 psia = 759.7 mm Hg.

Benzene's vapor pressure at 25°C is 95.2 mm Hg.

For Denver:

$$VHR = 95.2 \times 10^6/0.1 \times 625.2 = 152,271 \text{ (dimensionless)}$$

For Atlantic City:

$$VHR = 95.2 \times 10^6/0.1 \times 759.7 = 125,313 \text{ (dimensionless)}$$

All other things being equal, we see VHR is greater in Denver than in Atlantic City by a factor of 152,271/125,313 = 1.215, or 21.5% greater. This is not surprising because the lower atmospheric pressure in Denver promotes faster evaporation of liquids and volatile solids.

This can be applied as well to preventing heat strain at higher altitudes. All other things being equal, an "average" heat-strained worker in Denver will require 21.5% more hydration per unit time than a comparable worker at sea level. Dehydration occurs more quickly at high elevations. Moreover, with thinner atmosphere, more infrared radiation penetrates at high elevations enhancing heat gain of the body.

More easily, we can simply compare each city's respective atmospheric pressure: 759.7/625.2 = 1.215. Such is helpful if extrapolating a vapor hazard ratio or index, from one altitude to another.

648. The methyl ethyl ketone vapor concentration vapor in an empty warehouse with interior dimensions of 42' × 36' × 14' is 378 ppm_v at 3:10 p.m. All supply air and exhaust fans are turned on at this time. At 4:40 p.m., MEK vapor concentration was 119 ppm_v. What is the rate of air exchange in this building? Assume 15% of warehouse's interior volume is occupied by objects. Assume excellent mixing of the outside air with the contaminated air.

$(42' \times 36' \times 14') \times 0.85 = 17{,}992.8$ ft³, the net volume of the building

$$\text{ach} \times (t_e - t_o) = \ln C_o - \ln C_e,$$

where

ach = air changes per hour
C_o = original concentration at time, t_o
C_e = concentration after elapsed time, t_e

Rearranging:

$$\text{ach} = \frac{\ln C_o - \ln C_e}{t_e - t_o} = \frac{\ln 378\ ppm_v - \ln 119\ ppm_v}{4{:}40 - 3{:}10 = 1.33\ \text{hours}}$$

$$= \frac{5.935 - 4.779}{1.33} = 0.869\ \text{ach}$$

649. Polychlorinated biphenyls (PCB) are a class of chemical compounds based upon various degrees of chlorination of biphenyl. The base molecular structure of PCB is

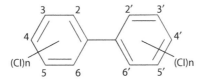

Anywhere between 1 and 10 chlorine atoms can attach to the indicated positions on the two biphenyl rings. For example, chlorine atoms attaching to form 2,4,3',5',6'-pentachlorobiphenyl is one of the 209 possible congeners of PCB. Arochlor® is one brand name. PCBs have a high chemical stability and, therefore, a very long environmental half-life. PCBs are highly soluble in lipids, oils, and fats. Therefore, they tend to be stored in fatty tissues where they persist. Their combined toxicity and long environmental and biological half-lives place them high on EPA's toxicant watch list. PCB production was banned in 1978. They are carefully regulated by EPA under the Toxic Substances Control Act and by the OSHA. Materials containing 50 ppm_m PCB or greater must be treated as hazardous waste per the EPA's Toxic Substances Control Act.

As the degree of chlorination increases, toxicity increases, and water solubility and vapor pressure decreases. For example, Arochlor® 1016 has a water solubility of 0.42 mg/L at 25°C, whereas Arochlor 1260's water solubility is 0.0027 mg/L at 25°C. Their respective vapor pressures are 4.0×10^{-4} and 4.05×10^{-5} mm Hg, respectively.

PCBs were used primarily as insulating fluids in electrical transformers, capacitors, and other electrical equipment and present inhalation and skin absorption risks to electricians and other workers who encounter them. PCBs exposed to fires can release even more toxic chemicals such as dioxins and various chlorinated furans:

TCDD

A PCB-containing cable leaked in a tunnel at an electricity generator's sub-station. The tunnel was not ventilated, and PCB had accumulated over months. Assume an average molecular mass of 341 grams/mole and an average vapor pressure of $4 \times 10^{-4.5}$ mm Hg, what is the PCB saturation in ppm_v?

Assume the sub-station is in New York City and has no water on the floor.

$$[(4 \times 10^{-4.5} \text{ mm Hg})/760 \text{ mm Hg}] \times 10^6 = 0.1664 \text{ ppm}_v$$

$$mg/m^3 = \frac{ppm_v \times \text{molecular weight}}{24.45 \text{ L/gram-mole}}$$

$$= \frac{0.1664 \text{ ppm}_v \times 341 \text{grams/gram-mole}}{24.45 \text{ L/gram-mole}} = 2.33$$

2.33 mg/m³ greatly exceeds NIOSH's REL of 0.001 mg/m³ as 8-hour time-weighted average exposure (2330). Moreover, PCB are absorbed through intact skin. Target organs are skin, eyes, liver, and reproductive system. The nursing infant at top of the human food chain is at great risk of future adverse health effects. As a probable occupational carcinogen, PCB cancer sites are the pituitary gland, liver, and bone marrow (leukemia) based on animal studies.

650. The OSHA defines an oxygen-deficient atmosphere as one with the oxygen level below 19.5% by volume (29 *CFR* 1910.134, *Respiratory Protection*). What is the partial pressure of O_2 at sea level? What is partial pressure of O_2 at a worksite at 12,000 feet altitude in the American Rocky Mountains above Denver (e.g., Loveland Pass on Interstate 70 at 11,990 feet altitude)? Compare these.

At sea level: 760 mm Hg $\times$ 0.20946 = 159.1896 mm Hg, the partial pressure of O_2

At sea level: 760 mm Hg $\times$ 0.195 = 148.2 mm Hg, the OSHA lower limit for an O_2 sufficient (not deficient) atmosphere

From physics and chemistry reference books, at 12,000 feet altitude the standard barometric pressure is 496 mm Hg. So, at 12,000 feet:

$$496 \text{ mm Hg} \times 0.20946 = 103.892 \text{ mm Hg}$$

$$(103.892/159.1896) \times 100 = 65.26\%$$

That is, at 12,000 ft altitude, the oxygen concentration is reduced $100 - 65.26 = 34.74\%$. Simply being there places one in an OSHA oxygen-deficient atmosphere if one is not acclimatized over several weeks to produce more erythrocytes. One can easily imagine a highway work crew from Los Angeles winning a contract to do repair work at Loveland, Colorado where Interstate 70 crosses Continental Divide and several workers doing heavy lifting collapsing without understanding why. The OSHA 19.5% O_2 "rule" is good only up to about 5000 ft. One must take special provisions beyond this elevation.

Using the volume percent of oxygen can be confusing. In the author's view, the partial pressure of oxygen in the inhaled air is more meaningful to assess risks of inhaling oxygen deficient atmospheres.

Otherwise, healthy workers who are not acclimated to working at this altitude are at great risk of O_2 insufficiency. Those with cardiac ischemia are also at risk of heart attack and death. Special industrial hygiene precautions are necessary. With an acclimatization to working at high altitudes, workers, to some extent, produce more erythrocytes with an increased hemoglobin content over several weeks. However, with this, blood is denser further compromising a weak heart. Medical screening of those with angina, frequent rest breaks, controlling heat strain, careful supervisory oversight, supplemental oxygen-supplying masks, and education and training are important components of the industrial hygiene program. Carbon monoxide, as a chemical asphyxiant gas, must be monitored regularly if any combustion process is nearby. Biomonitoring using digital oximetry should be considered.

651. A large building's ventilation maintenance engineer checks a main exhaust duct by cutting a 3" $\times$ 3" hole through the sheet metal. He measures the static pressure a few inches downstream of this hole at—3.5" water gauge. He forgets to patch and seal the hole after his inspection. How much air leaks through the

hole and shunts from the design exhaust point? The hole may be considered a sharp edge orifice with a coefficient of entry of approximately 0.72.

$$Q, \text{cfm} = 4005 \times \text{ft}^2 \times C_e \sqrt{SP, "H_2O}$$

$$= 4005 \times \frac{3" \times 3"}{144 \text{ in}^2/\text{ft}^2} \times 0.72 \times \sqrt{3.5"} = 337 \text{ cfm}$$

This is energy wasteful and reduces exhaust control volume where it is required. If the hole is reduced to 2" × 2", the volumetric flow rate through the hole is 150 cfm. Standard practices should be to make smallest hole possible using an inspection mirror and flashlight and carefully sealing the hole with gasket adhesive and duct tape after inspecting.

652. Refer to Problem 644 where we calculated that 811,803 cfm as dilution ventilation are required to dilute benzene vapor to acceptable levels—a practice that is highly unacceptable but performed here to demonstrate its unreasonableness. Consider a worker exposed to this vapor is at a fixed work location 7' high and 8' wide such as a work bench. What would be his cross-draft air velocity?

$$811,803 \text{ cfm}/(7' \times 8') = 14,496 \text{ fpm} = 164.7 \text{ mph!}$$

A category 5 hurricane is defined as one with sustained wind exceeding 155 mph. Such air velocity results in catastrophic damages. This worker certainly could not stand upright in such a wind speed.

653. A machinist worked 3 hours out of his 8-hour workday at a bench removing lapping compound from machinery valves. He used recycled mineral spirits that contained 0.0015 vol. % (15 ppm$_v$) benzene. This is hand intensive work to remove dried adherent and caked lapping compound at this open surface tank. What was his TWAE during actually cleaning dirty metal parts? What was his most likely 8-hour TWAE to benzene vapor?

Hemeon's Plant and Process Ventilation, 3rd Edition (D. Jeff Burton, editor, Lewis Publishers, 1999) provides the following table useful for industrial hygienists:

Organic Solvent Application Rates in Certain Typical Industrial Operations (Table 10-1, p. 187)

Work Operation	Pints/Minute/Worker
Manual, small brushing, cementing, "fussy" work	0.02–0.03
Manual, large-brush operations ("daubing")	0.2
Manual, gross applications, maximum use rate by hand	0.75–1.5
Mechanical coating operation	0.33–2
Spray finishing machinery	0.25–0.5

This table of experimental data assists industrial hygienists and product engineers to design appropriate control methods to protect the health of workers exposed to the emitted solvent vapors before the solvent cleaning process equipment becomes functional and is placed into commercial production for retail customers.

One can make reasonable assumptions to calculate the worker's most likely vapor exposure during the specific work period and his 8-hour average exposure:

• Select 0.1 pint/minute from the above table based on personal experience and description of work practices.
• Use 2 feet as the typical distance from open surface tank solvent vapor point source to the worker's breathing zone.
• Choose 20 feet/minute single-plane dilution ventilation based on the exposure authoritative reference (ibid) and personal experience.
• Vapor concentration decreases hemispherically from point source following inverse-square law and diffusion ventilation mass transfer kinetics. Mists that are most likely generated are not covered by this prediction model.
• The surface area of a hemisphere excluding the flat base is $2\pi r^2$ (r = radius).
• This plant is located near Detroit where the average atmospheric pressure is 740 mm Hg. 740/760 mm Hg = 0.97368 atmosphere.

3 hours/workday $\times$ 0.1 pint/minute = 180 minutes $\times$ 0.1 pint/minute = 18 pints of mineral spirits used during 3-hour cleaning operations

$$1 \text{ pint} = 473.176 \text{ mL}$$

$$18 \text{ pints} = 8517 \text{ mL}$$

8517 mL mineral spirits $\times$ 0.0015% = 12.78 mL benzene evaporated over a 3-hour cleaning period.

12.78 mL (0.430 fluid ounce = 2.58 teaspoons = 0.86 tablespoon). Moreover, the benzene is far more volatile than the mineral spirits.

We might ask what volume of benzene vapor is produced from the evaporation of 1 gallon of liquid benzene at 70°F and 740 mm Hg.

$$1 \text{ gallon} = 3785.412 \text{ mL}$$

$$3785.412 \text{ mL} \times 0.8765 \text{ g/mL} = 3317.9136 \text{ grams of benzene}$$

$$3317.9136 \text{ g}/78.112 \text{ g/mole} = 42.476 \text{ moles of benzene}$$

$$\text{Volume} = \frac{nRT}{P} = \frac{42.476 \text{ moles} \times 0.0821 \text{ L} - \text{atm/k} - \text{mole} \times 294.261 \text{ kelvin}}{740/760 \text{ mm Hg}}$$

$$= 1053.9047 \text{ L}$$

$$1053.9047 \text{ L} = 37.218 \text{ ft}^3$$

One gallon of liquid benzene at 70°F and 740 mm Hg atmospheric pressure yields 37.218 ft³ of benzene vapor after evaporation.

$$\frac{37.218 \text{ ft}^3}{3,785.412 \text{ mL}} = \frac{x \text{ ft}^3}{12.78 \text{ mL}}$$

$$x = 0.1257 \text{ ft}^3 \text{ of benzene vapor}$$

$$0.1257 \text{ ft}^3/180 \text{ minute} = 0.000698 \text{ ft}^3/\text{minute}$$

ppm_v near field (2') breathing zone benzene vapor exposure during parts cleaning

$$= \frac{2' \times 0.000698 \text{ ft}^3/\text{minute}}{20 \text{ fpm} \times 2\pi(2')^2} = 0.00000278 = 2.78 \text{ ppm}_v \text{ (ibid)}$$

Distance from Breathing Zone ($r = 2$ feet)	ppm_v Benzene Vapor $2(1/r^2)$
4 feet 2 r	1.4
6 feet 3 r	0.62
8 feet 4 r	0.35
10 feet 5 r	0.22
12 feet 6 r	0.15

Far-field breathing zone is reported at 8' to 12' from near field > 0.1 ppm_v.

$$8\text{-hour TWAE} = \frac{(2.78 \text{ppm}_v \times 3 \text{ hours}) + (\sim 0.1 \text{ppm}_v \times 5 \text{hours})}{8 \text{ hours}}$$

$$= \sim 1.1 \text{ppm}_v$$

The machinist performed this work for approximately $2\text{-}\frac{1}{2}$ years (5–6 days/week), so: cumulative, chronic dose $= 1.1$ $ppm_v \times 2.5$ years $= 2.75$ ppm_v-years, minimally.

The predicted exposure exceeds OSHA's PEL of 1.0 ppm_v and is 11 × the NIOSH REL (0.1 ppm_v 8-hour average) to prevent hematopoietic cancers in benzene-exposed workers. The worker's TWAE is more than twice the TLV.

This model is conservative because a "puff" exposure concentration which occurs briefly as the work station lid is opened is not included. Mist and airborne particles are not included in the model. Several similar work stations were located throughout the facility contributing to background benzene vapor levels. Moreover, benzene skin absorption contributes to worker's systemic dose. Regardless, there is sufficient health hazard assessment information from the model for one to design necessary ventilation engineering, work practice controls, and personal protective equipment.

Robust industrial hygiene control methods must be installed to conserve workers' health. Included in these are OSHA's *Hazard Communication*

Standard (29 *CFR* 1910.1200), *Open Surface Tank* local exhaust ventilation (29 *CFR* 1910.94, part d), and OSHA's *Benzene Standard* (29 *CFR*
1910.1028). The nature of this cleaning work requires careful selection of
protective equipment such as gloves and aprons because liquid benzene can
be absorbed through intact healthy skin and more so when skin is breached
by lacerations and nicks, sores, infections, dermatitis, and enhanced by the
presence of other skin defatting solvents. Chemical splash goggles and a
face shield are needed to protect against splashes. Use impervious aprons.

654. Stack exhaust gas (excluding particulates) from a combustion process was
11,670 ppm_v CO (1.167%$_v$). Water vapor was 28.3 vol. %. Concentrations
have been corrected to normal temperature and pressure. What was the
concentration of CO gas on a dry gas basis?
ppm_v, dry concentration = ppm_v, wet concentration/(1 – fraction wet concentration) = 11,670 ppm_v/(1 – 0.283) = 16,276 ppm_v CO.

655. What is the air exchange rate in a small building where homogenous concentration of *n*-heptane vapor is 76.8 ppm_v at 3:10 p.m. subsiding to 36.7 ppm_v
by 4:30 p.m.? Assume no further vapor generation and steady-state ventilation with good mixing.

$$ach = \frac{\ln C_o - \ln C_a}{t_a - t_o},$$

where
ach = air changes per hour
C_o = original concentration at time t_o
C_a = concentration after ventilation at elapsed time t_a

$\Delta t_a - t_o$ = 4:30 p.m. – 3:10 p.m. = 1 hour, 20 minutes = 1.33 hours

$$ach = \frac{\ln 76.8 - \ln 36.7}{1.33} = \frac{4.341 - 3.603}{1.33} = 0.55 \text{ ach}$$

656. Consider the previous problem (655). What is the projected air concentration of *n*-heptane vapor at 9:00 p.m.? Assume steady-state ventilation with
no further vapor generation and good mixing of solvent vapor-free air with
contaminated air.

$$C_{later} = C_o e^{(-ach \times t)}$$

where
C_{later} is the projected concentration after elapsed time t
C_o = original air concentration
ach = air changes per hour

elapsed time = 9:00 p.m. – 3:10 p.m. = 5 hours, 50 minutes = 5.833 hours

$$C_{later} = 76.8 \text{ ppm}_v \, e^{(-0.55 \times 5.833)} = 76.8 \text{ ppm}_v \, e^{-3.2082}$$
$$= 76.8 \text{ ppm}_v \times 0.0404 = 3.1 \text{ ppm}_v$$

657. The manager of a five-chair beauty parlor 64' long and 14' wide wants suggestions for outdoor air ventilation requirements to provide comfort and reasonable control of hair spray mists, perfumes, shampoo odors, powders, and so on. The five beauticians average 0.7 customer/h/each at this appointment-based salon. What do you review in your considerations, and what would you recommend to the manager?

A good starting point is the *Required Air Ventilation Rate per Unit Area from the International Code* (adopted from the American Society of Heating, Refrigeration, and Air-Conditioning Engineers Standard 62-2001:

ft³/minute/ft²	Occupancy Classification
0.05	Public corridors, retail or storage warehouse
0.10	Education corridors, retail florists, hardware, fabrics, supermarket
0.15	Library, retail shipping and receiving
0.20	Arcade or mall retail showroom
0.30	Basement or street-level retail showroom, clothier, furniture shop
0.50	Locker room, darkroom, duplicating or printing room
1.00	Commercial dry cleaner, conference room, elevators, pet shop
1.50	Automotive service station, repair garage, enclosed parking garage

Typically, that which is sought is not in the searched table or chart. So, judgment is required. The relatively small area has a high population density (0.7 customer/hour/beautician, five beauticians, waiting customers, others coming and going). 0.2 ft³/minute/ft² might be too low, whereas 0.50 ft³/minute/ft² is probably too high. 0.4 ft³/minute/ft² of floor area appears to be a reasonable compromise. However, the outdoor air supply fan and the shop's exhaust system should have reasonable flexibility to change the outdoor air volumetric flow rate as conditions present with, say, a three-speed fan or a variable air volume system (VAV).

Air should be supplied at about the 10' level so that it gently sweeps past waiting customers toward patrons and their beauticians. A balanced air exhaust plenum above the bench and behind where beauticians locate their tools and beauty aids would be ideal.

(0.4 ft³/minute/ft²) × 64' × 14' = 358.4 cfm. Arguably, one may subtract the area of mechanical room and lavatory from total area of shop because they have separate ventilation exhaust systems to control odors from the lavatory and capture CO from the combustion equipment (water heater and furnace). However, if the overall supply is applied for negative pressure to these rooms, use 358.4 cfm.

Provide temperature and humidity control of the mixed air. The lavatory and the furnace and water heater closet must have an independent air exhaust to prevent recirculation of CO and other tramp, fugitive exhaust gases. Design these spaces so they are under negative pressure with respect to the main parlor.

658. The author encountered a safety engineer who was not concerned when his meter measured 20.1 vol.% oxygen in a confined space work project. He said, "It's not below OSHA's 19.5% lower limit." We evacuated workers immediately. I explained he should be highly concerned at any oxygen measurement below 21% because (assuming meter is calibrated and properly functioning) 21% − 19.5% = 1.5% (15,000 ppm$_v$) of some gas and/or vapor which has intruded into the space. What is the gas/vapor? Will the intrusive leak increase? Could 1.5% exceed LEL of some gases and vapors? Be worried. Be very worried! Take all interventional steps immediately: evacuate workers, determine source(s), repair, ventilate, test.

But wait! Is it really 1.5% of an unknown gas or vapor intrusion? We are mistaken because (19.5%/21%) × 100 = 92.86%. 100% − 92.86% = 7.14% = 71,400 ppm$_v$. Our first mistaken calculation assumed we were just measuring "air." No, we were measuring oxygen. 71,400 ppm$_v$ of something mysterious is in this tank. Anybody strolling into this tank without a SCBA and an extraction harness might be taking their last saunter into never-never land.

659. Consider a previous problem (650). What is the equivalent percent oxygen in dry air at sea level?

From the American Conference of Governmental Industrial Hygienists' *Threshold Limit Values® for Chemical Substances and Physical Agents (2011), Appendix F: Minimal Oxygen Content:*

$$P_{\%O_2} = 20.948 \times e^{-(\text{altitude in feet}/25,970)} = 20.948 \times e^{-(12,000/25,970)}$$

$$= 13.2\% \text{ by volume}$$

Healthy workers would, sooner or later, experience abnormal fatigue with exertion, poor coordination, impaired judgment, emotional upset, and other symptoms and signs listed by ACGIH for lower partial pressure oxygen exposure levels. Workers with cardiac ischemia and a history of angina are at risk of a fatal heart attack.

660. An acetylene gas cylinder falls to the floor of a confined space rupturing its valve at an oxy-acetylene torch cutting operation. Workers evacuate. A subsequent air sample did not show acetylene gas concentrations above 10% of the LEL, but a mysterious analytical peak from the gas chromatograph suggested acetone later confirmed by GC-MS. Acetone was not used in the confined space tank. Explain.

Acetylene gas is dissolved under pressure into highly volatile liquid acetone in the gas cylinder. This helps to ensure steady, predictable release of acetylene gas at oxy-acetylene torch metal-cutting activities.

661. Industrial hygienists and chemical safety engineers regularly make decisions about the flammability and explosion potential of gases and organic solvent vapors. The following can assist them.

To determine the saturation concentration in air at 75°F, divide the vapor pressure of the organic solvent at this temperature by the barometric pressure and multiply by 10^6. Add vapor pressure to the atmospheric pressure if, for example, the vapor pressure is more than 20 mm Hg.

An example for toluene: $(21 \text{ mm Hg}/760 \text{ mm Hg}) \times 10^6 = 27{,}632 \text{ ppm}_v = 2.76_v\%$. Toluene vapor has a "built in match" because saturation concentration is between the LEL ($1.1_v\%$) and the UEL ($7.1_v\%$). All that is needed for an explosion is source of sufficiently sustained ignition of at least 8.82 eV.

An example for turpentine: $(4 \text{ mm Hg}/760 \text{ mm Hg}) \times 10^6 = 5263 \text{ ppm}_v = 0.53_v\%$. This is below the LEL of 0.8% for turpentine. One must not confuse explosion with combustibility. As we see with turpentine at room temperature and at nearly sea level barometric pressure, an explosive atmosphere will not form. However, this kerosene will obviously burn thereby heating the remaining liquid kerosene. Once the fire is extinguished, the warm kerosene's vapor pressure increases so that an explosive atmosphere can ensue with a subsequent ignition source.

Organic Solvent	Flashpoint (°F)	LEL (%)	UEL (%)	Evaporation Rate	Ignition Temperature (eV)	Vapor Pressure (mm Hg)
Diethyl ether	−49	1.9	36.0	1.0	9.53	440
Acetone	0	2.5	12.8	1.9	9.69	180
Toluene	40	1.1	7.1	4.5	8.82	21
Ethyl alcohol	55	3.3	19	7	10.47	44
2-Butoxyethanol	143	1.1	12.7	85	10.00	0.8
Turpentine	95	0.8	?	375	?	4
Gasoline[a]	−45	1.4	7.6	2.5	?	38–300
Carbon tetrachloride	None	None	None	2.6	Will not ignite	91

[a] Gasoline must never be used as a solvent, and none of the above or other organic solvents must be used as skin and clothing cleaners. Perchloroethylene is an exception in some commercial clothing dry cleaning establishments.

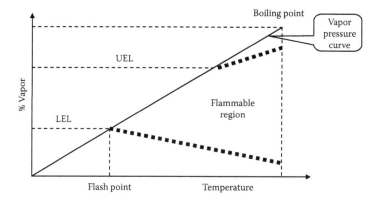

Note that chemicals have a range of LELs and UELs. This range expands as the liquid temperature increases because more energy is now in the liquid phase, and less energy is needed to shift a liquid-to-vapor-phase transition.

Also note how LEL decreases with increasing liquid temperature—nothing very surprising about this.

662. Half-lives (t) can be mathematically expressed several ways. Assuming first-order decay kinetics, the half-life can be helpful in toxicology, ventilation, and radioactive decay. State a few ways half-lives can be expressed.

$$\log C = \log C_o - \frac{k}{2.3}t$$

$$\frac{C_o}{2} = C_o\, e^{-kt}$$

Dividing both sides by C_o: $\quad e^{-kt} = \frac{1}{2}$

$$k = \frac{0.693}{t}$$

Consider the remaining concentration after, for example, nine half-lives:

$$C = C_o\, e^{-k \times 9t}$$

$$t = \frac{0.693}{k}$$

$$\frac{C}{C_o} = -\frac{1}{e^{kt \times 9}} = \frac{1}{e^{\ln 2 \times 9}} = \frac{1}{2^9} = \frac{1}{512}$$

663. An industrial hygienist studied solvent degreasing work stations where the workers cleaned dirty machinery parts with an organic solvent, A. Several breathing zone air samples revealed excessive exposures exceeding the TLV of 20 ppm$_v$. While mechanical local exhaust ventilation would be the best choice to control exposures, this little company was on the cusp of bankruptcy. He studied another solvent, B, with a TLV of 100 ppm$_v$ finding it was as effective as A for cleaning, however, it cost 10% more. Management was reluctant to purchase the costlier solvent. The parts cleaning time was the same for both solvents. The target organ for both A and B is the same (brain; CNS effects of dizziness, incoordination, and vertigo).

The industrial hygienist did another study by placing equal volumes of A and B in two small Pyrex® Petri dishes. Both dishes were placed side-by-side in a gentle air stream of 50 fpm. The liquid solvent and the air temperature were identical to this workplace, 70°F. After 4 minutes and 32 seconds, all of solvent A evaporated, whereas after 30 minutes and 12 seconds, all of solvent B evaporated. Using this information, he justified to the management to replace solvent A with solvent B.

4 minutes, 32 seconds = 272 seconds

30 minutes, 12 seconds = 1812 seconds

1812 seconds/272 seconds = 6.66

That is, evaporative losses of A would be about 6-⅔ times more than B. A costs $5.36/gallon, and B costs $5.90/gallon. Plant purchase records revealed that, on average, workers used 18.9 gallons of A daily, or a cost of $101.30 each day. The industrial hygienist's research demonstrated that only about 6.3 gallons of B would be used daily [(1 − 0.666) × 18.9 gallons$_A$)].

6.3 gallons$_B$/day × $5.90/gallon = $37.17/day

($101.30 − $37.17) × 230 work days/year = $14,749.90 saved annually

Subsequent air samples demonstrated that exposures were well below applicable TLV for B. Moreover, solvent A is skin absorbed but solvent B is not appreciably absorbed through intact, healthy skin. Special, pricey gloves were not needed.

Plan B might consider warm soapy water cleaning with ambient air drying.

664. For stoichiometric combustion of one cubic foot of natural gas (83–99% CH_4), how many cubic feet of exhaust gases are created?

One cubic foot of natural gas (CH_4) needs 9.6 cubic feet of air (21% O_2) for perfect (100%) stoichiometric combustion. The total gas volume, therefore is 10.6 cubic feet that expands to a larger volume when heated. Charles' law applies when we know initial mixed gas temperature and the exhaust gas temperature (absolute): $V_1/T_1 = V_2/T_2$. Assume both the initial natural gas and air temperature are 70°F, and the exhaust gas temperature is 1100°F. Further assume both air and natural gas contains no moisture.

70°F = 21.11°C

21.11°C + 273.16 = 294.27 kelvin

1100°F = 593.33°C   593.33°C + 273.16 = 866.49 kelvin

$V_2 = T_2V_1/T_1 = [(866.49 × (1\ ft^3 + 9.6\ ft^3)]/294.27 = 31.21\ ft^3$

10.6 ft³ of mixed gas expands (31.21 ft³/10.6 ft³) 2.94 times under these stipulated conditions. After the expansion, of course, the gas contains less energy releasing energy to the heating process.

665. Downdraft ventilation at a conveyor line where workers glued plastic parts using a solvent (methyl ethyl ketone) was adequate to control workers' exposures to MEK vapor. Glued parts dried in the open to which the local fire marshal objected. He insisted on installing a drying tunnel with sufficient ventilation to control MEK vapor to no more than 0.01% by volume

(100 ppm$_v$) inside the tunnel. This fire marshal is overly conservative because the LEL of MEK in air is 1.4$_v$% or 14,000 ppm$_v$. Regardless, management agreed. Calculate the volumetric rate of dilution air for the drying tunnel if purchasing records show that an average of 1.5 pints of MEK are used each hour by four employees gluing parts. Assume mixing factor of 2 inside the tunnel. The tunnel is 2' high × 2.5' wide × 24' long.

$$1.5 \text{ pints/hour/employee} \times 4 \text{ employees} = 6 \text{ pints MEK/hour}$$

$$6 \text{ pints/hour} \times 1 \text{ hour/60 minutes} = 0.1 \text{ pint of MEK/minute}$$

Specific gravity of MEK is 0.805 g/mL. MEK's molecular weight is 72.1 grams per mole.

$$\frac{\text{ft}^3}{\text{minute}} = \frac{403 \times 0.1\,\text{pint/minute} \times 0.805\,\text{g/mL} \times 2 \times 10^6}{72.1\,\text{grams/mole} \times 100\,\text{ppm}_v} = 8999\,\text{cfm}$$

If the exhaust duct is located at the center (12') of the tunnel, what is the capture velocity at each end of the tunnel?

$$V = Q/A = [8999 \text{ cfm/2'} \times 2.5']/2 = 900 \text{ fpm}$$

The author maintains this is energy wasteful and would argue for 10% of the LEL inside the tunnel, or 1400 ppm$_v$ thereby conserving the energy costs for tempering make-up air and fan operating costs. If this volumetric rate was successful to dry parts, the volumetric flow rate would reduce to 633 cfm, and the capture rate at the tunnel inlet and outlet reduces to 63 fpm—satisfactory with control of cross drafts. Give some thought to reducing the 2' × 2.5' tunnel openings consistent with size of glued parts to increase capture velocity and rate of airflow through a drying tunnel.

666. The concentration of carbon dioxide gas inside an empty beer brewing vat at 2:00 p.m. is 468,500 ppm$_v$—residual headspace gas from yeast fermentation of barley, hops, water, and other ingredients. The vat's volume is 52,800 gallons. What is the expected CO_2 gas concentration at 3:10 p.m. with an exhaust fan with a 2300 cfm capacity. Estimate the air-gas mixing factor at 5. Use calculations to assist in determining if it would be safe for a worker to enter without respiratory protection such as a SCBA and other elements of OSHA's permit required Confined Space Entry standard (29 *CFR* 1910.146).

$$t_2 - t_1 = 3:20 \text{ p.m.} - 2:00 \text{ p.m.} = 1.33 \text{ hours} \times 60 \text{ minutes/hour}$$
$$= 79.8 \text{ minutes}$$

$$52,800 \text{ gallons} = 7058.33 \text{ ft}^3 = V$$

$$C_o = 468,500 \text{ ppm}_v$$

$$\text{Mixing factor, } M = 5$$

$$Q = 2300 \text{ cfm}$$

$$2300 \text{ cfm} \times 79.8 \text{ minutes} = 183,540 \text{ ft}^3$$

$$C_{after} = C_o \times e^{-[Q(t_2 - t_1)/MV]}$$

$$C_{after} = 468,500 \text{ ppm}_v \times e^{-(2300 \text{ cfm} \times 79.8 \text{ minutes})/(5 \times 7058.33 \text{ ft3})} = 2583 \text{ ppm}_v \text{ CO}_2$$

While this gas concentration does not exceed occupational exposure limits, the ventilation should continue until CO_2 gas level in the tank approaches ambient levels of approximately 360–400 ppm$_v$. The site industrial hygienist or chemical safety engineer should ensure that the exhaust duct vents in a manner that does not reintroduce the CO_2 into spaces that could be occupied. A stenching agent could be added to odorless CO_2 to alert people, but the brewers and customers, I am sure, would object.

667. Industrial hygienists, toxicologists, and risk managers are involved in calculating a person's total daily dose to one or more toxicants. This is no easy feat and relies on making several assumptions and calculating various iterations in a stochastic manner. One must consider all routes of absorption: percutaneous, trans-ocular, inhalation, and food and beverage ingestion. An example follows for chloroform.

Our hypothetical male weighs 185 pounds; is 5 feet, 10 inches tall, is 35 years old; drinks 1.5 liters of beverages daily; eats 6 pounds of food daily; takes one 15-minutes shower daily; and washes his hands 4 times daily at 20 seconds per washing. His totally glass-enclosed shower is 6 feet × 5 feet × 8 feet high. He adjusts the shower water flow rate to 4 gallons/min with temperature of 100°F. Assume 90% of chloroform evaporates from the shower stream into air in the shower. Assume chloroform ($CHCl_3$) in all beverages is 30 parts per billion (mass/volume). Assume $CHCl_3$ is absorbed through intact healthy wet skin at 10 mcg/cm^2/h. Assume chloroform in food is 40 mcg/kg. $CHCl_3$ in ambient air is 2 parts per billion by volume. Assume the following absorption factors: skin and eyes at 100%, inhalation at 90%, ingestion of food at 60%, and 70% from all beverages. His average alveolar ventilation rate is 7 L/min. Also assume he is not exposed to chloroform vapor or liquid in his occupation as a high school economics teacher. His only hobbies are golfing and trombone playing in a community orchestra. Absorption factors vary greatly, but by assigning different values one can calculate iterations within a broad range. These were selected on judgment and reasonableness and are presented here simply to demonstrate the calculation processes.

First, convert units and measurements to metric. A very helpful online conversion tool for placing as an icon on your computer screen is www. onlineconversion.com.

$$185 \text{ lb } \male = 83.915 \text{ kg} = 83,915 \text{ g}$$

$$5' - 10'' = 70'' = 1.778 \text{ m} = 177.8 \text{ cm}$$

$$30 \text{ ppb} = 0.03 \text{ ppm} = 0.03 \text{ mg/L} = 30 \text{ mcg/L } H_2O$$

$$6 \text{ lb} = 2.722 \text{ kg}$$

$$40 \text{ mcg} = 0.04 \text{ mg}$$

$$4 \text{ gpm} = 15.142 \text{ Lpm}$$

$$6' \times 5' \times 8' = 240 \text{ ft}^3 = 6.796 \text{ m}^3 = 6796 \text{ L}$$

$$4 \text{ times/day} \times 20 \text{ seconds/time} = 80 \text{ seconds} = 0.0222 \text{ hour}$$

23 hours, 45 minutes = 23.75 hours = 1425 minutes (nonshower time)

Beverages: 1.5 L/day $\times$ 0.03 mg/L $\times$ 0.7 = 0.0315 mg/day

Food: 2.722 kg/day $\times$ 0.04 mg/kg $\times$ 0.6 = 0.065 mg/day

Ambient air: 1425 minutes $\times$ 7 L/min = 9975 L = 9.975 m^3

Ambient air less shower time: 2 ppb$_v$ $\times$ 0.9 = 1.8 ppb$_v$ = 0.0018 ppm$_v$

$$\text{mg/m}^3 = (0.0018 \times 119.4)/24.45 = 0.00879$$

0.00879 mg/m^3 $\times$ 9.975 m^3 = 0.0877 mg/day

Hand washing: 810 cm^2 $\times$ 10 mcg/cm^2/hour $\times$ 0.0222 hour = 179.8 mcg
$$= 0.180 \text{ mg/day}$$

Ambient air in shower: 15 minutes $\times$ 7 L/min = 84 liters = 0.084 m^3

0.00877 mg/m^3 $\times$ 0.084 m^3 = (0.0007 mg/day)

Chloroform vapor in shower air: 15.142 L H_2O/min $\times$ 12 minutes = 181.7 liters

181.7 liters $\times$ 0.015 mg/L $\times$ 0.9 = 2.18 mg volatilizes

2.18 mg/6796 L = 0.0003 mg/L

0.0003 mg/L $\times$ 7 L/min $\times$ 12 min = (0.0252 mg/day)

Total inhalation (0.0023 mg/day + 0.84 mg/day) = 0.0259 mg/day from shower
plus ambient air

$$\text{Skin absorption} = (83.915 \text{ kg})^{0.425} \times (177.8)^{0.725} \times 0.007184$$
$$= \text{skin in shower (DuBois formula)}$$

$$\text{Skin surface area} = 2.02 \text{ m}^2 = 20{,}200 \text{ cm}^2$$

$$20{,}200 \text{ cm}^2 \times 6 \text{ mcg/cm}^2 \times 0.2 \text{ hour} = 24{,}200 \text{ mcg} = 24.2 \text{ mg/day}$$

$$0.01575 + 0.049 + 0.0879 + 0.1079 + 24.2 = 24.46 \text{ mg/day}$$

$$\text{Total absorbed chloroform} = 24.46 \text{ mg/day/83.95 kg}$$
$$= 0.29 \text{ mg/kg/day (way high)}$$

$$23 \text{ hours, } 48 \text{ minutes} = 1428 \text{ minutes (nonshower time)}$$

$$\text{Water: } 1.5 \text{ L/day} \times 0.015 \text{ mg/L} \times 0.7 = 0.01575 \text{ mg/day}$$

$$\text{Food: } 2.722 \text{ kg/day} \times 0.03 \text{ mg/kg} \times 0.6 = 0.049 \text{ mg/day}$$

$$\text{Ambient air: } 1428 \text{ minutes} \times 7 \text{ L/min} = 9996 \text{ L} = 9.996 \text{ m}^3$$

$$\text{(less shower time) } 2 \text{ ppb} \times 0.9 = 1.8 \text{ ppb} = 0.0018 \text{ ppm}$$

$$\text{mg/m}^3 = (0.0018 \times 119.4)/24.45 = 0.00879$$

$$0.00879 \text{ mg/m}^3 \times 9.996 \text{ m}^3 = 0.0879 \text{ mg/day}$$

$$\text{Hand washing: } 810 \text{ cm}^2 \times 0.006 \text{ mg/cm}^2/\text{hour} \times 0.0222 \text{ hour}$$

$$= 0.1079 \text{ mg/day* } (0.1079 \text{ mg/day}/10^6 = 0.00000018 \text{ mg/day})$$

$$\text{Ambient air in shower: } 12 \text{ minutes} \times 7 \text{ L/min} = 84 \text{ liters} = 0.084 \text{ m}^3$$

$$0.00879 \text{ mg/m}^3 \times 0.084 \text{ m}^3 = (0.0007 \text{ mg/day})$$

$$\text{Chloroform vapor in shower air: } 15.142 \text{ L } H_2O/\text{min} \times 12 \text{ minutes} = 181.7 \text{ liters}$$

$$181.7 \text{ liters} \times 0.015 \text{ mg/L} \times 0.8 = 2.18 \text{ mg volatilizes}$$

$$2.18 \text{ mg}/6796 \text{ L} = 0.0003 \text{ mg/L}$$

$$0.0003 \text{ mg/L} \times 7 \text{ L/min} \times 12 \text{ min} = (0.0252 \text{ mg/day})$$

Total inhalation from shower plus ambient air: 0.0023 mg/day + 0.84 mg/day
= 0.0259 mg/day

Skin absorption in shower: $(83.915 \text{ kg})^{0.425} \times (177.8)^{0.725} \times 0.007184 = \text{skin}$

Surface area $= 2.02 \text{ m}^2 = 20{,}200 \text{ cm}^2$

$20{,}200 \text{ cm}^2 \times 6 \text{ mcg/cm}^2 \times 0.2 \text{ hour} = 24{,}200 \text{ mcg} = 24.2 \text{ mg/day}$

$0.01575 + 0.049 + 0.0879 + 0.1079 + 24.2 = 24.46 \text{ mg/day}$

Total absorbed chloroform $= 24.46 \text{ mg/day}/83.95 \text{ kg} = 0.29 \text{ mg/kg/day}$
(way high)

Preceding calculations rely on many assumptions. Risk management toxicologists typically state their assumptions and calculate most likely exposures through many iterations selecting the most plausible exposure scenario when done.

668. Air pollution engineers must regularly report losses of organic vapors and mists to the atmosphere. There are many ways to do this. For example, we can apply the total amount into the mathematical box from the purchase records. Knowing how much is incorporated into the product(s) and amounts lost to plant drainage water, we can rightly assume the difference is transferred to community air. Another way is to calculate atmospheric emissions from existing analytical data. For example, an estimate of acetone vapor released to ambient air from wood pulp digested by caustic soda ($NaOH$) in a paper mill is known to be 1.08 lb/ton of paper pulp that is produced. This mill produces 100,000 tons of pulp per year. Acetone vapor is fed to an incinerator at steady 99.9% combustion efficiency. What mass of acetone is released annually?

$$\frac{1.08 \text{ lb}}{\text{ton}} \times \frac{100{,}000 \text{ tons}}{\text{year}} \times \left(1 - \frac{99.9}{100}\right) = 108 \text{ pounds of acetone/year}$$

669. A tank truck delivered 4000 gallons 10% w/v (in H_2O, pH 8) sodium hydrosulfide solution to their customer. The driver, having done this monthly for several years, connected the tank's discharge pipe to the client's hose fitting not knowing that the client added 6500 gallons of 17% (v/v) sulfuric acid to their 20,000 gallon storage tank the day before. The ensuing chemical reaction released hydrogen sulfide gas into the surrounding air killing the truck driver and two workers in the plant. Assuming quantitative stoichiometric conversion, what mass of H_2S was released to the air?

$$2NaHS + xs\ H_2SO_4 \xrightarrow{H_2O} 2\ H_2S + (Na)_2\ SO_4 + \text{exotherm}$$

$$4000 \text{ gallons} = 15{,}142 \text{ liters}$$

$$10\%\ \text{m/v solution} = 100 \text{ g } H_2SO_4/L \text{ of solution}$$

$$100 \text{ g/L} \times 15{,}142 \text{ liters} = 1{,}514{,}200 \text{ grams NaHS}$$

Molecular weight NaHS = 56.06 grams/mole

1,514,200 grams NaHS/56.06 grams/mole = 27,010 moles NaHS

Thus, from the balanced chemical reaction equation above, 27,010 mol of H_2S were released to the air disregarding solubility of H_2S in the sulfuric acid solution. This exothermic reaction would encourage rapid release of dissolved gas into the surrounding air. Note in the following equilibrium reaction, H_2S ionizes as a weak acid, and the presence of excess sulfuric acid shifts the reaction to the left. This, plus the highly exothermic reaction, drives the dissolved H_2S into gaseous phase. Moreover, as H_2S escapes from the liquid, reaction is driven rapidly to the left.
H_2S is a diprotic acid. *Ergo*, ionization is

$$H_2S \xrightleftharpoons{\ H_2O\ } H^+ + HS^- \xrightleftharpoons{\ H_2O\ } S^= + 2\, H^+$$

(Note how adding acid shifts reaction to left.)

Molecular weight H_2S = 34.08 grams/mole

27,010 moles × 34.08 grams/moles = 920,501 grams = 920.5 kilograms of H_2S gas were released to the ambient air.

This rapid generation and atmospheric release of a potent chemical asphyxiant is another example of a *de novo* toxicant sprinkled throughout this book. Others are household chlorine bleach plus liquid ammonia releases considerably more toxic chloramine gases [(NH_2Cl, $NHCl_2$, NCl_3), soluble cyanide salts plus acids or water release HCN, other soluble sulfides plus acids also release H_2S, household bleach plus acids release Cl_2 and ClO_2, a reducing agent plus inorganic arsenic releases arsine (similar with phosphine and stibine), and thermal decomposition of chlorine-containing solvents produces phosgene gas].

One might wonder why the author selected this problem (and others similar to it for this book). The problem is not a mere exercise in mathematics-engineering equity. There are at least three good practical reasons.

First, the calculations permit us to provide a worst-case estimate of the total mass of a gaseous strongly toxicant released into the community. Regulatory agencies often require this after a release (e.g., Coast Guard, EPA, and state departments of environmental quality). Indeed, even irate community residents demand data.

Second, the calculations can be one of several iterations of theoretical worst-case estimates by industrial hygienists predicting mass emissions and emission rates in potential problem analyses, "What if ...?" engineering calculations, and necessary due diligence practiced by responsible chemical and nuclear material processing facilities and their risk management engineers.

Third, such calculations can help one to predict what exposures of workers and those in the community most likely experienced—or could experience—if process upset and failures occur. The scope of such modeling calculations is beyond this book. The equations to do this can be daunting, however, modeling software is available.

For workers, in the absence of blood specimen biomarkers, one needs proximity of worker to the source, emission rate, mass emitted, ventilation, perturbances such as walls and equipment, respiration rate, and the duration of exposure. For those in the community, one needs time of day, emission rate, distance from emission source (x, y, and z coordinates), height of emission point above ground, duration of exposure, topography, meteorological status (precipitation, wind direction, wind velocity, cloud cover, temperature).

Also note 6500 gallons of 17% (v/v) sulfuric acid = 6500 gallons × 0.17 = 1,105 gallons sulfuric acid. 1105 gallons = 4182.88 L. Density of H_2SO_4 is 1.85 g/mL. So, on hand, chemical facility had at least 4,182,880 mL × 1.85 g/mL = 7,738,328 g = 17,060.1 pounds of sulfuric acid. This greatly exceeds the 1000 pounds threshold limit for reporting under OSHA's Instruction CPL 2–2.45 (*Systems Safety Evaluation of Operations with Catastrophic Potential*). This instruction lists H_2SO_4 as an "Extremely Hazardous Substance."

Surprisingly, sodium hydrosulfide is not listed, and, apparently, reporting does not appear to be required. Yet, H_2S is also an "extremely hazardous substance" with a reporting threshold of 500 pounds. 920.5 kg = 2029.36 lb. Since the H_2S was generated *in situ* and not initially present, technically, the facility appears to comply with the OSHA instruction. The author, however, believes NaHS should be listed as either an "extremely hazardous substance" or as a "hazardous chemical." In so doing, a responsible facility in compliance with the OSHA instructions most likely would have averted the three-death catastrophe. Since more than the threshold quantity of H_2S could have been foreseen by a diligent industrial hygienist, the OSHA should provide for *in situ* generation of *de novo* toxicants. Regardless of OSHA's requirements, the facility failed to provide a reasonably safe workplace.

670. What elements of a risk management plan must have been in place to prevent the horrific catastrophe described in the previous problem (669)? Of all, rank the top three with the greatest degree of prevention.

   a. There must be advance communications between all parties and full agreement on nature of tasks to be performed, when to be performed, how to be performed, where to be performed, steps to be taken in emergency, who will perform, their education and training regarding task including OSHA's hazard communication training, and lock out—tag out (LOTO), and so on.

   b. Dedicated and unique pipe threads for NaHS pipeline should be installed in the tank truck for the pipe conveying NaHS solution into client plant's storage tank. Unique external pipe diameters should be used to hinder pipefitters and others from interchanging valves, fittings, and hoses and pipes.

   c. Truck driver and chemical plant receiving operator must each have dedicated, personally assigned keys for all valves and valve locks.

d. Clear, large warning signs that exhort the H_2S gas inhalation hazards must be installed at all transfer points and possible gas release points. Color-coded pipe helps to alert, but is never a substitute for the unique pipe fitting flanges and external diameters.

e. Inspect the receiving tank before dispensing NaHS solution. Both parties must verify that the tank is empty and clean. Double block and bleed systems must be used if inadvertent introduction of incompatible chemicals into the tank could occur.

f. Insist on a two-person team with trained and equipped standby personnel.

g. Emergency audio notification system must be in place (cell phone, megaphone, police whistle, bull horn, klaxon horn) with a supporting action plan—that is, who does what when?

h. SCBAs and escape respirators must be available with every element of OSHA's respiratory protection program in place (29 *CFR* 1910.134).

i. There should be ample community and residential set-back distances from this hazardous chemical transfer point.

j. A wind sock or a weather vane should be used to ensure that if a release occurs the greatest numbers of those at greatest risk are not located in harm's way.

k. Mechanical local exhaust ventilation should be used at all pipe fittings where gas release could occur.

l. The receiving tank should have mechanical local exhaust ventilation to capture fugitive emissions of H_2S gas displaced as the tank is filled.

m. Only daytime transfers should be permitted.

n. A spill containment system to collect overflow splashes and leaks must be used. This must include neutralization by alkaline materials (e.g., sodium bicarbonate).

o. A public notification system through collaboration with local public safety officers must be in place.

p. H_2S gas alarms at sites where maximum gas leaks could occur or accumulate should be electrically connected to the plant security office who, in turn, notify on advice about community evacuation plans and procedures.

q. Antidote kits containing amyl nitrite inhalation and sodium nitrite infusion hasten forming sulfmethemoglobin thereby preventing methemoglobinemia. Personnel must be trained in cardiopulmonary resuscitation. Oxygen must be nearby.

r. Elements of OSHA's Process Safety Management regulation (29 *CFR* 1910.119) must have been in place. This powerful standard requires responsible parties to address all "management of changes" that could result in a chemical catastrophe before such occurs.

s. Workers at greatest risk of H_2S gas inhalation exposure should be provided with personal gas monitors.

t. Local public safety agencies must be notified under EPA's "Community Right-to-Know" regulations of amounts of hazardous chemicals on site that exceed the threshold reporting quantities including NaHS.

Answer: b, c, and a

671. A homeowner intended to refinish a bathtub by dispensing estimated six ounces of a stripping chemical. He brushed this on the surface of the tub's floor returning 15 minutes later to finish the job. His wife found him kneeling and slumped over tub's edge face down in stripping material an hour later. He was dead. The decedent did not take special efforts to provide local exhaust ventilation. He did not wear an air-supplied or an organic vapor cartridge respirator. This product contained 80% methylene chloride by volume. Forensic reconstruction of the most likely CH_2Cl_2 vapor concentration indicated about 123,900 ppm_v vapor existed 14 in above tub's floor when he returned. Assume NTP (20–25°C, 760 mm Hg). Comment.

   This concentration of methylene chloride vapor is IDLH. This solvent is highly volatile with a boiling point of only 104°F at sea level. Compared to clean air (1.00), methylene chloride has a relative density of 2.93. A narcotic solvent vapor, it has the potential to sensitize the myocardium to a fatal arrhythmia when there is coincident release of endogenous adrenalin such as in a flight or fight response. It would be easy to "blame the victim," but were the product warnings sufficiently compelling, and did manufacturer consider a product formulation with a less volatile and toxic solvent?

   123,900 ppm_v = 12.39%$_v$. We might ask what the density of a 12.39%$_v$ CH_2Cl_2 in air vapor mixture is.

$$100\% \text{ air} - 12.39\% \text{ } CH_2Cl_2 = 87.61\% \text{ air}$$

For CH_2Cl_2:	$0.1239 \times 2.93 = 0.363$
For air:	$0.8761 \times 1.00 = \underline{0.876}$
For air and vapor mixture:	1.239

That is, this vapor mixture is 23.9% denser than air. No wonder the atmosphere in this bathtub was immediately capacitating. With little air movement, this mixture is so dense it hovers inside the tub with little dissipation into ambient bathroom air.

   With this much dense solvent vapor in air, we can also ask how much oxygen is in this resultant air and vapor mixture.

$$20.95\%_v \times 0.8761 = 18.35\%_v \text{ } O_2$$

18.35%$_v$ O_2 is an oxygen-deficient atmosphere per OSHA (<19.5%$_v$) that, in itself, is unremarkable for healthy people with strong hearts. But, when combined with a very high CNS toxicant solvent vapor, this becomes an additive toxicity mixture by only a few inhalations.

   What happens when one is in an oxygen-deficient atmosphere? One will breathe faster (tachypnea), and with the higher respiratory rate, the victim inhales a greater amount of solvent vapor. One can reasonably state the root cause of death of the victim was brief inhalation of an excessively high level of methylene chloride vapor augmented by atmosphere with decreased oxygen content. This will tax anyone's heart, brain, lungs. One kneeling and leaning in this space would be immediately overcome compounded by oxygen deficiency. Falling forward, their nostrils and mouth are now in the no escape zone. Rapid breathing leads to panting and quick unconsciousness and death, perhaps, in only a few minutes.

672. The chemical processing industry is replete with mishaps where tanks overflow leading to explosions, environmental pollution, property damages, atmospheric emissions, fires, and personal injuries and deaths. Root-cause analyses of such mishaps often reveal the absence of tank fill capacity alarms or their malfunction, other process equipment failures, dispensing materials into the tank with an un-calibrated flow measurement meter, corroded and weak tank walls and pipe walls, leaky valve gaskets, not considering preexisting tank volume contents, insufficient written safety work procedures, missing or nonlabeled tanks, inattentive process operators (e.g., dispensing into wrong tank, or the correct tank at the wrong time), and poor communications among all involved parties.

For example, consider a 1000 gallon steel tank containing 200 gallons of 23% w/v sodium hydroxide in water. The same concentration of lye solution is continuously added at 100 gallons/hour. Another process removes this solution concurrently at 50 gallons/hour. How many gallons are in this tank after 3 hours? When, if ever, will this tank begin to overflow?

The mass balance equation is: accumulation = material$_{in}$ − material$_{out}$.

$$\text{Material}_{in} = 3 \text{ hours} \times 100 \text{ gallons/hour} = 300 \text{ gallons}$$

$$\text{Material}_{out} = 3 \text{ hours} \times 50 \text{ gallons/hour} = 150 \text{ gallons}$$

$$\text{Accumulation} = 300 \text{ gallons} - 150 \text{ gallons} = 150 \text{ gallons}$$

$$\text{Accumulation} = 1000 \text{ gallon tank} - 200 \text{ gallons initially contained}$$
$$= 800 \text{ gallons}$$

$$(100 \text{ gph} - 50 \text{ gph}) \times t = 800 \text{ gallons}$$

$$t = 800 \text{ gallons/50 gallons/hour} = 16 \text{ hours}$$

Such overflow might be unnoticed, for example, if filling commences at noon at a plant unoccupied after 5:00 p.m., and the security guard fails to notice spillage and overflow at 4:00 a.m. or after. From the data, one can easily calculate how much NaOH (not including the water portion) was released to the environment. Try it!

All such systems must have an overflow sensor that is interlocked with supply line and active audio-visual alarms. For extremely hazardous materials, moats must be present to collect and scavenge overflow. As important is good communication between all involved in transfer process as the following example demonstrates (from a manual written by the author of medical education fact sheets for physicians who practice occupational and environmental medicine):

OEM MEDUCATION FACT SHEET 97

CLEAR COMMUNICATIONS TO PREVENT DEATHS

The title of this *Fact Sheet* speaks volumes and indicates in the absence of clear communications between two or more parties, there can be deaths, life-altering injuries, and property damages. The editor investigated four fatalities from gas inhalation (NH_3, Cl_2, and two H_2S) and numerous others short of lethal (chemical pneumonitis, bronchiolitis obliterans, RADS, and so on) where there was the lack of understanding between those shipping a hazardous material and those receiving it. SHE and OEM personnel must ensure that clear communications are established before the actual shipment (or any other hazardous activity). Here are opportunities for industrial hygienist and safety engineer to prepare an advance script covering all safety aspects involved in receiving hazardous materials. The time that it takes to prepare a well-written script is invaluable because it allows all parties to apply various hazard metrics (e.g., What if?, HAZOP, FMEA). The following is a basic example.

Simple Communication Failures and Poor Safety Coordination
Can Cause Life-Altering Injuries, Deaths, and Property Damages

Railroad: "Hi. This is Bob Adams with Central Michigan Railroad on behalf of Ammonia Chemicals. We're planning to spot three rail cars of ammonia next to your plant in Flint between 8:00 a.m. and 9:00 a.m. on Monday, October 30, 2000. Are you unloading any anhydrous ammonia cars then that could expose our guys to gas?" (*Note*: This communication should be initiated by chemical plant to railroad as well. Both parties have safety engineering coordination duties.)

Plant: "No, we're not. Everything should be OK then. Go ahead and spot your cars. We'll let you know if our plans change. Thanks for asking." *or*: "Great minds think alike. We were just about to call you and let you know our plans. Yes, we are. We'll be unloading one ammonia car at that time. It's a jumbo that should take about three hours. Can you reschedule your train to be there some time after 11:30 a.m. when we'll be done?"

Railroad: "Yup, we can. How's noon that day?" *or*: Nope, we can't. We have to switch and set out cars at that time. We're on a tight schedule to make Detroit and the connection to St. Louis."

Plant: "OK. Then we'll have our safety man on site to watch out for your people. The anhydrous ammonia car will be 'blue flagged.' He'll have a powered megaphone to warn you about any upset conditions. He will assess the conditions at the time to determine boundaries of the setback lines and danger zone. This might be 50 to 100 feet or more from the anhydrous car. Your people must not enter this zone without wearing self-contained breathing apparatus. There will be a secondary zone from 50–100 feet to 300 feet or more where your workers must have a five-minute escape air pack on them or a SCBA. You won't be able to miss him. He'll be on the unloading ramp about 12 feet up wearing a SCBA and bright yellow turnout gear. Look for the red flag and red windsock. Your people must not enter either zone if they do not see him. We assume your people have good hearing and have been trained in SCBA and escape pack use and in the hazards of ammonia gas. Right? Also, your people must understand that conditions can change quickly and place them in harm's way. They must remain in the sight of our safety engineer. He is in the best position to know if there is a release of gas. Have your engineer signal him with three short whistle blasts when you're done, and the train is about to leave. We'll let him know when you are arriving. By the way, the anhydrous car is a large black GATX, Serial No. 7350194. By the time you get there, it will be spotted and braked at the unloading rack. As you know a derail is in place and must be removed after replacing our empties with your loads. After, replace the derailer. Our guy will wave to you as you arrive. Again, thanks for calling. We're in this together."

Railroad: "Thanks. The work plan sounds good. We'll let you know if there are any changes. Please get back and let us know that you informed your safety engineer and anybody who might replace him that day. He should know that, besides the engineer who remains in the locomotive, there will be a brakeman and the conductor—John and Marty—on the ground. We'll pass this safety information onto the train crew verbally and in writing. They'll have two-way radios. Their band frequency is 1234.5 Hertz in case your safety guy has to talk to them. It's good to work with you. Any questions? In the future, let's have a check list so that we can pre-arrange all the safety items, and so we don't overlook any. OK?"

Plant: "You bet. I'll get cracking on it and fax my draft to you so that we agree on everything."

673. Refer to the previous problem (672). Assume a process operator noticed the overflow condition and stopped it promptly at 9:30 a.m. How many pounds of NaOH were released to the environment and, in this case, reported to the Coast Guard and the State Department of Environmental Quality?

$$9:30 \text{ a.m.} - 4:00 \text{ a.m.} = 5.5 \text{ hours}$$

$$5.5 \text{ hours} \times 50 \text{ gallons/hour } 275 \text{ gallons}$$

$$275 \text{ gallons} \times 0.23 = 63.25 \text{ pounds}$$

While this amount of NaOH is relatively small, an area in the receiving water could have substantial increase in pH depending on dilution rate from water inflow with a resultant fish and plankton kill.

674. The Carboniferous and other periods in Earth's early history had huge time spans of many millions of years with an atmosphere considerably richer in carbon dioxide gas than today. Plants were abundant sequestering tremendous amounts of CO_2 from the warm atmosphere into their tissues ultimately forming today's fossil fuels of natural gas, shale oil, mineral petroleum oil, and coal. In only a few centuries, we will reverse this long

natural process by combusting these fuels returning the CO_2 to the atmosphere. Global warming ensues with horrible sequelae.

The author teaches a graduate school course (Principles of Environmental Health) in which robust discussion of global warming ensues. A student assignment is to suggest a plausible engineering method to capture CO_2 from fossil fuel electricity generating power plants. I encourage time-tested engineering and crazy notions as well. I love it when I see student pods "brain storming" the assigned problems.

One student suggested that CO_2 in stack gases could be bubbled through milk-of-lime scrubbers [$Ca(OH)_2$] to capture the gas as calcium carbonate ($CaCO_3$). She claimed the $CaCO_3$ precipitate could be safely disposed in a landfill and/or used in chemical manufacturing processes. She presented the following cogent chemical equation:

$$Ca(OH)_2 + CO_2 \rightarrow CaCO_3 \text{ (precipitate)} + H_2O$$

Close, but no cigar, because when I asked her where she would obtain the milk-of-lime, she said, "Well, purchase lime (CaO) from a chemical supplier and mix it with water," and offered this equation:

$$CaO + H_2O \rightarrow Ca(OH)_2 \text{ (slurry)}$$

What she failed to consider is that lime is obtained by calcining limestone (calcite) in a lime kiln:

$$CaCO_3 \underset{\Delta}{\rightarrow} CaO + CO_2$$

The energy for this is $\Delta H_{1200-1300°C} = +4.25$ million Btu/ton of limestone produced assuming perfect heat transfer and residence baking/calcining time in the kiln.

So, as you see, every mole of CO_2 removed from the power plant exhaust gases evenly balances 1 mole of CO_2 released into the atmosphere from the lime kiln. Her proposal does not change the CO_2 balance in the atmosphere. Regardless, I gave her a good grade for her out-of-the-box thinking. Depending on combustion efficiency at the power plant, her idea should be given consideration because the energy required to calcine limestone might be substantially less than the energy gained from combusting the fossil fuel. That is, although CO_2 is not removed from the atmosphere, the net energy generation balance might be socially economical.

But wait! Later, she came exclaiming, "Professor, our discussion gave me pause to my whimsical idea. You said my proposal is CO_2 neutral because for each mole collected in powerhouse stack gases one mole is released at the lime kiln." "Yes, I replied." Wondering why she brought this up again, she said, "We are both wrong because fossil fuel energy is required to roast and calcine the limestone to lime at the kiln. The overall process releases even more CO_2 gas to the atmosphere than we first believed." I uttered something like, "Ouch! You got me on that one, and a power of two engineering a solution surely beats solitary engineering." I was toast, and humility and collective thinking saved my day. We laughed and thanked each other from which action I coined a new hybrid word from humor and humility for my vocabulary: "humorility." Humorility goes a long way and will often carry the day.

675. As graduate students of industrial hygiene, our professor, the late Dr. Ralph Smith, provided co-students and me his challenge: Search chemical processing industry literature to find a single manufacturing process from raw materials to the finished single chemical product that has the largest number of starting chemicals with highest overall toxicity and hazards—a daunting assignment. Teachers of IH take note!

My selection was the synthesis of toluene 2,4-diisocyanate from carbon monoxide, chlorine, and 2,4-dinitrotoluene. The simplified reactions are:

$$CO + Cl_2 \rightarrow COCl_2 \text{ (phosgene)}$$

$$2H_2 + 2,4\text{-dinitrotoluene} \rightarrow 2,4\text{-diaminotoluene}$$

$$2,4\text{-diaminotoluene} + COCl_2 \rightarrow \text{toluene } 2,4\text{-diisocyanate} + 2HCl$$

This batch of materials contains a chemical asphyxiant, a highly explosive gas, a methemoglobin forming agent, an upper respiratory irritant gas, a lower respiratory irritant gas, a former battlefield war gas, a cardiotoxicant, and a potent respiratory sensitizer. The challenge to the reader is to match these chemical toxicants to the potential adverse health effect.

Eight nasty chemicals earned an "A." Dr. Smith's assignment was a wise choice because industrial hygiene practitioners must think outside of their box. That is, if the industrial hygienist, for example, only considered phosgene evaluation in this process, much would be missed in the overall exposure assessment process and implementation of solid risk management plans.

676. Toxicological terms of additive, synergistic, potentiating, and antagonistic are used by industrial hygienists to describe effects of toxicants, toxins, and physical agents on worker health. Express the toxicity concept for each term using whole numbers with a few examples.

Additive: $2 + 3 = 5$

CO + HCN: Both gases are systemic chemical asphyxiants. Their mechanisms of toxicity are different, but both hinder the body's oxidative metabolic pathways.

CO + heat strain: Heat-strained worker shunts a large blood volume to peripheral tissues to promote evaporation cooling and radiation heat transfer. Tachycardia ensues. Meanwhile, his brain, heart, and retinas—highly oxygen dependent—become oxygen-starved by carboxyhemoglobin formation.

Pb + certain organic solvents: Both are ototoxic.

Pb + *n*-hexane: Both are peripheral neurotoxicants.

Synergistic: 2 + 3 = 12

Inhalation of tobacco smoke and asbestos fibers is a classic example.

Ingestion of ethyl alcohol and inhalation of trichloroethylene promotes "degreaser's flush" while neither alone does.

Potentiation: 0 + 3 = 8

A cardiologist controls his patient's coronary ischemia with medications (0) with no knowledge his patient is exposed to carbon disulfide (3), a cardiac ischemia agent. Improved dialogue between the cardiologist, nurse, worker-patient, and industrial hygienist would identify this hazard with attendant risks of heart attack or stroke.

Antagonism: 2 + 3 = 1, or 3 + (−3) = 0, or 3 + 0 = 1 (where 0 is a pharmaceutical)

The author encountered a maintenance worker of a pharmaceutical client cleaning an air pollution control bag house who simultaneously inhaled minute amounts of dusts of testosterone and estrogens—pharmacological agonists. He said, "This is better than working in the androgens or estrogens departments—a (physiological) conflict between beards or breasts."

677. Previous problems demonstrated how to calculate vapor saturation concentrations in the head space above the volatile liquid or solid for a sealed container. That is, simply divide the vapor pressure for the chemical (liquid or solid) by the barometric pressure and multiply the result by 10^6 to get the saturation concentration in ppm_v. This method is correct for chemicals with low volatility because the vapor does not add very much to the system's pressure. Recall molecules migrate back and forth between the liquid or solid phase and the vapor phase. When this migration is essentially zero, we may say the head space is vapor saturated. Use the following formula for high volatility chemicals:

$$\frac{\text{vapor pressure at a temperature}}{\text{barometric pressure} + \text{vapor pressure}} \times 10^6 = ppm_v \text{ of vapor}$$

For example, naphthalene has a low vapor pressure at 25°C of 0.082 mm Hg. We know naphthalene has a vapor pressure because of its unique sharp odor. Substituting

$$\frac{0.082 \text{ mm Hg}}{760 \text{ mm Hg} + 0.082 \text{ mm Hg}} \times 10^6 = 107.88 \text{ ppm}_v$$

The vapor saturation concentration is 107.89 ppm_v if we do not account for the slight back pressure of naphthalene vapor on the solid naphthalene. Practically speaking, one normally does not calculate for this variance for chemicals of low volatility.

Now let us see what happens with a highly volatile chemical such as methyl ethyl ketone with a vapor pressure of 95.1 mm Hg at 25°C.

$$\frac{95.1 \text{ mm Hg}}{760 \text{ mm Hg} + 95.1 \text{ mm Hg}} \times 10^6 = 111,215 \text{ ppm}_v$$

Compare the above result to the following where we did not account for high back pressure exerted by methyl ethyl ketone at saturation:

$$\frac{95.1 \text{ mm Hg}}{760 \text{ mm Hg}} \times 10^6 = 125,132 \text{ ppm}_v \text{ of vapor}$$

This result is incorrect. The industrial hygienist and chemical safety engineer must be mindful of the increased pressure inside, for example, 55-gallon drums of such volatile solvents. We see in the first example the pressure in the drum is 760 mm Hg + 95.1 mm Hg = 855.1 mm Hg. This is substantially greater than the atmospheric pressure of 760 mm Hg. An unprotected worker opening the drum is enveloped in a cloud of MEK vapor as the pressure is released. The vapor cloud is larger for a partially filled drum than from a full drum. Mechanical local exhaust ventilation at the bung hole, personal protective clothing, and a SCBA are required. A chemical cartridge respirator is unacceptable because these cartridges are limited to 1000 ppm$_v$ total organic vapors among other restrictions.

678. The target tissue of benzene (as a hematopoietic carcinogen) is the bone marrow. Calculate the average number of benzene molecules per bone marrow cell for a worker at the end of his work shift inhaling benzene vapor at OSHA's PEL of 1 ppm$_v$. A 70 kg adult male individual has an estimated 100 trillion cells in his body. Yellow and red bone marrow for the standard man is 3000 grams (6.61 pounds, or 4.28% of body weight). Assume an alveolar ventilation rate of 6.5 Lpm and 50% absorption of benzene vapor in inhaled air. Further assume this worker has no skin contact with benzene. Molecular weight of benzene is 78.11 g mole^{-1}.

$$\frac{mg}{m^3} = \frac{1 \text{ ppm}_v \times 78.11}{24.45 \text{ L}} = \frac{3.195 \text{ mg}}{m^3} = \frac{3.195 \text{ mcg benzene}}{\text{liter of inhaled air}}$$

480 minutes $\times$ 6.5 Lpm $\times$ 3.195 mcg/L $\times$ 0.5 = 4984.2 mcg benzene absorbed

78.11 grams of benzene contain 6.023×10^{23} molecules. 78.11 micrograms of benzene contain 6.023×10^{17} molecules. $(6.023 \times 10^{17}$ molecules/78.11 mcg) $\times$ 4984.2 mcg = 3.843×10^{19} molecules of benzene. 100 trillion cells $(10^{14}) \times 0.0428 = 4.28 \times 10^{12}$ cells in red and yellow bone marrow. 3.843×10^{19} molecules/4.28 $\times 10^{12}$ cells = 8,978,972 benzene molecules per bone marrow cell. This does not mean 8,978,972 benzene molecules are in or on each bone marrow cell at the end of his work shift. Some of these are in interstitial fluid and might never contact a cell. What it means is that by end of his work shift, this large number of benzene molecules in perfusing blood could contact, on average, every bone marrow cell. All would have to be catabolized or excreted before the start of the next work shift to ensure benzene accumulation is not occurring from one day to the next. Our bone marrow is metabolically highly active, and, for example, an average of 2,500,000 erythrocytes are produced each second in healthy adults. How's that for some roaring biochemistry?

Obviously, this large number of benzene molecules per bone marrow cell is neither highly accurate nor precise because of the stated assumptions and the total number (10^{14}) of body cells. The author places the number in a range of 7–11 million. However, by any measure, results are undisputedly large and provide a reasonable frame of reference for occupational toxicologists.

679. A wash cloth containing 34.5 mL H_2O hangs on a hook inside an empty closet $3.5' \times 3.5' \times 8'$. Closet air initially had 37% relative humidity at 72°F, is tightly sealed, and has no ventilation. What is the relative humidity after this water evaporates?

$$3.5' \times 3.5' \times 8' = 98 \text{ ft}^3$$

Air at 37% relative humidity and 72°F contains approximately 0.0064 pound of water per pound of dry air (obtain from psychrometric chart).

$$98 \text{ ft}^3 \times 0.075 \text{ lb/ft}^3 = 7.35 \text{ pounds of air in closet}$$

$$7.35 \text{ lb air} \times 0.0064 \text{ lb } H_2O/\text{lb dry air} = 0.04704 \text{ lb moisture in air}$$
$$\text{before wash cloth}$$

$$34.5 \text{ mL } H_2O = 34.5 \text{ g} = 0.07606 \text{ lb}$$

$$0.04704 \text{ lb} + 0.07606 \text{ lb} = 0.1231 \text{ lb } H_2O \text{ vapor after evaporation}$$

$$0.1231 \text{ lb } H_2O/7.35 \text{ lb air} = 0.01675 \text{ lb } H_2O \text{ vapor/pound dry air}$$

From psychrometric chart, this is equivalent to 100% relative humidity—saturation. In time, mold growth can be expected on vulnerable substrates. However, as the mold amplifies, water vapor is removed from the air. If the relative humidity drops below 60–70%, growth stops. With mold death, water vapor not incorporated into cells will again enter the atmosphere, and the cycle continues until most water is bound in dead mold molecules.

680. A fan installed in a fixed air-handling system operates at 10,000 cfm, 1.5" water gauge static pressure, 5.0 brake horsepower, and 1000 rpm. What fan rpm is required to handle 25% more air (12,500 cfm) through this system? This example can be viewed as either the present installation requires more air than designed, or the air balancing report demonstrates 25% less air than specified in design plans.

Rearrange the fan laws:

$$\text{rpm}_2 = (\text{cfm}_2/\text{cfm}_1) \times \text{rpm}_1$$

$$\text{rpm}_2 = (1250 \text{ cfm}/1000 \text{ cfm}) \times 1000 \text{ rpm} = 1250 \text{ rpm}$$

$$\text{sp}_2 = \text{sp}_1 \times (\text{rpm}_2/\text{rpm}_1)^2$$

$$\text{sp}_2 = 1.5'' \text{ w.g.} \times (1250 \text{ rpm}/1000 \text{ rpm})^2 = 2.34'' \text{ w.g.}$$

$$\text{bhp}_2 = \text{bhp}_1 \times (\text{rpm}_2/\text{rpm}_1)^3$$

$$\text{bhp}_2 = 5.0 \times (1250 \text{ rpm}/1000 \text{ rpm})^3 = 9.77 \text{ bhp}$$

According to the fan laws, in order to use the original fan, the fan speed must be increased from 1000 to 1250 rpm, and the fan motor must be changed from a 5 hp to a 10 hp. The energy requirements almost double for only a 25% gain in system air volume.

It is important to ensure that the new fan rpm do not exceed the maximum allowable fan tip speed for the existing fan. Maximum fan rpm are listed in fan catalogs. One may always consult fan manufacturers for engineering specifications if you would like them to review the application.

681. Assume gasoline is entirely octane (2,2,4-trimethylpentane). It is not, but we may assume this simply when asking how much carbon dioxide gas can be produced from burning about 6 pounds of this hydrocarbon. Well, how much?

Gasoline weighs 6.25 pounds per gallon at NTP. Each molecule of C_8H_{18} produces eight molecules of CO_2 if perfectly combusted. The molecular weights of octane and CO_2 are 114 and 44 grams/gram-mol, respectively.

$$44 \text{ pounds mole}^{-1} \times 8 \text{ moles} = 352 \text{ pounds}$$

$$6.25 \text{ pounds} \times (352/114) = 19.3 \text{ pounds}$$

That is, combusting 1 gallon of gasoline produces 19.3 pounds of atmospheric CO_2. It has been calculated that 70 million tons of CO_2 enter our Earth's atmosphere daily. The official *World Resources Institute* adopted conversion of 19.564 pounds of CO_2 released per gallon of gasoline.

Every combusted gallon of gasoline adds roughly 8 pounds of water vapor to our atmosphere that ultimately returns to our hydrosphere.

682. Industrial hygienists and forensic toxicologists are asked now and then to estimate the uptake of an inhaled material into a person's body. What one must know, or state as assumptions, are the concentration of the air contaminant inhaled, nature of the contaminant (particulate and gas or vapor), respiration rate, absorption through the alveoli (as a fraction), duration of exposure, additional routes of exposure, and the pulmonary function parameters (tidal volume, dead space, minute ventilation, etc.).

The alveolar ventilation must be determined. This is defined as the volume of gas per unit time that reaches alveoli (respiratory portion of lungs where gas exchange occurs). This can be calculated by the equation $A_v =$ (tidal volume − dead space) × respiratory rate. The following table provides example values for an adult male with two levels of activity:

Respiration Metric	Resting	Moderate Exercising
Tidal volume	0.5 L	1.8 L
Respiratory rate	15/minute	30/minute
Minute ventilation	7.5 L/minute	50 L/minute
Dead space	0.1667 L	0.1667 L
Alveolar ventilation	5.0 L/minute	49 L/minute

For this gentleman, calculate his alveolar ventilation rate while he walks inhaling air at 1.1 Lpm with a respiratory rate of 20 inspirations–exhalations per minute.

683. In this business of industrial hygiene risk management, we encounter the terms of hazard and risk frequently. Some wrongly use the terms interchangeably. Hazard ≠ risk. A hazard is an object and/or energy that can cause harm. That is, hazard is the *potential* for harm. The author uses the six Ds (discomfort, reversible health disorders or dysfunctions, irreversible diseases, deaths, or significant personal or property damages when identifying hazards. Risks, on the other hand, are the *potential* outcomes of hazards. Risk has two major components: *probability and severity.* One can never manage risks without first carefully identifying hazards. *To wit*, a sealed drum of 97% sulfuric acid is an incredible hazard, but it will never be a risk as long as the drum remains sealed and intact. Once the drum is breached in any manner, deliberately for use or inadvertently (e.g., impaled by blades of fork lift truck), numerous risks present to those nearby and the environment.

Most would agree that, in the modern world, the automobile and truck are hazards to us either as drivers, passengers, or pedestrians. Few recognize that in the United States 2% (i.e., 1/50) of adults die from an accident involving a car or a truck. Did you know that—this probability? Severity of such an event ranges from minor to life-altering injuries to multiple deaths and property damages.

As practice, ponder the likelihood (probability) of risks from the following examples of hazards:

Air Contaminants(s)	Risks (Typically Multiple, >3)
CO + heat strain	
9.1% oxygen by volume in inhaled air	
Certain isomers of PCBs + intense fire	
Unmarked 55-gallon drum of some green liquid	
α-Quartz dust + tobacco smoke	
4% Phenol in 10% caustic soda (NaOH) solution	
Pb + intense noise	
Simultaneous inhalation exposure to Pb, Mn, Hg	

Immediately, you see challenges to "fill the blank spaces" because we ask: acute exposure?, chronic exposures?, skin contact?, preexisting medical issues?, work practices?, personal protective equipment?, education and training?, failures of existing controls?, special circumstances?, or other pertinent (or spurious) issues. It is not easy, and the chances of overlooking major concerns abound due to nonprofessional conduct, errors and omissions, out-and-out sloppy carelessness, and failure to recognize one's limitations and seeking advice of others more skilled.

684. Industrial hygienists and plant engineers often toss out phrases like "mechanical local exhaust ventilation," "dilution ventilation," "thermal head ventilation," and other arcane terms pertaining to air movement. For those responsible for protecting the health of the worker from toxic air contaminants and otherwise providing for their comfort, an understanding of terms is essential. The following diagram—although crude—attempts to clarify, in a general way, what is meant by these various ventilation engineering phrases. Please excuse the author's highly diagrammatic conceptual sketch; an artist I am not, but I strive to be a good HVAC engineer.

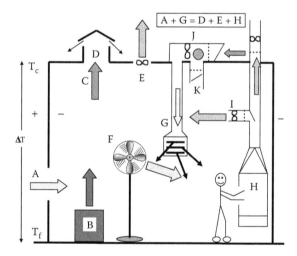

A = natural ventilation ("God given" wind, humid, dry unpredictable, unreliable, cold, hot, polluted)

B = hot process with convective heat air currents

C = thermal gradient ventilation ("chimney effect")

D = natural (thermal exhaust becoming rare)

E = general mechanical exhaust (ceiling and/or wall exhaust fans)

F = dilution ventilation

G = air supply ventilation

H = mechanical local exhaust

I = recirculation ventilation

J = make-up air unit (air filtered and tempered)

K = shunt ventilation

685. The following are guidelines for indoor air comfort for up to 80% of the occupants recognizing that about 10% of the occupants will claim "It is too cold," and another 10% will say "It's too hot in here." The author calls this the

"Goldilocks Principle" where MaMa Bear found her porridge too hot, PaPa Bear claimed it too cold, and Baby Bear uttered, "My porridge is just right."

Indoor Comfort for Offices, Commercial Buildings, Schools, and so on where Worker Heat Stress and Heat Strain are Not an Issue

Acceptable Ranges of Indoor Temperatures and Relative Humidity during Summer and Winter[a]		
Relative Humidity (%)	Dry Bulb, Winter (°F)	Dry Bulb, Summer (°F)
30	68.5–76	74–80
40	68.5–75.5	73.5–79.5
50	68.5–74.5	73.0–79
60	68–74	72.5–78

Source: Adapted from American Society of Heating, Refrigeration, Air-Conditioning Engineers (ASHRAE) Standard 55-1981: *Thermal Environmental Conditions for Human Occupancy* (see later editions as indicated).

[a] For people dressed in typical summer and winter clothing in geographic regions with significant differences between the summer and winter climates and who are mainly at light, sedentary activity. Values should be adjusted as metabolic activity and radiant temperatures increase, and airflow rate over occupants significantly departs from 50 fpm (0.57 mile/hour). The outside wall temperatures should approximate dry bulb air temperature to eliminate excessively cold drafts resulting from temperature differences between the floor and the ceiling. Rooms with high ceilings have a large gradient in air temperatures from floor to ceiling (e.g., >30°F). Judicious placement of ceiling fans in such rooms promotes occupant comfort if the air velocities passing over persons are not excessive (e.g., >100–200 fpm).

There is considerable debate among researchers, building ventilation engineers, indoor air quality specialists, industrial hygienists, and public health professionals concerning the optimum levels of relative humidity. In general, the broad range of relative humidity levels recommended by different organizations is 30–60%. Relative humidities below 30% produce discomfort by promoting drying of skin, eyes, and mucous membranes. Low humidity results in accumulating electrostatic charges on objects including people and clothing. Maintaining relative humidity at lowest possible level greatly restricts mold and mildew growth. Rhinoviruses and influenza virus do not thrive in the air when humidity is below 50%. Concerns (comfort for most part) associated with dry air must be balanced against risks (enhanced microbial growth) associated with over-humidification. If dry bulb temperatures are maintained at lower end of comfort range (68–70°F) during heating periods, relative humidity in most climates will not fall much below 30% (also within comfort range) for most occupied buildings.

Industrial hygienists investigating worker complaints of chilly air must be mindful of the complainant's clothing. That is, one who wears light summer clothing in a cold climate building with low humidity during the winter should expect goose bumps.

686. List the industrial hygiene control methods within their philosophical framework.

Industrial Hygiene Control Methods

1° Eliminate > 2° Engineering > 3° Administrative > lastly PPE

Hierarchy of Approaches (Combinations are often required.)
1. Engineering methods (normally the best; OSHA requirements when feasible)
2. Administrative methods and changes in work practices
3. Personal protective equipment (generally the last choice; often temporary)

Application Points (Combinations are often required.)
1. At the source (best; e.g., mechanical local exhaust ventilation, process enclosure), but only after elimination has been shown impossible or costly.
2. In the path between hazard source and worker (e.g., pedestal fan, shield, distance).
3. At the worker (worst, i.e., use of personal protective equipment, e.g., respirator)

18 Basic Industrial Hygiene Control Methods
1. Hazard evaluation, analysis, and risk assessment; risk ranking and priorities
2. Hazard education, risk communication, and training (necessary for all hazards, workers)
3. Eliminate > substitute with less hazardous materials, processes, components
4. Process isolation; guards, barriers, and shields
5. Process enclosure; interlocks
6. Wet methods of dust suppression
7. Time, distance, and shielding (inverse square law, half-thickness shields)
8. Ventilation (mechanical local exhaust, dilution, general supply, and exhaust)
9. Warnings, labels, posting, and written safety instructions and work practices
10. Housekeeping and plant sanitation and hygiene
11. Changes in work practices and applying administrative controls
12. Medical placement, screening, and routine medical surveillance
13. Biological monitoring, biomarkers
14. Regular workplace surveillance, employee monitoring, reporting, and record keeping (e.g., air sampling and analysis, noise monitoring, radiation testing, wipe samples)

15. Emergency and hazard contingency plans and procedures, spill response plans
16. Prescription of personal protective equipment (PPE: respirators, gloves, etc.)
17. Regular inspection and preventative maintenance of equipment, tools, devices
18. Modify management cultures and supervisor and employee attitudes.

687. Display some data characteristic of environments with suspended airborne dust particles.

The following chart displays data obtained from four distinct environments. It is offered only to give an idea of the nature of data and how it is typically displayed. It must not be construed as typical average or range—only to illustrate diversity. Note how the particles are log-normally distributed. This is typical because sub-micron particles tend to remain airborne by numerous aerodynamic physical factors. For example, 3% of airborne particles by mass could comprise over 95% of the particles by count. One milligram of silica dust in the size range reported for the atmosphere in an iron foundry or quarry typically contains 200–300 million particles. A single particle of this silica dust, therefore, weighs only 4 picograms (4×10^{-12} gram). Tiny!

Particle Size, (microns, μ)	Particles per liter of air			
	Outside Air	New Home (<5 Years)	Old Home with Visible Mold	Office[a]
0.3	814,908	315,298	2,651,469	113,899
0.5	94,721	101,875	291,193	21,896
1	16,530	61,879	70,852	9283
2	7264	45,519	36,837	5934
5	2926	28,105	17,993	3285
10	145	2607	1979	617

Source: Data from *Evaluating Indoor Air Quality*, Fluke Corporation, 2005.

[a] The office make-up air unit had a good, but non-HEPA, dust filter system to remove many, but not all, ambient air particles. The HVAC system was carefully maintained.

688. List the risk factors for workers exposed to airborne particles (dusts, mists, fibers, fog, smoke). Several risk factors must be evaluated in evaluating a worker's dose to particles suspended in their breathing zone air. Failure to account for these can over or underestimate the worker's dose with risks of adverse health outcomes—or deaths.

Workplace Airborne Particle Inhalation Considerations

- 1Breathing zone concentration (mg/m^3, ppm_v, fibers/cc, etc.)
- Particulate toxicity
- Duration of exposure (daily, weekly, yearly, career)
- Acute or chronic exposures
- Exposure limits (RELs, TLVs®, PELs, WEELs, EPA air quality, etc.)

- Particle size distribution, length-to-width aspect ratios
- Respiration rate (driven by metabolic rate)
- Individual risk factors, biological variability
- Industrial hygiene control methods (eliminate $> 1° > 2° > 3°$)
- Medical screening; medical surveillance, biomarkers
- Life style factors
- Other air contaminants, gas adsorption on particles
- Age, health, mouth breathing, and so on
- Co-toxicity of other air contaminants (gases, vapors) and adsorption
- Co-toxicity of over-the-counter and prescription medications
- Biological particles (e.g., mold, fungi, bacteria, viruses, biogenic fragments)
- Overtime work
- Signs and/or symptoms concordant with toxicants, toxins
- Target organs
- Hand-to-mouth activities
- Education and training
- Management culture
- Respirator use and practices
- Applicable regulations
- Heat strain

689. See preceding problem (688) to define risk factors for skin-absorbed chemicals.

    Skin Absorption Risk Factors (supplemented by risk factors in Problem 688):

- Skin absorption (perfusion) rate of chemical (mcg per cm^2 of skin per minute)—obtained from *in vivo* and *in vitro* research
- Surface area covered
- Glove or clothing occlusion
- Duration of contact; evaporation
- Skin washing efficiency, personal hygiene
- Integrity of skin; any breaches (sores, rashes, diseases, lacerations, eczema, psoriasis, etc.)
- Type of skin (area of body, thickness, hairiness, pores)
- Biological variability, individual susceptibility; medical screening
- Cofactors (heat strain, abrasion, other chemicals, medications, friction, pressure, perspiration rate, etc.)
- Metabolic rate
- Blood perfusion rate through dermal contact area
- Acute (one time) or chronic (e.g., regular daily) exposures
- Personal protective equipment permeability

690. List some clinical tips for physicians, nurses, and dentists that can help them see if there is nexus between their patient's presentation and possible work-relatedness.

Industrial Hygiene Clinical Tips

The worker presenting clinically often demonstrates, explains, or exhibits numerous stigmata of occupational disorders and diseases. These diagnostic clues include:

- The appearance and condition of a worker's hands speaks volumes on what they do for a living and their health status (e.g., clubbing, blanching, dermatitis, nail status, injuries, calluses, fissures). Other diagnostic clues elicited clinically are:
- Do you, or others, have to raise your voice to be heard at work?
- Are sounds muffled and faint after you leave work? Do you have to turn up the volume on your car radio after you leave work?
- Are there odors? Smells like? When? Duration and frequency?
- Do your eyes burn, tear? When? Where? Doing what?
- Can you see your work clearly? How is the lighting?
- Burning nostrils, throat, lungs? When? Where? Frequency? Duration?
- Productive cough? Dark sputum? What do you think is causing this?
- Dizzy, nauseous, vomiting? Headaches? When, where, duration, frequency?
- Settled dust? Process emissions? Dust, fumes in the air? Your proximity?
- Extreme fatigue? When, where, duration, frequency?
- Chemical spills? When, where, duration, frequency?
- Dermatitis? Itching? When, where, duration, frequency?
- Elaborate ventilation systems in plant? Local exhaust ventilation?
- Labels, placards, warning signs, and posters? For what? MSDS? Training?
- Production or maintenance work?
- Nasal discharge? Epistaxis? Dirty residue in nostrils?
- Work and medical histories?
- Dirty work clothing? How soon into shift?
- Prior industrial hygiene studies? When? For what? Findings?
- Anything unique or distinctive in workplace since you noticed health problems?
- Special work clothing provided?
- Respirators provided? For what? What type? Do you use them? Other PPE?
- Other workers affected? How many? Clusters? Locations? Who?
- Community complaints? Stack emissions?
- Weight loss of more than 1.5% in any 2–8-hour period (dehydration).
- What do you think is the major cause of your health problems?

691. Discuss the important statistical underpinnings of the NIOSH/OSHA action level.

There are exposure limits for about 700 occupational air contaminants—a paltry few of the 4,000,000 chemicals so far identified and about one-third of 2000 chemicals that comprise 95% of chemical commerce in our nation. Perhaps fewer than 50 or so have substantial toxicological, epidemiological, and a foundation of clinical case reports. The remainder has much less than ideal research data and experience to support the limits. Animal data helps, but can mislead.

Since occupational toxicology is a highly inferential undertaking, there can be great risks to some workers if limits are applied using a binary attitude: Exposures above the limit are *per se* toxic, and exposures below the limit are acceptable. No! Most seasoned industrial hygienists would hope that it was that simple. How often have toxicologists, safety engineers, physicians, and industrial hygienists been asked: "Is it toxic?" Dose, dose rate, individual susceptibility, comorbidity factors, and everything Dr. Paracelsus taught us come into play. Although this question is simplistic, that, for the most part, is how most see our world: toxic or nontoxic? Good or bad? OK or not OK? Compliant or noncompliant?

The limits for the chemicals are decreasing as more is learned about their adverse effects on human health. For example, the limit for CO gas was 100 parts per million of air (ppm_v) when the author began his professional career. The limit for CO was reduced to 50 ppm_v, then 35 ppm_v, and it now is 25 ppm_v. Another chemical believed to be a simple CNS narcotic is now an aggressive liver and lung carcinogen. Still another treated as a simple upper respiratory irritant is now known to be a human teratogen capable of inducing severe cardiac malformations *in utero*.

Enter the action level! Considered by some to be a safety factor, it has much more far-reaching significance. The eminent NIOSH statisticians, Busch and Leidell did elegant research showing, for the most part, a one-day full-shift air sample collected in John's breathing zone cannot be used to infer John's exposure yesterday or what it would be tomorrow. We could infer a little about other workers near John when the air sample was taken, but little else. To deduce anything about John's exposure in the future, we need at least 2 days of air sampling. The action level, for all but lead and cotton dust, is defined as 50% of the exposure limit whether it is an OSHA PEL, an ACGIH TLV, or a NIOSH REL.

The statistical power of the action level is this: If a worker's TWAEs to an air contaminant or to a mixture of air contaminants are less than the action level on two days of sampling, it can be concluded at 95% confidence that future exposures of the worker will be below PEL, TLV, or REL if conditions in the worker's environment do not change. One day of air sampling does not cut it.

What can change? Production rates, ventilation, overtime, chemical concentrations, work practices, process changes, and so on, but if these are carefully documented during the initial study, future air samplings to ensure compliance are normally not required unless change has occurred.

692. On your first day on the job as an OEM and SHE professional, you are introduced to your coworkers, are shown your office and coffee pot, and the lavatory. You set up your files and books, adjust your chair, boot your computer, and study the company organization charts and telephone book. You then go to lunch with some of your new colleagues. That afternoon, you decide to conduct a preliminary walk-through audit of one nearby plant in your company. Through education and training, you learned that situations

arise that do not require air sampling, but prompt intervention is imperative. The six examples set out below require resolution—some with much greater urgency than others. Let us see how you do.

1. A worker is enveloped in clouds of silica flour dust as he slits 20 75-pound bags and dumps the powder into a slurry mixing tank. He does this once per workday.
2. A pipe fitter is exposed to cough-producing concentrations of chlorine gas while making adjustments and repairs to a high-pressure chemical plant process line.
3. Refrigeration plant workers are exposed to ammonia gas that irregularly causes throat irritation and lacrimation.
4. Employees are periodically exposed to offensive odors that intrude from a nearby butcher scrap meat rendering plant. The odors come and go. They smell offal.
5. Extensive mold is growing on a few ceiling tiles in the plant manager's office.
6. A worker without a SCBA is about to enter an empty confined space process tank to retrieve a man who collapsed from whatever unknown cause.

Obvious problems require intervention stat, not air samples!

Here's your quiz: Rank the six from highest priority to the lowest. Compare your set with the correct answers 13, 9, 10, 8, 11, and 12 by subtracting 7 from each of these. Great! You are an interventional OEM—SHE professional unsurpassed at triage!

Can you see how the amateur industrial hygienist or other OEM—SHE "professional" could dally about taking air samples before standing up to be counted to intervene on behalf of these workers? Obvious war problems require obvious prompt inserts.

As General George Patton bellowed:
"Lead, follow, or get out of my way!"

693. Discuss industrial hygiene ramifications of pregnant workers.

There are several industrial hygiene risk factors in workplaces that can be significant to pregnant employees. Of course, in every case, dose and dose rate are important. Exposures in early pregnancy for certain chemical and physical agents can be more significant than in late pregnancy. Conversely, musculoskeletal issues assume more importance toward the end of pregnancy. These hazards include:

• Excessive *heat strain*, especially in the first trimester, can cause developmental defects in the embryo/fetus neural tube. Heat stress is a teratogenic hazard.
• Heavy metals, especially *lead and mercury*, can cause birth defects, miscarriages, neurological impairments, and other adverse medical outcomes.
• *Ionizing radiation*, such as x-rays and radioactive materials including hazardous waste containing radionuclides, can cause abnormal fetal

development, cancer, mutations, miscarriage, and other dysfunctions—developmental and metabolic.

- Excessive exposure to workplace noise may adversely affect the child's hearing.
- Research animals exposed to *pharmaceutical agents* and *chemotherapeutic drugs* have demonstrated low birth weights, miscarriages, and deformations.
- *Infectious microbes* (e.g., rubella, hepatitis) are risk factors for both the mother and the fetus. Pregnant health care workers, in general, are a high-risk work group.
- Prolonged standing or sitting, heavy lifting, and walking are special *ergonomic factors* for pregnant workers. Job stresses, multitasking, circadian rhythms, and variable work shifts including prolonged overtime work can have adverse risks.
- Some *solvents*—inhaled and/or skin-absorbed—can produce fetal deformities (e.g., heart, brain) and other adverse developmental outcomes.
- *Chemical asphyxiant gases* such as CO, H_2S, and HCN can impede high oxygen demands of the developing embryo/fetus.
- While at work, and off the job, the pregnant woman should not smoke or chew *tobacco* or consume *alcohol*. While her obstetrician or family physician may have advised her of these risk factors, the OEM physician, OEM nurse, and industrial hygienist should reinforce these prudent practices with the worker at her job site.

The American Medical Association recommends the following for pregnant workers:

- Take a rest break every few hours.
- Take a longer meal break every four hours.
- Drink plenty of fluids while on the job.
- Vary work positions continuously, from sitting to standing to walking.
- Minimize heavy lifting and bending.

Heavy lifting becomes increasingly difficult during pregnancy. The gain in weight strains the back. This is compounded by lifting heavy objects. Frequent lifting and awkward postures are additional risk factors for back injuries. OEM professionals should educate and train pregnant workers the following ergonomic practices:

- Stand with both feet apart separated to the outer width of your shoulders.
- Tuck in your buttocks.
- Slowly bend at your knees—not at the back. Keep your back as straight as possible.
- Lift with the legs and the arms—not with your back.
- Limit the weight, size, and frequency of lift of objects. If in doubt about weight, enlist other workers to do the lifting or supply mechanical lifting devices.

- Back braces and supports are not recommended. This applies not just to women who are pregnant, but to all workers—male and female—as well.

694. Discuss toxicant absorption and excretion vis-á-vis first-order kinetics.

The following diagram depicts absorption and excretion (actual toxicant or one of its metabolites) of a toxicant following typical first-order kinetics. Most toxicants follow this pattern in a healthy person.

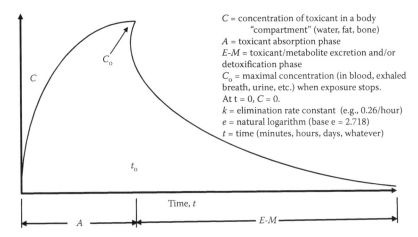

C = concentration of toxicant in a body "compartment" (water, fat, bone)
A = toxicant absorption phase
E-M = toxicant/metabolite excretion and/or detoxification phase
C_0 = maximal concentration (in blood, exhaled breath, urine, etc.) when exposure stops. At $t = 0$, $C = 0$.
k = elimination rate constant (e.g., 0.26/hour)
e = natural logarithm (base e = 2.718)
t = time (minutes, hours, days, whatever)

In toxicology, the body is divided into three "compartments": aqueous (blood, organs, muscles, tissues, lymph, etc.), lipids and fats, and mineralizing (bones, teeth). The curves shown above vary from toxicant to toxicant and person to person. However, they show absorption is rapid at first and later slows (saturation, tolerance, impaired enzymatic pathways, cofactors, etc.). Elimination is the inverse with rapid loss when exposure stops. One becomes "drunk" quickly. "De-tox hangovers" linger for hours.

For elimination: $C = C_0\, e^{-kt}$. This is the simplest mathematical model by which a toxic substance is eliminated from the body. The negative sign indicates loss or decay.

The elimination half-life is: $0.5\, C$ at $T_{1/2} = C_0\, e^{-kT}$. Note this is T, not t ($T = 0.693/k$).

For example, what fraction of the inhaled, absorbed dose remains in the body after a time $= 9\ T$? Remember that detoxification is occurring as intoxication proceeds.

$$C = C_0\, e^{-k9T}$$

$$C/C_0 = -1/e^{kT9} = 1/e^{\ln 2 \times 9} = 1/2^9 = 1/512 = 0.002\ C_0.$$

How much toxicant is in the body 50 hours after exposure stops if $T = 14.3$ hours?

$$50/14.3 = 3.5 \quad C/C_0 = 1/2^{3.5} = 1/11.3 = 9\%\ \text{of}\ C_0.$$

695. Provide a reasonable solvent selection guide policy for your industrial client.

SHE and OEM personnel must be vigilant and mindful of organic solvents used in their production and maintenance operations. The proper selection of a solvent is a robust, sophisticated undertaking. A choice to use a particular solvent over another can be daunting especially when chlorinated solvents are considered. Solvents are usually selected by their solvency power, past experience with the solvent in other operations, and, sometimes, costs. Rarely are inhalation toxicity, dermal absorption, evaporation rate, flammability, and chemical stability considered. For example, a highly volatile and relatively nontoxic, inexpensive solvent can be more costly and hazardous to workers' health than a more toxic solvent with a low evaporation rate. The use of highly volatile solvents in a confined space is particularly hazardous.

The following guidelines should be applied when selecting a solvent:

1. Try tap water at room temperature first. Next, consider cool water at high pressure.
2. Failing the above, try warm tap water. Next, consider warm water at high pressure.
3. Failing the above, try hot water not to exceed 120°F (instantly scalding).
4. Failing the above, consider steam using special burn and scald prevention methods.
5. Failing the above, try tap water at room temperature with a mild detergent.
6. Failing the above, use the same detergent with increasing water temperature not to exceed 120°F. Contact times cooler and longer are better than hot and brief.
7. Failing the above, consider selecting other cleaning methods, for example, dry ice pellets or liquid nitrogen blasting with good local exhaust ventilation and hypothermic protective clothing, ultrasonic cleaning, presoaking before cleaning, oven baking.
8. Failing the above, consider selecting a solvent, or a solvent blend, that meets the following criteria: flashpoint >120°F, highest published occupational exposure limit >100 ppm_v, not absorbed through intact skin, noncorrosive, chemically inert, free of analytically detectable benzene and other recognized carcinogens and teratogens, noncontributive to destruction of ozone layer in the stratosphere, vapor pressure below room temperature water, and compatible with the work process and materials.
9. Failing the above, consult an experienced industrial hygienist.
10. The following organic solvents, in the author's view, should never be used for any cleaning activities: benzene, mono-chloroacetic acid, chloroform, carbon tetrachloride, methylene chloride, trichloroethylene, enthers of any kind, some glycol ethes, chlorofluorobromo hydrocarbons (*Freons*®), n-hexane, and carbon disulfide.

11. Strong acids and alkalis are often inappropriately considered solvents. Their use should be reviewed by an industrial hygienist or chemical safety engineer.

696. Workers, more often than not, are potentially exposed to multiple stressors as they go about earning their living. OEM and SHE professionals must always be mindful of this because it is very easy to lose sight if our primary focus at the time is, for example, evaluating workers' exposures to welding fumes and gases. We must constantly ask: Besides fumes, "Is John exposed to other stressors in the major industrial hygiene categories: chemicals, physical agents, ergonomics, biological factors"? Multiple exposures (not always concurrent) impacts workers' well-being by additive, synergistic, and potentiating mechanisms or otherwise exacerbates occupational and nonoccupational health conditions. Use a well-designed check list at the job site will help us to be mindful of all risk factors impacting our clients.

This photograph of an electric arc ("stick") welder in a fabrication shop at one of author's clients reveals multiple industrial hygiene stressors listed below (before interventions).

Although we cannot "see" the noise, this gentleman is also exposed to welding fumes, smoke, and gases; airborne dust; ergonomic and postural issues; thermal skin burns; poor ambient lighting; radiation to eyes and skin; electric shock; heat strain, injuries by tools; and fires (requires a "hot work" permit). Can you identify other stressors? Human factors? Bully boss? Production demands? Note the objects on his tool belt: wire brush, meat cleaver, axe, ball peen hammer. Hand and power tools can be a clue to types of air contaminants the worker generates. Does this welder use the wire brush to remove surface materials such as lead, zinc, cadmium, and so on before welding? Does such dust contribute to his exposure?

697. There is increasing evidence that several industrial chemicals can induce hearing loss. Comment on this.

Ototoxic Chemicals + Noise = Whudyasay?

Toxicologists, audiologists, and others are increasingly finding a relationship between hearing loss and occupational and/or avocational exposures to certain chemicals. We can expect additional linkages in the future. In the interim, it would be prudent to control inhalation of, and dermal contact with, organic chemicals—particularly low molecular weight solvents—and metals known to be neurotoxic. Chemical asphyxiant gases and vapors— besides carbon monoxide—such as hydrogen sulfide, cyanide salts and hydrogen cyanide, aniline and homologues, and some pharmaceuticals should also alert SHE professionals.

It is unlikely that MSDS—in the near future—will declare possibility for products that contain known or reasonably suspected ototoxicants. Therefore, this topic should be covered in HazCom training sessions, with patients in the clinic, and one-on-one by SHE professionals as they interact with workers and supervisors. Those workers with exposures to high noise levels and ototoxic agents have—presumably—the risks of additive hearing loss. That is, noise + solvent vapor inhalation + dermal absorption of solvents = hearing loss greater than would be expected from independent exposures.

Effects on hearing most likely do not require concurrent exposures. For example, a paint sprayer inhales toluene and n-hexane for 10 years. He loses hearing, but never worked near high noise sources. For the next 10 years, he worked as a mechanic where he inhaled carbon monoxide gas and Stoddard solvent vapors. Without wearing impervious gloves, he washes oily parts in Stoddard solvent and gasoline (contains n-hexane). Noise exposures were intermittent. He is now 40 years old with the hearing of a 75-year-old man without a documented history of exposures to solvents, metals, noise, and carbon monoxide gas. In noise-induced hearing loss training programs, the author has shared this wake-up call: "When you reach 55, 65, or 75 and choose to retire because you finally have cash money for that dream stereo system, you won't be able to enjoy it."

The following chemicals are ototoxic (* = high priority ototoxicants per NIOSH):

Carbon monoxide*	Carbon disulfide
Arsenic	Mercury and mercury compounds
Lead and lead compounds*	Stoddard solvent (mineral spirits)
Cyanides	Styrene
Manganese	Toluene*
n-Hexane	Xylene isomers
Some aminoglycoside antibiotics	Organic tin compounds
Trichloroethylene*	

698. List six key points of occupational toxicology that influence human dose response.

Chemicals and radioactive materials are potential toxicants, whereas materials biologically derived are potential toxins. Toxicant will be used interchangeably for each type of poison here. One may be intoxicated by both a toxicant and a toxin via different biomechanisms.

Toxicology is the science and study of the effects of toxicants on living systems (plants, animals, and microbes). Occupational and environmental toxicology focuses primarily upon the adverse effects of toxicants on workers and people outside of workplaces. Since workers go home, they can, for example, be exposed to carbon monoxide gas in both environments: CO as a fork lift truck driver at work, CO from their active tobacco smoking, CO from passive smoke of others, CO in ambient air, and CO from a faulty gas hot water heater, fireplace, car, and furnace. When using a paint stripper that contains methylene chloride solvent (CH_2Cl_2), the inhaled and absorbed vapors are metabolically converted into CO *in vivo*. If this worker then inhales hydrogen sulfide gas (H_2S) while repairing his home septic tank, his ischemic high oxygen demand myocardium is stressed by both asphyxiants. Death could be at hand.

There are six key factors affecting responses to toxicants. Carefully consider all in your health hazard assessments.

1. The *toxicant itself.* Some produce immediate and dramatic adverse biological effects. Others may produce no discernible adverse effects—short or long term. Others may produce adverse effects delayed as long as 10–40 or more years later (carcinogens). Chromium^{+6}, for example, does both.

2. The *type and mode of contact.* Inhalation, percutaneous absorption, ocular absorption, direct injury to the skin and eyes, ingestion. Some toxicants are harmless after one mode of application (e.g., skin), but have serious effects when contacted another way (e.g., lungs). Brief dermal contact with silica dust is unremarkable, but chronic inhalation of high amounts of respirable silica dust (quartz) can produce a progressive, nodular pulmonary fibrosis that leads to an early death.

3. *Dose of toxicant.* The amount actually absorbed into the body or in direct contact with tissues that interface with our environments: respiratory tract, skin, eyes, and gastrointestinal tract.

4. *Duration of exposure.* Some toxicants produce signs and symptoms after only one exposure (acute), some only after exposures of long time periods (chronic), and some produce adverse health effects from both acute and chronic exposures. For example, a one-time excessive dose of inhaled hydrogen chloride gas can be lethal or cause coughing and a reversible, treatable chemical pneumonitis, but inhalation of this gas at sub-acute levels in the air for many years can cause severe corrosion of tooth enamel, bronchial asthma, and laryngeal cancer. The dose of a toxicant is, then, a function of concentration and duration of exposure—sometimes referred to as the dose rate. Equal dose rates in John, Mary, and Bob may produce three different responses.

5. *Individual sensitivity.* Humans vary considerably in their adverse responses from exposures to toxicants. These can range from no

discernible adverse effects to mild signs and symptoms to serious illness to death. Different effects can occur in the same person at different exposures. Certain preexisting medical conditions can influence and promote exaggerated response to toxicants.

6. *Interaction with other toxicants or agents.* Toxicants in combination with other toxicants or agents can produce different or exacerbated adverse biological responses observed when the exposure is to one toxicant alone. There may be comorbidity between inhaled chemical at work and prescribed oral medication or an over-the-counter drug. Adverse toxic effects can be additive, antagonistic, synergistic, or potentiating. The example given in the preface of a worker inhaling CO, H_2S, and CH_2Cl_2 is a simple example of additive toxicity. These can be sneaky.

699. By now, readers have encountered several ways to calculate the air contaminant concentration in a space when the volume of space is known, the generation rate of air contaminant is known, the initial concentration is known, and the ventilation rate of space is given or assumed. American Society of Heating, Refrigeration, and Air-Conditioning Engineers provides another formula for these calculations:

$$t = (v/V)\ln\left[\frac{(C_i)V/(N-1)}{X}\right]$$

where
t = time required to lower the concentration to a fraction X above the final steady-state concentration (X should, in turn, be a fraction of the most restrictive limit),
v = room volume,
V = ventilation rate,
N = contaminant generation rate,
C_i = initial concentration

This equation is elegant, however, the author recommends adding a coefficient, K, to the right-hand side of equation from 1 to 5 where 1 = excellent ventilation air distribution, and 5 = extremely poor air mixing—both judgment calls of the industrial hygienist.

Assume an initial concentration of 200 ppm_v carbon monoxide, a small garage 20' × 30' × 10' high, ventilation rate of 2.5 cfm/ft² floor area, CO generation rate of 32 ft³ per hour, X selected at 0.1 (= 20 ppm_v), and air mixing factor of 3. The garage is empty but for an unattended small gasoline-fueled electricity generator.

$$V = (20' \times 30') \times 2.5 \text{ cfm/ft}^2 \times 60 \text{ minutes} = 90,000 \text{ ft}^3/\text{hour}$$

$$v = 20' \times 30' \times 10' = 6000 \text{ ft}^3$$

$$t = 3\,(6000\,\text{ft}^3/90{,}000\,\text{cfh})\,\ln\left[\frac{(200\,\text{ppm}_\text{v}) \times 90{,}000\,\text{ft}^3\ \text{hour}/32\,\text{ft}^3/\text{hour} - 1)}{0.1}\right]$$

$$= 15.6\,\text{hours}.$$

In other words, 15.6 hours/0.1 = 156 hours (assuming generator does not run out of fuel, ventilation remains constant, ambient CO level does not increase, etc.). What this calculation tells us is the eventual steady state of 20 ppm$_\text{v}$ is reached in 156 hours (6.5 days!). Clearly, this is way too long because a worker, believing the fan operated long enough for 5 hours, might wrongly assume a safe atmosphere.

Other than turning the generator off, the ventilation must be greatly increased, and mixing clean outdoor air with indoor air must be improved. One cannot reasonably allow the generator to run in this open space and expect dilution to 20 ppm$_\text{v}$ in a short period of time. One must not enter until air quality has been quantitatively determined "safe," or those who do wear self-contained breathing apparatus while under observation by others remaining outside.

700. Analytical chemists are now able to detect some environmental toxicants to as low as 1 part per trillion. Most people, including the author, find the concept of such a tiny amount difficult to grasp. Using an example, try to make this meaningful to people with the need to know—for example, a group of workers for whom you are conducting chemical hazard communication training classes per OSHA's 29 *CFR* 1910.1200.

   31,688 years have 1 trillion seconds. That is, 1 second in 31,688 years is 1 part per trillion. Oops! I forgot to include leap years. A part per billion is 1 second in 31.688 years. The chemist can search and identify that single second. And we say, "Where did the time go?"

701. When plotting data consider which regression trend line has the best fit: normal or logarithmic as depicted below:

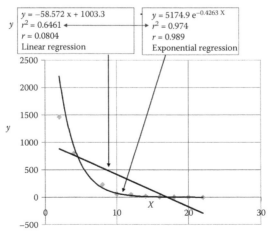

Clearly, the exponential regression has the best fit (r^2: $0.974 > 0.6461$) providing superior predictive value particularly for higher values of y and lower values of x.

As seen here, the familiar algebraic equation $y = mx + b$ (equation for a straight line) has limitations when paired data sets are not linear correlates. Empirical data for the above graph are:

X	Y
2	1460
4	800
8	230
10	60
12	50
14	20
16	5
18	4
20	1
22	0.2

702. Refer to Problem 544. In this modification, ducts bring in outside air to this building at 4000 cfm. So, instead of calculations based on x air changes per hour, use the following formula:

$$\frac{240,000\,\text{ft}^3/\text{hour} \times (1\,\text{air change/hour}) \times 47\,\text{grains}}{13.5\,\text{ft}^3/\text{lb of air} \times 7000\,\text{grains/ft}^3} = 119\,\text{lb}\,H_2O\ \text{per hour}$$

where $240,000\ \text{ft}^3/\text{hour} = 4000\ \text{cfm} \times 60$ minutes/hour

703. The author, early in his career, did not have access to most of the sophisticated instruments available in today's modern industrial hygiene laboratories. He only had a visible wavelength spectrophotometer, a crude balance worthless for most gravimetric methods for airborne dust, pH meter, a few direct reading devices for combustible gases and mercury, an oxygen meter, microscope, combustible gas indicator, sound level meter, an electrostatic precipitator, sling psychrometer, and numerous detector tubes for common gases and vapors. Basic tools, yes, but he had to improvise devices and methods in many situations.

Colorimetric methods were used for mercury, lead, cadmium, nickel, manganese, a few other metals, and SO_2, H_2S, Cl_2, and TDI. Many of the chemicals made and used by his employer were highly acidic or alkaline. For these, he relied upon his knowledge of chemistry and acid–base reactions in what he termed "gas phase air titration." A midget impinger was used in place of, and analogous to, a burette.

For example, to measure NH_3 and organic amine vapors, he prepared a very dilute solution of sulfuric acid (0.0001 N) adding bromocresol purple as an indicator. 10 mL were added to impingers. Breathing zone air was bubbled through the solution until there was a distinct change from yellow to purple. Based on calibrating visual color changes in the calibration chamber for several concentrations, he now could measure alkaline air contaminants directly in the field without a need for laboratory analysis. The time to a color change was the average vapor or gas concentration. Field results were typically ±10% of true calibration value. Practical, satisfactory.

The advantages of this method are immediate results, no costly laboratory bench analyses, ability to discuss results immediately with workers and their supervisors, better accuracy than most detector tubes, air-sampling times (depending on acid or base strength of impinger solution) of approximately 15 minutes or less based on air concentration, and low cost. A disadvantage is lengthy, necessary one-time calibration protocols. Another disadvantage was the presence of other acidic or alkaline air contaminants competing for the indicator. Besides, it is fun to "titrate" the air. The worker, see the color change, is impressed and now teachable.

A list of various indicators follows. Select one near neutral pH with abrupt, narrow change of color and narrow range. The author used those in italics with success for alkaline and acid gases and vapors. Others were not evaluated.

Indicator	Low pH Color	pH Range	High pH Color
Gentian violet	Yellow	0.0–2.0	Blue-violet
Leucomalachite yellow	Yellow	0.0–2.0	Green
Leucomalachite green	Green	11.6–14	Colorless
Thymol blue B	Red	1.2–2.8	Yellow
Thymol blue	Yellow	8.0–9.6	Blue
Methyl yellow	Red	2.9–4.0	Yellow
Bromophenol blue	Yellow	3.0–4.6	Purple
Congo red	Blue-violet	3.0–5.0	Red
Methyl orange	Red	3.1–4.4	Orange
Bromocresol green	Yellow	3.8–5.4	Blue
Methyl red B	Red	4.4–6.2	Yellow
Methyl red	Red	4.5–5.2	Green
Azolitmin	Red	4.5–8.3	Blue
Bromocresol purple	Yellow	5.2–6.8	Purple
Bromothymol blue	Yellow	6.0–7.6	Blue
Phenol red	Yellow	6.4–8.0	Red
Naphthol red	Red	6.8–8.0	Yellow
Naphtholphthalein	Colorless to reddish	7.3–8.7	Greenish to blue
Cresol red	Yellow	7.2–8.8	Reddish-purple
Phenolphthalein	Colorless	8.3–10.0	Fuchsia
Thymolphthalein	Colorless	9.3–10.5	Blue
Alazarine red	Yellow	10.2–12.0	Red

The equation for a sampling rate of 0.1 cfm (2.832 Lpm) using a 0.01 N reagent is

$$\frac{0.2445 \times 10^6 \times \text{mL reagent}}{2832 \times \text{minutes sampled}} = \frac{86 \times \text{mL reagent}}{\text{minutes sampled}}$$

1 mL 0.01 N reagent is equivalent to 0.2445 mL monovalent gas or vapor at 25°C and 760 mm Hg pressure. The industrial hygienist records the pump's sampling time to observe a color change in the impinger solution. See Problem 429.

Air titration works best for alkaline air contaminants. Always use non-volatile acids such as H_2SO_4 in the impinger or fritted gas bubbler because the concentration of volatile acids (e.g., HCl, acetic, HNO_3) reduces as air sampling proceeds. Results would be unreliable. The use of dilute alkaline solution for acidic air contaminants is unreliable because atmospheric CO_2 competes for alkali reactant. For example, if one selected 0.0001 N NaOH to sample HCl in air, the reaction:

$$NaOH + CO_2 \rightarrow NaHCO_3$$

interferes because sodium bicarbonate results from CO_2 scavenging NaOH in the impinger.

704. Morpholine (not morphine!) is used for, among other applications, an emulsifier for waxes to coat fruits such as apples and pears and as a boiler corrosion inhibitor in steam-powered electricity generation plants. Its vapor pressure is 7 mm at 20°C. The molecular weight is 87.1 g/mole. A 329.4 liter air sample obtained at 722 mm Hg and 17.5°C contained 17.6 mg. Calculate the concentration in parts per million by volume for this worker applying the NIOSH coefficient of variation.

$$ppm_v = \frac{17.6\,mg}{0.3294\,M^3} \times \frac{22.4}{87.1} \times \frac{273.16\,K + 17.5°C}{273.16\,K} \times \frac{760\,mm\,Hg}{722\,mm\,Hg}$$
$$= 15.39\,ppm_v$$

NIOSH reports combined correlation coefficient (air sampling, laboratory analysis) for morpholine at 0.06. Therefore, the true value at the 95% confidence level for this test result is

$$15.39\,ppm_v \times 0.06 = \pm 0.92\,ppm_v$$

$$15.39\,ppm_v \pm 0.92\,ppm_v = 14.47\,ppm_v - 16.31\,ppm_v\ (i.e.,\ 14.5\,ppm_v$$
$$to\ 16.3\,ppm_v)$$

This range is below the OSHA's PEL and the ACGIH TLV of 20 ppm_v. However, it exceeds the NIOSH action level. Prompt industrial hygiene intervention is required.

705. Consider the previous problem (704). Another chemical safety engineer evaluated the same worker on the same day using colorimetric detector tubes. Several were used throughout the workday, and the average breathing zone concentration was 17 ppm_v. Again, using the NIOSH coefficient

of variation for detector tubes (0.14), calculate the 95% confidence level range. With both methods, the lower ends of the 95% confidence levels are essentially identical (14.47 ppm_v vs. 14.6 ppm_v).

$$17 \ ppm_v \times 0.14 = \pm2.38 \ ppm_v$$

$$17 \ ppm_v \pm 2.38 \ ppm_v = 14.6 \ ppm_v - 19.4 \ ppm_v$$

Both air sampling and analysis methods demonstrate reasonably good agreement between results. Sometimes it is not the means but the ends. Regardless of the evaluation method used, intervention is required. Slightly higher readings obtained with the detector tubes could be attributed to bias in determining the length of stain readings. The reader, by now, will have recognized that preservation of employee health is paramount—not necessarily the method selected to evaluate exposures when there is slight disagreement between the results.

NIOSH published combined (air sampling and/or laboratory analysis) coefficients of variation for several methods as listed below:

Colorimetric detector tubes	0.14
Rotameter on personal pumps (sampling only)	0.05
Charcoal tubes	0.10
Asbestos (sampling, counting)	0.24–0.38
Respirable dust, except coal mine dust (sampling, weighing)	0.09
Total airborne dust	0.05

The coefficients are not surprising for seasoned practitioners with field experience in these instruments and subsequent analysis of the laboratory samples. See the NIOSH's *Occupational Exposure Sampling Strategy Manual*, N.A. Leidel, K. Busch, J. Lynch (1977).

706. The exhaust gas velocity in a 14" internal diameter chemical plant stack at NTP is 2260 fpm. The concentration of sulfur dioxide gas in these exhaust gases is 31.7 ppm_v. How many pounds of SO_2 are emitted per hour from this stack?

$$\text{Stack cross-sectional area} = \pi r^2 = \pi(7")^2/144/in^2/ft^2 = 1.0724 \ ft^2$$

$$\text{"molecular weight" of air} = 28.966 \ g \ mole^{-1}$$

$$\text{Molecular weight of } SO_2 = 64.066 \ g/mole$$

$$Q = AV = 1.0724 \ ft^2 \times 2260 \ fpm = 2423.6 \ \text{standard cubic feet per minute}$$

$$\text{lbs/hour} = \text{scfm} \times \text{molecular weight} \times \frac{(60 \ minutes/hour) \times 0.07639}{(28.966 \ g/mole) \times 10^6}$$

$$\text{lbs/hour} = 2423.6 \text{ scfm} \times 64.066 \text{ g/mole} \times \frac{(60 \text{ minute/hour}) \times 0.07639}{(28.966 \text{ g/mole}) \times 10^6}$$

$$= 0.02457$$

707. The architectural design plans for a new school stipulate an air change every 20 minutes for each classroom. Each classroom is 35' × 25' × 10'. How many cubic feet of tempered, clean air are required per minute per classroom?

$$35' \times 25' \times 10' = 8750 \text{ ft}^3$$

Base your calculations on an empty classroom:

$$\text{cfm required} = \frac{\text{room volume}}{\text{minutes per air change}} = \frac{8750 \text{ ft}^3}{20 \text{ acm}} = 437.5 \text{ ft}^3/\text{minute}$$

How many cfm/ft^2 of floor area is this?

$$437.5 \text{ cfm}/35' \times 25' = 0.5 \text{ cfm/ft}^2$$

This air exchange rate might be low. Reconsider 1.0 cfm/ft^2. Consider occupants' comfort, the air flow distribution patterns, climate and energy costs, tempering and humidifying, and students' physical activities (e.g., music, dance vs. mathematics, history).

708. Automotive service garages commonly have flexible hoses that can be attached to the exhaust pipes of operating car and truck engines. These typically vent through a hole in garage service doors. Exhaust gas flow is not mechanically assisted with a fan. Problems with this approach to protect occupants' health by accumulation of CO include kinked hoses, hoses too short, back pressure from long hoses and condensed water vapor to liquid water, holes in the hose, and wind forcing exhaust gases back into the garage through cracks under doors. Workers sometimes fail to make the connections or to ensure a secure connection.

An example (with permission) from the 26th edition of *Industrial Ventilation ... A Manual of Recommended Practice* published by the American Conference of Governmental Industrial Hygienists (2007), demonstrates design requirements for mechanically exhausted tailpipe system: A 15-liter engine operating at 1000 rpm displaces 530 scfm. Under heavy load, corrected for a 1300°F exhaust temperature, the actual exhaust is 1758 acfm. This example applies a 20% safety factor in the equation:

$$Q, \text{acfm} = (1.2)\left[15 \text{ L} \times \frac{0.0353 \text{ ft}^3}{\text{L}} \times 1000\right] \times \left[\frac{460°F + 1300°F}{530°F}\right]$$

$$= 2110 \text{ acfm}$$

This equation can be used for any internal combustion engine as long as engine rpm are known (design for maximum) and the exhaust gas temperature is known. It is far safer to vent exhaust gases through a high stack instead of through the door. The "safety factor" can be eliminated with excellent design. The author suggests installation of CO gas alarms at strategic locations in the shop and education and training of workers in the hazards of CO gas inhalation. A common myth is that engines with catalytic converters generate less CO. This is true if the converter is not fouled and is hot. Cold starting generates much CO until convertor acquires a high operating temperature.

These calculations also apply to engine dynamometer actual operating conditions testing.

709. A ranch house built on a concrete slab above a crawl space has wood, linoleum, and tile floors for all floor surfaces. Area dimensions for rooms are living room (22' × 26'), two bedrooms (10' × 12'/each), dining area (10' × 8'), kitchen (8' × 10'), total hall area (3.5' × 18'), four closets (3' × 3.5'/each), and a bathroom (6' × 8'). A virgin HEPA dust filter was used on the vacuum machine to double sweep all floor areas. Laboratory analysis of filter determined 2267 milligrams of inorganic lead (spalled paint, window well chips, chipped door casings), outdoor soil tracked inside, etc.).

What is the minimum average Pb dust concentration in micrograms per square foot?

Living room	22' × 26' =	572 ft^2
Bedrooms	2 (10' × 12') =	240 ft^2
Dining room	8' × 10' =	80 ft^2
Kitchen	8' × 10' =	80 ft^2 (including counter areas)
Hall areas	3.5' × 18' =	63 ft^2
Closets	4 (3' × 3.5') =	42 ft^2
Bathroom	6' × 8' =	48 ft^2 (including bathtub, counters)
	Total house floor area =	1125 ft^2

$$2{,}267{,}000 \text{ mcg Pb}/1125 \text{ ft}^2 = 2015 \text{ mcg/ft}^2$$

This is over 10 times the current Housing and Urban Development standard of a maximum of 200 micrograms of lead per square foot of interior floor surface area. Occupancy must be curtailed immediately and not allowed until a robust cleaning and abatement program demonstrates acceptable lead-free surfaces. Some key applicable regulations are OSHA's (29 *CFR* 1910.1025)—for all lead abatement workers, HUD's *Lead-Based Paint Poisoning Prevention Act*, Title X *Residential Lead-Based Paint Hazard Reduction Act*—for residential occupants, and EPA's hazardous waste disposal, and EPA's *Clean Air Act*—both for our environment.

One, of course, must be aware that vacuum sweepers have different efficiencies. So, the result in the example problem is low. Canister vacuums are the best for hard surface floors. Those with power brushes could be up to 6 times more efficient than those without. Shag carpets and deep plush rugs are the most difficult to clean. Upright vacuum sweepers will remove 35–55% of recently deposited dust. Older dust is often more difficult to vacuum clean.

710. Until now, we have not applied Henry's law: At constant temperature, the amount of gas (or as vapor) that dissolves in a given type and volume of a liquid is directly proportional to the partial pressure of that gas in equilibrium with that liquid. Think of an un-opened bottle of soda or a sparkling wine where the head space is rich in CO_2. After opening, head space loses pressure from the CO_2. *Ergo*, so will the carbonated liquid and, in time, "goes flat."

The following example is modified with permission from a problem in professional development course attended by the author (*Mathematical Modeling of Exposures*, given at American Industrial Hygiene Conference and Exhibition, Toronto, Course 707, 2009).

Consider a 1 cubic meter (264 gallons) open surface tank of water that contains 10 mg benzene/liter (10 $ppm_{m/v}$). The empty room volume is 100 m^3 (3531.5 ft^3), and there is no ventilation of the room with outside air. A 3531.5 ft^3 room is, for example, 20' × 20' × 8.83'. For reference, the solubility of benzene in water at 15°C is 1.8 g (1800 mg) per liter of water.

10 mg benzene/L H_2O = 0.001%—is much less than the 0.1% stipulated by OSHA to be reported on Material Safety Data Sheets for carcinogens in product mixtures.

Henry's law constant for benzene in water is 0.22 (no units). This constant is the water-to-air partition coefficient for benzene for a set volume and temperature. Or, 0.22 = C_{air}(benzene in air)/C_{liquid}(benzene in water).

Apply a mass balance at equilibrium: initial mass of benzene in water = final mass of benzene in water + benzene vapor in air. That is, benzene that was dissolved in water is now in air as vapor, and some benzene remains dissolved in the water in this two-phase system at equilibrium.

$$2m^3 \text{ tank} = 1000 \text{ L}$$

$$1000 \text{ L} \times 10 \text{ mg/L} = 10,000 \text{ mg total benzene in water tank}$$

$$0.22 \times (C_{\text{benzene in air}} \times 99 \text{ m}^3) + (C_{\text{benzene in water}} \times 1 \text{ m}^3) = 10,000 \text{ mg benzene}$$

$$C_{\text{benzene in water}} = 459 \text{ mg/m}^3$$

$$C_{\text{benzene in air}} = 0.22 \times C_{\text{benzene in water}}$$

$$0.22 \times 459 \text{ mg/m}^3 = 101 \text{ mg/m}^3 = 31.6 \text{ ppm}_v$$

31.6 ppm_v is over 30 times the OSHA's PEL for benzene and 316 times NIOSH REL for benzene. Both limits are worst acceptable vapor concentrations for this genotoxic hematopoietic carcinogen.

711. Industrial hygienists investigate workplace health disorder outbreaks to determine if there is any work-relatedness in groups of those of concern. The following is an analytical tool they use to this end.

Two-By-Two Tables			
Present	A	B	A + B
Absent	C	D	C + D
Total	A + C	B + D	A + B + C + D

Calculations and comparisons of disease or injury incidence rates are helpful in determining strengths of association between risk factors and a health problem.

The data are summarized and distributed into four cells: A, B, C, and D. These data are used to compare occurrence and exposure. The standard format is

<u>Health Event, Injury, or Disease</u>
present, absent, total

exposure

A = those with disease and exposure   A + C = those with disease
B = those exposed without disease   A + B = those with exposure
C = those with disease, but not exposed   B + D = those without disease
D = those neither exposed nor with   C + D = those without exposure
    disease   A + B + C + D = those at possible
                                                  risk

Apply this to a chemical plant where some workers present with chronic peripheral neuropathy of upper extremities—bilateral in all cases. The plant manufactures a neurotoxic pesticide. Research on the disease incidence and populations reveals: A = 43, B = 312, C = 2, and D = 401.

$$\text{Cumulative incidence of exposed group} = \frac{A}{A + B} = \frac{43}{43 + 312}$$
$$= 0.122 = 12.2\%$$

$$\text{Cumulative incidence in unexposed group} = \frac{C}{C + D} = \frac{2}{2 + 401}$$
$$= 0.005 = 0.5\%$$

$$\text{Relative risk} = \frac{A/A + B}{C/C + D} = \frac{43/355}{2/403} = 24.4$$

Group A was initially seven workers. Plant physician was suspicious of this apparently high incidence of affected workers and evaluated all at possible risk. This uncovered 36 more workers with bilateral peripheral neuropathy. Group A expanded to 43 employees.

These calculations help to compare risks that a worker in this plant will develop a nerve disease of this type. This, of course, depends on the worker's inhalation, dermal, and the ingestion dose of this pesticide, its chemical intermediates, raw materials, and other factors. Ratio of rates in the unexposed and exposed work groups are then compared—the relative risk (RR). If RR = 1.0, there is identical risk between the exposed and unexposed workers and is unremarkable. If RR exceeds 1.0, this suggests there could be association between the disease and exposure. This RR of 24.4, in the example, is extremely high. The 95% and 99% confidence intervals for RR can be determined from standard statistical tables. If RR is substantially below 1.0, one might tentatively conclude exposure presumably protects against a particular health issue: disease or injury.

This high RR of 24.4 requires a prompt investigation, reporting the data to public health agencies, and immediate industrial hygiene interventions. Investigation is worthy of reporting in the scientific literature. An alerting and prompt report to the EPA is required as a matter of law under the Toxic Substances Control Act. State OSHA and public health agencies require reporting of known or suspected disease originating in workplaces.

When considering medical statistics, we should be mindful of the late Sir Austin Bradford Hill who reminded us, "Health statistics represent people with the tears wiped off."

Last, J. M. and R. B. Wallace, *Public Health and Preventive Medicine*, 13th Edition, Appleton & Lange, 1992—an excellent textbook for practitioners.

712. Nonoccupational benzene vapor exposures of residents in urban and suburban areas of the United States are typically 5.76 mcg/m^3 (1.8 ppb$_v$). By how much will one's exposure increase if he or she has daily 8-hour occupational exposures at NIOSH's occupational REL of 0.1 ppm$_v$ (100 ppb$_v$)?

EPA (*Exposure Factors Handbook*, 1997) gives the following inhalation rates and daily air volumes inhaled for a 70-kilogram adult male:

Sedentary	0.62 m^3/hour × 8 hours (sleeping)	= 4.96 m^3
Light activity	1.40 m^3/hour × 8 hours (leisure time)	= 11.2 m^3
Moderate activity	1.78 m^3/hour × 8 hours (working)	= 14.24 m^3
	Total inhaled volume/24 hours	= 30.4 m^3/day

Ambient benzene vapor exposure (no occupational exposure):

$$5.76 \text{ mcg/m}^3 \times 30.4 \text{ m}^3 = 175.1 \text{ micrograms benzene}$$
$$\text{per day (ambient inhalation).}$$

To one's ambient exposure we must add occupational exposure at NIOSH limit:

$$\text{mg/m}^3 = \frac{\text{ppm}_v \times \text{molecular weight}}{24.45 \text{ L/gram-mole}} = \frac{0.1 \text{ ppm}_v \times 78.11 \text{ grams/mole}}{24.45 \text{ L/g-mole}}$$
$$= 0.3195 \text{ mg/m}^3$$

$$0.3195 \text{ mg/m}^3 = 319.5 \text{ mcg/m}^3$$

$319.5 \text{ mcg/m}^3 \times 14.24 \text{ m}^3 = 4549.7$ micrograms of benzene from occupational exposure vapor inhalation

worker's total inhalation exposure/day = ambient air + workplace air
= 175.1 mcg + 4549.7 mcg = 4724.8 micrograms

contribution from occupational exposure to total daily exposure
= (4549.7 mcg/4724.8 mcg) × 100 = 96.3%

That is, for this worker, ambient benzene exposure contributes only 3.7% of daily dose.

For workers exposed at OSHA's PEL of 1 ppm$_v$, occupational workplace air contributes:

$$\text{mg/m}^3 = \frac{\text{ppm}_v \times \text{molecular weight}}{24.45 \text{ L/gram-mole}} = \frac{1.0 \text{ ppm}_v \times 78.11 \text{ grams/mole}}{24.45 \text{ L/g-mole}}$$
$$= 3.195 \text{ mg/m}^3$$

$$3195 \text{ mcg/m}^3 \times 14.24 \text{ m}^3 = 45{,}496.8 \text{ micrograms of benzene}$$

$$45{,}496.8 \text{ mcg} + 175.1 \text{ mcg} = 45{,}671.9 \text{ micrograms}$$

Occupational = 99.6%, and nonoccupational = 0.4% of benzene vapor daily dose excluding weekends.

Skin contact will increase workers' absorbed systemic benzene dose. Calculate this worker's weekly exposure at the OSHA's PEL:

$$1 \text{ week} = 168 \text{ hours}$$

40 hours working: 5 days × 45,496.8 mcg = 227,484 micrograms benzene

Sleeping: 7 days × 8 hours/day × 0.62 m^3/hour × 5.76 mcg/m^3
= 200 mcg benzene

Light activity: 5 days × 8 hours/day × 1.40 m^3/hour × 5.76 mcg/m^3
= 322.6 mcg

Light activity: 2 days × 16 hours/day × 1.40 m³/hour × 5.76 mcg/m³
= 258 mcg

Weekly dose = 227,484 mcg + 200 mcg + 322.6 mcg + 258 mcg
= 228,264.6 mcg

Workplace contribution = (227,484 mcg/228,264.6 mcg) × 100 = 99.66%

713. The CO_2 gas concentration in a beer brewing plant's fermentation area is 33,800 ppm$_v$. What is the oxygen gas level?

$$33,800 \text{ ppm}_v = 3.38\%_v$$

$$100\% \text{ air} - 3.38\% \text{ } CO_{2v} = 96.62\%_v \text{ air remaining}$$

Air is 20.95% oxygen by volume.

$$96.62\%_v \times 0.2095 = 20.24\% \text{ by volume } O_2$$

Although the oxygen gas concentration complies with OSHA's minimum level of 19.5% oxygen by volume, the CO_2 gas concentration is excessive. Preventing CO_2 gas from leaking from fermentation tanks and piping and increasing dilution ventilation must be done. Of these intervention methods, eliminating gas leaks is the least expensive and most effective. Workers with cardiac issues might be at risk because elevated CO_2 levels stimulate the aorta's carotid body. Heart and breathing rates increase, and diminished oxygen in the air can act in combination to stress an already compromised, ischemic heart. Protracted inhalation of CO_2 could increase arterial carbonic acid resulting in acidosis. Inform workers. Fix it now.

The TLV and STEL for CO_2 are 5000 and 30,000 ppm$_v$, respectively.

714. Industrial hygienists and toxicologists are asked what was the person's exposure to carbon monoxide gas when his carboxyhemoglobin concentration is known. Or, conversely, at what time can we expect a certain COHgB level when the CO gas level is known or assumed. Or, more commonly, when we know carbon monoxide gas concentration and duration of exposure, what is the expected COHgB level? The following equation can provide answers to these questions (Wallace, L. A. et al., A linear model relating breath concentrations to environmental exposures: Application to a chamber study of four volunteers exposed to volatile organic chemicals, *Journal of Exposure Analysis and Environmental Epidemiology*, 7(2): 141–163 (1993):

$$[COHgB]_{total} = [COHgB]_{base} + [COHgB]_{exog} \times e^{-\gamma t} + \frac{\beta}{\gamma}[CO](1 - e^{-\gamma t}),$$

where
COHgB = percent hemoglobin combined with exogenous CO in the blood
CO = concentration of inhaled carbon monoxide gas, ppm$_v$

β = rate constant for exogenous CO uptake by blood (0.06% COHgB ppm$_v^{-1}$/hr^{-1})

γ = rate constant for return of exogenous blood CO to the lungs (0.402 hr^{-1})

$[COHgB]_{total}$ = baseline (endogenous) CO level in blood, constant at 0.5% (background carboxyhemoglobin level derived from catabolism of heme), %

$[COHgB]_{exog}$ = exogenous CO level in blood at time $t = 0$, %

For example, what is COHgB level in a one pack per day cigarette smoker ($[COHgB]_{exog} = 5.0\%$) exposed to 350 ppm$_v$ CO for 90 minutes?

$$90 \text{ minutes} = 1.5 \text{ hours}$$

$$[COHgB]_{total} = 0.5\% + 5\% \times e^{-(0.402/\text{hour} \times 1.5 \text{hours})}$$

$$+ \frac{0.06}{0.402}[CO](1 - e^{-(0.402/\text{hour} \times 1.5 \text{ hours})})$$

$$= 0.5\% + (5\% \times 0.547) + 0.149 \times 350 \text{ ppm}_v \times (1 - 0.547)$$

$$= 26.9\% \text{ COHgB}$$

715. A pollutant source releases 238 mg CO every minute into an empty non-ventilated building (30' × 20' × 12') that has 0.5 air change each hour. What is the maximum CO gas concentration that will be achieved when the CO emission rate is balanced with the fresh air dilution ventilation?

$$30' \times 20' \times 12' = 7200 \text{ ft}^3 = 203.88 \text{ m}^3$$

$$\frac{238 \text{ mg/minute}}{[0.5 \text{ air change/hour}] \times [1 \text{ hour/60 minutes}] \times [203.88 \text{ m}^3/\text{air change}]}$$

$$= 140.1 \text{ mg/m}^3$$

Assume this pollutant source is repaired after 3 hours. What is the CO gas concentration at this time?

$$3 \text{ hours} = 180 \text{ minutes}$$

$$\frac{238 \text{ mg/minute}}{[0.5 \text{ air change/hour}] \times [1\text{hour/60minutes}] \times [203.88 \text{ m}^3/\text{air change}]}$$

$$\times (1 - e^{-(0.5)(1/60)(180\text{minutes})}) = 140.1 \text{ mg/m}^3 \times 0.777 = 108.9 \text{mg CO/m}^3$$

$$\text{ppm}_v = \frac{\text{mg/m}^3 \times 24.45\text{L/gram-mole}^{-1}}{\text{molecular weight}}$$

$$= \frac{108.9 \text{ mg/m}^3 \times 24.45 \text{ L/gram-mole}}{28.01 \text{grams/mole}} = 95.1 \text{ ppm}_v$$

716. An industrial hygiene chemist prepares a dilute concentration of hydrogen bromide gas in a Tedlar® bag for a direct reading instrument calibration. She injects 1.0 mL HBr into an air stream delivering 7.65 Lpm for 23.5 minutes into the bag. What is the final HBr concentration in the bag?

$$7.65 \text{ Lpm} \times 23.5 \text{ minutes} = 179.775 \text{ L}$$

$$C_{HBRv}, \%_v = 100 \times \frac{V_{HBr}, \text{mL}}{1000(V_{air}) + (V_{HBr})}$$

$$= 100 \times \frac{1.0 \text{ mL}}{[(1000 \text{ mL/L}) \times (179.775 \text{ L})] + 0.1 \text{ mL}}$$

$$= 0.0005563\% \text{ by volume}$$

$$0.0005563\%_v = 5.563 \text{ ppm}_v \text{ HBr}$$

Hydrogen bromide does not have a TLV for an 8-hour TWAE. However, the TLV ceiling level is 2 ppm$_v$. A set of serial dilutions of the mixed gas from this bag must be made to make instrument calibration bags of, for example, 0.1, 0.5, 1, 2, and 5.563 ppm$_v$. This can be done easily by setting up arithmetic dilution ratios. Take instrument readings from each bag, and plot the six (one on "zero" air) results on normal graph paper (ppm$_v$ HBr vs. meter reading at the calibration pressure and absolute temperature).

Contact the Tedlar® bag manufacturer or the NIOSH laboratory to ensure that HBr will not react with the material resulting in low, false-negative readings.

717. The concentration of carbon monoxide gas in an empty room is 573 ppm$_v$. We will not permit occupancy until the CO gas concentration is reduced to 5 ppm$_v$. The room is 30' × 25' × 12'. The ventilation of the room with fresh air is 620 scfm, and air mixing is excellent ($K = 1$). How long will it take?

$$30' \times 25' \times 12' = 9000 \text{ ft}^3$$

$$t_{req=} \frac{V}{q} \times \ln\left[\frac{C_{initial}}{C_{final}}\right] = \frac{9000 \text{ ft}^3}{620 \text{ scfm}} \times \ln\left[\frac{573 \text{ ppm}_v}{5 \text{ ppm}_v}\right] = 19.26 \text{ minutes}$$

On a scale from 1 to 5 (no units), if we believed air mixing was greater than dismal and less than excellent, we might assign a mixing factor of 3. We will then multiply 19.26 minutes × 3 = 57.78 minutes and only allow occupancy after 1 hour and several air samples confirm the desired 5 ppm$_v$ CO gas occupancy concentration. Determine source of CO and fix it to assure no recurrence. Obtain blood samples from those who occupied the room before ventilation intervention to determine their COHgB concentrations and to treat medically as appropriate.

718. Calculate the mass of EPA air pollutant SO_2 represented by a fritted gas bubbler containing 36.5 mL of SO_2 collection solution. One milliliter was taken for analysis. The sample contained 13.2 mg, and the blank contained <0.01 mg. The collection efficiency of the bubbler is 0.87 (87%, no units). The sampling rate was 5.2 Lpm, and the stack exhaust gases were sampled for 19.5 minutes (corrected to NTP).

$$mg/L = \frac{(36.5 \text{ mL}/1 \text{ mL}) \times (13.2 \text{ mg} - 0.009 \text{ mg})}{0.87 \times 5.2 \text{ Lpm} \times 19.5 \text{ minutes}} = 5.46$$

$$5.46 \text{ mg/L} \times 1000 \text{ L/m}^3 = 5460 \text{ mg/m}^3$$

$$ppm_v = \frac{(5460 \text{ mg/m}^3) \times 24.45}{64.06 \text{ grams/gram-mole}^{-1}} = 2083.9 \text{ ppm}_v = 0.2084\%_v$$

719. The Agency for Toxic Substances and Disease Registry offers the following equation to calculate a reasonably acceptable risk for carcinogens:

$$C_m = \frac{RWL}{PIA(ED)},$$

where
    C_m = the action level, for example, air concentration of carcinogen above which remedial action must be taken,
    R = acceptable risk, or the probability of developing cancer,
    W = body mass,
    L = assumed lifetime,
    P = potency factor (unitless),
    I = intake rate (total from inhalation, dermal, ingestion, trans-ocular, injection),
    A = body absorption factor, the fraction of carcinogen absorbed by each route of exposure (unitless), and
    ED = exposure duration.

Assume a 70-kg male individual with life expectancy of 70 years exposed to benzene vapor for 17 years. Further assume an acceptable risk one cancer per million persons, or 10^{-6}. Benzene's established potency factor is 1.8 mg/kg-day. Assume his lung absorption is 60% (0.6). By the inhalation route:

$$C_m = \frac{(10^{-6})(70 \text{ kg})(70 \text{ years})}{(1.8 \text{ mg/kg/day})(30.4 \text{ m}^3/\text{day})(0.6)(17 \text{ years})(\text{mg}/1000 \text{ mcg})}$$

$$= 0.088 \text{ mcg/m}^3$$

Risk is unacceptable if this person's 17-year inhalation exposure exceeds 0.088 mcg/m³ = 0.000088 mg/m³.

$$\text{ppm}_v = \frac{(0.000088 \text{ mg/m}^3)(24.45 \text{ L/g-mole})}{78.11 \text{ grams/mole}} = 0.0000275 \text{ppm}_v$$

$$= 0.0275 \text{ppb}_v$$

This concentration is less than benzene vapor levels reported for ambient urban air. Moreover, because the equation contains several fuzzy variables, the utility of these calculations, in the author's view, has severe limitations. The value of these calculations is to determine relative risks between several carcinogenic chemicals in the atmosphere. In time, with more research and refinements, the calculations and iterations will have greater predictive power and public health utility.

720. A general area ambient air sample was obtained for respirable dust near a large stone-crushing operation. Process enclosures, wet methods of dust suppression, and mechanical local exhaust ventilation were not used to prevent and capture fugitive dust that contained silica (α-quartz) at 23%. The filter used in respirable dust cyclone weighed 41.22 milligrams initially and 44.08 milligrams after air sampling at a rate of 1.7 Lpm for 7 hours, 42 minutes. What was the average respirable dust concentration in the air over this sampling period? What was the ambient silica dust level?

7 hours, 42 minutes = 420 minutes + 42 minutes = 462 minutes

462 minutes × 1.7 Lpm = 785.4 liters = 0.7854 m³

44.08 mg – 41.22 mg = 2.86 mg respirable dust

2.86 mg/0.7854 m³ = 3.64 mg total respirable dust/m³

3.64 mg total respirable dust × 0.23 = 0.84 mg respirable silica/m³

Prompt intervention by installing dust controls are necessary to protect the workers and people living nearby. Dispense appropriate respirators to educated, trained, and medically approved workers. Inform neighbors to remain inside and to close all windows and doors. Respirable silica dust causes pulmonary fibrosis (silicosis) and lung cancer.

721. Subsequent analysis of the respirable dust in the previous problem (720) revealed, in addition to quartz, the presence of cristobalite (1.1%) and tridymite (0.6%). What is the PEL for this three-mineral component respirable dust?

$$\text{PEL, respirable mg/m}^3 = \frac{10 \text{ mg/m}^3}{[\% \text{quartz} + (\% \text{cristobalite} \times 2) + (\% \text{tridymite} \times 2)] + 2}$$

$$= \frac{10 \text{ mg/m}^3}{[23\% + (1.1\% \times 2) + (0.6\% \times 2)] + 2]}$$

$$= \frac{10 \text{ mg/m}^3}{28.4} = 0.352 \text{ mg/m}^3$$

This PEL assists in prevention of silicosis. However, it might not be sufficiently low to prevent lung cancer. The formula is based on the 1968 TLV adopted by OSHA to prevent silica pneumoconiosis. Silica has since been identified as a pulmonary carcinogen. Industrial hygienists must be forever mindful to exposures below OSHA PELs and other guidelines might be regulatory compliant but remain as inhalation health hazards based on any current clinical, toxicological, and/or epidemiological evidence.

We must also consider what the remaining respirable dust is: 23% + 1.1% + 0.6% = 24.7%. 100% − 24.7% = 75.3% (silicates? asbestos? heavy metals?, and so on). And do not forget we must also consider qualitatively and quantitatively what total airborne dust is.

722. Stack exhaust gases from hot combustion processes typically contain much water vapor. As the exhaust gas cools, water will condense as a fog and a liquid film on cool surfaces. We easily notice this during the winter in cold climates where water in a car or truck exhaust forms a plume of visible water fog and mist. The vehicle's exhaust is just as hot in summer, but since warmer ambient summer air can hold more moisture than in cold winter air, a visible water vapor cloud does not form.

The amount of water vapor can be highly variable between different combustion processes and other moisture-containing gases. Failure to account for the water vapor content can lead to significant reporting errors. Therefore, air-sampling data from various degrees of "wet air" are converted into a standard "dry air" as the following calculations will demonstrate. Breathing zone ambient air samples obtained for industrial hygiene purposes are not typically corrected for water vapor content because the water content is much less than from a combustion process. An air-sampling device placed extremely close to the nostrils and mouth could mislead because (1) exhaled breath at body temperature is 100% saturated by water vapor, and (2) exhaled air could be sampled along with contaminated air diluting mixture and giving false-negative results.

350 mL of water was condensed from 50.0 cubic feet of flue gas (as indicated at the dry gas meter conditions) drawn from a duct through a water condensation train. The gas temperature at both the meter and the last condensing stage was 70°F. Dry gas meter vacuum was −5.0 inches of mercury. Barometric pressure was 29.92 inches of mercury. Calculate the moisture content of the flue gas.

$$\text{Observed: } T_m = 70° + 460°R = 530°R$$

$$P_m = 29.92'' - 5.00'' = 24.92\text{: Hg}$$

$$(P_w)_m = 0.39'' \text{ Hg at } 70°F)$$

$$\text{Condensed water volume: } V_c = 2.67 \times 10^{-3} \times \left[\frac{T_m}{P_m}\right] \times L_c$$

$$= 2.67 \times 10^{-3} \times \left[\frac{530}{24.92}\right] \times 350 = 19.85 \text{ ft}^3$$

$$\text{Water vapor volume} = V_c = \frac{(P_w)_m}{P_m} \times V_m = \frac{0.739}{24.92} \times 50.0 = 1.48 \text{ ft}^3$$

$$\text{Total water volume} = V_c + V_w = 19.85 + 1.48 = 21.33 \text{ ft}^3$$

$$\text{Total volume} = V_T = V_c + V_m = 19.85 + 50.0 = 69.85 \text{ ft}^3$$

$$\text{Moisture content} = \frac{V_w}{V_T} \times 100 = \frac{21.33}{69.85} \times 100$$

$$= 30.5\% \text{ H}_2\text{O vapor by volume}$$

723. A chemical plant maintenance worker had the following TWAEs to organic solvent vapors (acceptable exposure limits within parentheses): 9 ppm$_v$ A (50 ppm$_v$), 45 ppm$_v$ B (200 ppm$_v$), 2 ppm$_v$ C (75 ppm$_v$), and 13 ppm$_v$ D (100 ppm$_v$). Calculate his additive mixture exposure and comment appropriately.

$$8 \text{ hour TWAE ppm}_v = \left[\frac{9 \text{ ppm}_v}{50 \text{ ppm}_v} + \frac{45 \text{ ppm}_v}{200 \text{ ppm}_v} + \frac{2 \text{ ppm}_v}{75 \text{ ppm}_v} + \frac{13 \text{ ppm}_v}{100 \text{ ppm}_v} \right]$$

$$= 0.18 + 0.23 + 0.03 + 0.13 = 0.58 \text{ (unitless)}$$

This exposure does not exceed 1.0 unity—the TLV for the solvent vapor mixture. However, 0.58 exceeds a 0.5 action level. Moreover, this worker, in maintenance activities, would most likely have highly variable exposures from day to day. More thorough investigation of his most likely highest (not low or average!) exposures is required. During interim, after being medically qualified, require use of acceptable respirator with fitting training. Ask the worker and his supervisor what contributes to his greatest vapor exposure, and focus on control of highest peak exposures as indicated.

724. The worker as in the previous problem (723) was not exposed to the same solvent vapors on another day. He was, however, exposed (8-hour time-weighted averages) to methyl cellosolve at 1.6 ppm$_v$ (limit = 25 ppm$_v$) and methyl cellosolve acetate at 1.1 ppm$_v$ (limit = 25 ppm$_v$). Calculate his additive mixture exposure.

$$8\text{-hour TWAE ppm}_v = \left[\frac{1.6 \text{ ppm}_v}{25 \text{ ppm}_v} + \frac{1.1 \text{ ppm}_v}{25 \text{ ppm}_v} \right] = 0.064 + 0.044$$

$$= 0.108 \text{ (unitless)}$$

Exposure is 10.8% of the worst acceptable exposure and (0.108/0.5) 21.6% of the action level. However, because he had frequent unprotected hand contact with these skin-absorbed solvents, the industrial hygienist immediately intervened by providing acceptable impervious gloves. Regular skin protection, improving the local exhaust ventilation, hazard communication education and training, enclosure of vapor-emitting processes, and medical surveillance were implemented. Follow-up studies revealed nondetectable vapor exposures and regular compliance with industrial hygiene recommendations—especially prevention of all skin contact.

725. The industrial hygienist becomes involved in the education and training of workers. This, by nature, involves several cryptic terms and unfamiliar concepts for trainees. Now and then, the author encounters some workers who have a hard time coming to grips with notions of TWAEs and short-term exposure limits.

The following example has been used as a reasonably apt analogy. Consider a person who travels nonstop on a highway for eight straight hours. Four highway sections have a speed limit of 60 mph which, at this speed, covers 15 minutes of driving for each section—comparable to four STEL exposures separated by an hour or so for each. The remainder of the highway has a speed limit of 50 mph.

The driver races at 80 mph through the 60 mph sections. He drives at 35 mph for the remainder of the highway. What is his time-weighted average velocity? On an average, is he compliant with the posted speed limits?

$$80 \text{ mph} \times 1 \text{ hours} = 80 \text{ mph-hours}$$

$$35 \text{ mph} \times 7 \text{ hours} = 245 \text{ mph-hours}$$

$$(80 \text{ mph-hours} + 245 \text{ mph-hours})/8 \text{ hours} = 40.625 \text{ mph}$$
average driving speed

So, we see this driver is speed limit-compliant—on average, but his high velocity bursts place him over the STEL. He's at risk of harm (ticket or disease). Do you believe the Federales Polizia will buy his average velocity argument if he is radar zapped at 80 mph anywhere on his trek? "But, officer, on average, I am a very careful driver." Close, but no cigar. "Driver's license, registration, and insurance record, please."

Index to Problems

Milton Keynes UK
Ingram Content Group UK Ltd.
UKHW031139141024
449569UK00024B/1216